Physics and Chemistry in Space Vol. 13

Edited by L. J. Lanzerotti, Murray Hill

Vladimir A. Krasnopolsky

Photochemistry of the Atmospheres of Mars and Venus

Technical Editor
Ulf von Zahn

With 209 Figures

Springer-Verlag
Berlin Heidelberg New York Tokyo

Professor Dr. VLADIMIR A. KRASNOPOLSKY
Space Research Institute
Academy of Sciences
Profsojusnaja, 84/32
117810 Moscow, USSR

Technical Editor:

Professor Dr. ULF VON ZAHN
Rheinische Friedrich-Wilhelms-Universität Bonn
Physikalisches Institut
Nußallee 12
5300 Bonn, FRG

Title of the original Russian edition:
Fotokhimia atmosfer marsa i veneri

ISBN-13: 978-3-642-70403-1 e-ISBN-13: 978-3-642-70401-7

DOI: 10.1007/978-3-642-70401-7

Library of Congress Cataloging-in-Publication Data. Krasnopol'skiĭ, V. A. (Vladimir Anatol'evich), 1938– Photochemistry of the atmosphere of Mars and Venus. (Physics and chemistry in space ; v. 13) Translation of: Fotokhimiía atmosfer Marsa i Venery. Bibliography: p. Includes index. 1. Mars (Planet) – Atmosphere. 2. Venus (Planet) – Atmosphere. 3. Photochemistry. 4. Cosmochemistry. I. Title. II. Series. QC801.P46 vol. 13 530'.0919 s 85-25008 [QB641] [551.5'0999'23]

Softcover reprint of the hardcover 1st edition 1986

2131/3130-543210

Preface

Spacecraft study of the Solar system is one of humanity's most outstanding achievements. Thanks to this study, our present knowledge of properties of and conditions on the planets exceeds many-fold that of 20 years ago: planets have been rediscovered. This is especially the case for planetary atmospheres, whose properties were for the most part either not at all or only erroneously known.

Much research has been invested in the study of the atmospheres of Mars and Venus, and their chemical composition and photochemistry are basic problems in these studies. In the present publication I have tried to summarize all findings in this field.

The English version of the book includes new data in the field from the last 3 years since the book was published in Russian.

I wish to thank U. von Zahn, who initiated my talks with Springer-Verlag and acted as technical editor.

December 2, 1985 V. A. KRASNOPOLSKY

Contents

Introduction

The main aim of this book is to present experimental and theoretical results achieved over the last decade on the chemical composition and structure of the atmospheres of Mars and Venus, processes defining the properties of the atmospheres, and other related phenomena.

Three basic aims for studies of an atmosphere exist: chemical composition, thermal regime, and dynamics. The chemical composition of an atmosphere is determined by its origin and evolution, processes of gas exchange between the atmosphere and the planetary body and space, and by chemical and physical processes inside the atmosphere. The latter processes are defined in the science of aeronomy, and this book is concerned primarily with findings in this field. However, a planetary atmosphere is a single entity and all its components are interrelated, e.g., its chemical composition determines thermal regime and the latter, in turn, affects its dynamics. Therefore, this book provides a brief description of these problems as well.

Ample space is given to experimental results and their interpretation. For theoretical findings, I will try to give an idea of a method or an interpretation, followed by the results obtained. The advantages or speculative aspects of some approaches are also considered. Unfortunately, in many cases I had no possibility to discuss the papers described with their authors, thus, this may have led sometimes to errors in my evaluations. Therefore, I would greatly appreciate receiving some response from readers.

Of similar importance to experimental results are theoretical models which describe the transformations of atmospheric constituents by photochemical processes in the presence of transport. If these models are in accord with the scope of experimental data, then they demonstrate the adequacy of our knowledge of the intimate properties of the atmosphere. A photochemical approach may be of use in the case of controversial experimental data. Besides, model calculations yield the amounts and vertical profiles of components which have been undetected or measured in a limited altitude range.

Important aspects of the space programs of the USSR and USA over the last two decades include investigations of Mars and Venus. The structures of these programs were different. The Soviet program from the beginning of the 1960's was directed toward the designing of space vehicles with descent probes, while in the United States spacecraft with more and more complicated functions were developed: flyby systems for short-term observations from a distance of $\sim 10^4$ km, orbiters, and descent probes.

The first successful experiments for the study of Mars were carried out by the American flybys Mariner 4 (1965) and Mariner 6, 7 (1969). The first orbiters were in 1971: Soviet Mars 2 and 3 and U.S. Mariner 9. That year, the first descent probe, Mars 3, was launched into the Martian atmosphere to measure its properties. Though the descent and landing proceeded normally, information transmission ceased soon after it started, probably as a result of extremely strong winds of a global dust storm on the planet which could have led to the disturbance of the antenna's operation. In 1974, four Soviet probes, i.e., two flybys (Mars 4 and 7), one orbiter (Mars 5), and one descent probe (Mars 6), investigated the planet. The most extensive research program of Mars was carried out by Viking 1 and 2 (1976 until 1981), which incorporated long-lived landers and orbiters.

The first space studies of Venus were made by the American flybys Mariner 2 (1962) and Mariner 5 (1967). The latter was accompanied by the successful descent of the Soviet Venera 4. Then, a series of first generation Soviet descent probes followed (Veneras 5 to 8, 1969–1972). In 1974, Mariner 10 flew by Venus and Mercury. The second generation of Soviet spacecraft to study Venus was started in 1975 by Veneras 9 and 10, each including a descent probe and an orbiter. At the end of 1978 the U.S. Pioneer Venus spacecraft arrived at Venus, which incorporated four descent probes intended to operate at different locations (low and high latitudes, day and night sides), an entry probe (bus) for measurements down to 130 km, and an orbiter with a pericenter at 150 km. At the same time, Soviet Veneras 11 and 12 operated. Since then, studies of Venus were continued by Veneras 13, 14 lander probes (1982) and Veneras 15, 16 orbiters (1983–1984).

Results obtained by the spacecraft, supplemented by some ground-based observations, especially high resolution spectroscopy in infrared and microwave regions, form the foundation of our modern knowledge of both planets.

Studies of the solar system, which may seem at first glance to be far removed from the needs and concerns of humanity, are of great cognitive and philosophical value, especially the problem of the origin and evolution of the solar system. However, it is our deep belief that atmospheric photochemistry will find ever increasing use. Human activity influences the environment negatively, in particular, leads to air pollution. Effective methods to maintain the existing properties of the atmosphere might be developed in the future. One significant problem is to study the atmospheric resilience to influences which might result in some irreversible consequences in definite conditions. For example, the atmospheric temperature depends on abundances of water vapor and carbon dioxide and the area of clouds reflecting solar radiation. The situation when high atmospheric temperature could be maintained by the amount of increased atmospheric water vapor does not seem totally improbable. Another apt example widely discussed in the last decade is the negative consequences of reduced amounts of ozone in the atmosphere as a result of enhanced delivery of chlorfluormethanes and nitrogen oxides, which act as catalysts in the process of destroying the ozone layer.

The mathematical background of photochemical modeling is a system of continuity equations for every component under consideration:

$$\frac{\partial n_i}{\partial t} + \nabla \bar{\phi}_i = P_i - n_i L_i$$

where n_i is the number density, $\nabla = i\frac{\partial}{\partial x} + j\frac{\partial}{\partial y} + k\frac{\partial}{\partial z}$ is the differential operator, P_i and L_i are the production and loss rates by chemical reactions, $\bar{\phi}_i$ is the flux of the component which can be given by

$$\phi_i = -(K + D_i)\left[\frac{\partial n_i}{\partial z} + \frac{n_i}{T}(1 + \alpha_i)\frac{\partial T}{\partial z}\right] - n_i\left(\frac{K}{H} + \frac{D_i}{H_i}\right)$$

for one-dimensional models and in the approximation of the minor component $n_i \ll \sum_j n_j$. Here, K is the eddy mixing coefficient, D_i is the molecular diffusion coefficient, H and H_i are the mean atmospheric scale heights and that of the i component, T is the temperature, and α_i is the thermal diffusion factor. In a physical sense, the continuity equation implies that changes in the density of a component number are due to chemical reactions or transport from or to other atmospheric regions.

Since altitude variations of atmospheric composition as a rule greatly exceed horizontal variations, derivatives with respect to horizontal coordinates are much less than those with respect to height and may be neglected. This approximation gives one-dimensional models. When diurnal and other temporal variations are not taken into consideration, time independent models are used which simplify the system of partial differential equations to that of ordinary differential equations. In some cases, e.g., for ionospheric calculations, when photochemical lifetimes, $\tau_c = L^{-1}$, are much less than transport time constants, $\tau_m = H^2/(K+D)$, transport effects can be neglected, and one gets a system of algebraic equations, $P_i = n_i L_i$.

In most cases these systems of equations need a special computation technique, the description of which is beyond the scope of our book.

Boundary and (in the case of time-dependent problems) initial conditions should be introduced. The usual form of boundary conditions is number densities or fluxes at both (upper and lower) boundaries. Below, some widespread cases are given: $\phi_i = 0$ at the upper boundary implies diffusive equilibrium of the component n_i, its altitude distribution having a scale height $H_i = \frac{kT}{m_i g}$ above the upper boundary. The same at the lower boundary implies that interactions with the low lying atmosphere or surface are negligible (chemically passive surface). A condition of photochemical equilibrium at the boundary, $n_i = P_i/L_i$, is a partial case of the boundary condition for a component number density.

This book is based on data published up to the beginning of 1984.

1 Chemical Composition and Structure of the Martian Atmosphere

1.1 Carbon Dioxide and Atmospheric Pressure

Carbon dioxide is the most abundant volatile matter of Mars and Venus. A great number of vibration-rotation bands in the infrared region facilitate CO_2 spectroscopic detection. Kuiper was the first to discover CO_2 on Mars from 1.57 and 1.6 μm bands in 1947. Figure 1 shows a detailed spectrum of Mars in this region measured by Moroz (1967). The bands in the Martian spectrum are much more intense than telluric bands (caused by absorption in the Earth's atmosphere) in the solar spectrum.

Quantitative interpretation of spectroscopic observations presents some difficulties (see Moroz 1967). Although this problem for CO_2 bands in the spectrum of Mars has been successfully solved, it is still of interest for other spectroscopic measurements. The idea of the interpretation is as follows. Observations for a band or rotational line in the case of fairly high resolution give a so-called equivalent width

$$W = \int_0^\infty \frac{I_0 - I_\lambda}{I_0} d_\lambda$$

where I_λ, I_0 are the intensities in- and outside the line, respectively. With the absorption weak even in the line center (unsaturated line) the equivalent width is given by $W = \sigma N$, where N is the number of absorbing molecules per unit

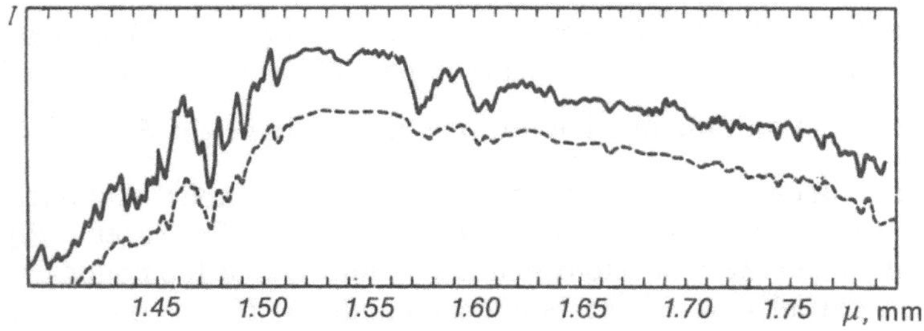

Fig. 1. Martian (*solid line*) and solar (*dashed line*) spectra in the range of 1.4 to 1.8 μm. CO_2 bands at 1.57 and 1.6 μm are more prominent in the Martian spectrum than those caused by telluric absorption in the solar spectrum (Moroz 1967)

square (column density), σ is their integral cross-section $\sigma = \int_0^\infty \sigma_\lambda d_\lambda$ or the line strength. For $\sigma_\lambda N > 1$ the line is saturated and its equivalent width dependence on σN, the so-called curve of growth, is governed by a process forming a line shape. At low pressure (lower than 100 mbar) the line broadening is due to the Doppler effect caused by thermal motion of molecules, and the Doppler line width is

$$\Delta\lambda_D = \lambda_0 V/C = 4.3 \times 10^{-7} (T/\mu)^{1/2} \lambda_0 ,$$

where $V = (2kT/m)^{1/2}$ is the most probable molecular velocity and μ is the molecular mass in atomic units. The line shape is

$$\sigma_\lambda = \frac{\sigma}{\pi^{1/2} \Delta\lambda_D} e^{-[(\lambda-\lambda_0)/(\Delta\lambda_D)]^2}$$

and a full line width at half maximum can be written as

$$\mathrm{FWHM} = 7.2 \times 10^{-7} (T/\mu)^{1/2} \lambda_0 .$$

At rather high pressure the line shape is collisionally induced (Lorentz or pressure-broadened line shape) and the line width in frequency scale is

$$\Delta\nu_c = \frac{1}{2\pi\tau_c} = \frac{P\delta}{\pi(kTm)^{1/2}}$$

with τ_c being the mean time between collisions and δ the collision cross-section. In the wavelength scale $\Delta\lambda_c$ is given by

$$\Delta\lambda_c = \frac{\lambda^2}{c} \Delta\nu_c .$$

The Lorentz line shape is

$$\sigma_\lambda = \frac{\sigma\Delta\lambda_c}{\pi[(\lambda-\lambda_c)^2 + \Delta\lambda_c^2]} .$$

Given a comparable contribution of both effects, a Voigt line shape forms

$$\sigma_\lambda = \frac{\sigma\Delta\lambda_c \pi^{-3/2}}{(\Delta\lambda_D)^2} \int_{-\infty}^{\infty} \frac{e^{-y^2} dy}{a^2 + (p-y)^2} \simeq \frac{\sigma}{\pi^{1/2}\Delta\lambda_D} \left[e^{-p^2} + \frac{a}{\pi^{1/2} p^2} \right]$$

with $p = \dfrac{\lambda - \lambda_0}{\Delta\lambda_D}$, $a = \dfrac{\Delta\lambda_c}{\Delta\lambda_D} = \dfrac{P\delta_\lambda}{\pi 2^{1/2} kT}$, where a equals unity at $P \simeq 100$ mbar for $\lambda \simeq 1$ µm and $T \simeq 300$ K.

Figure 2 shows typical curves of growth for different a. For low σN the curve of growth is a straight line with a slope of $\mathrm{tg}\,\alpha = 1$ in the logarithmic scale, i.e., it is proportional to the first power of σN. For the larger σN it depends on a. At $a = 0$ (the Doppler line shape) the equivalent width increases very little with increasing σN due to steep exponential decrease in the line wings. The Lorentz line shape is much less steep, and the equivalent width for $\sigma N \gg 1$ is given by

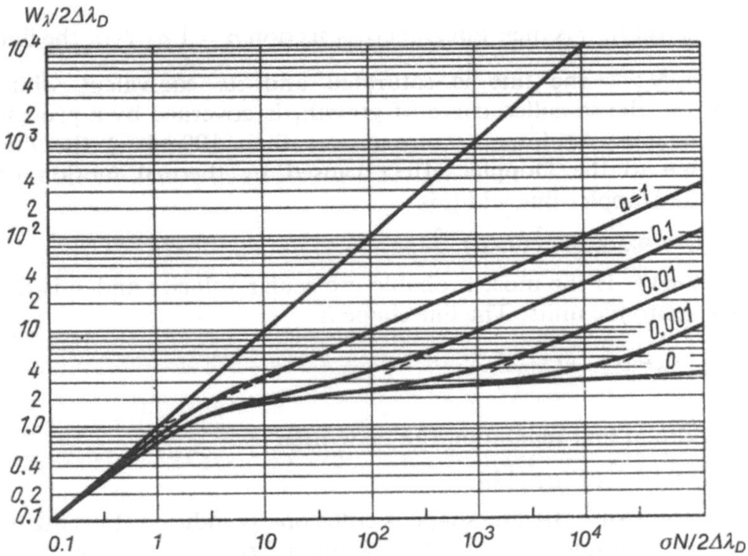

Fig. 2. Curves of line growth for different $a = \Delta\lambda_C/\Delta\lambda_D$

$$W = 2(\sigma N \Delta\lambda_c)^{1/2}$$

having a slope of tg $\alpha = \frac{1}{2}$.

Thus, from an unsaturated line one can find N. If it is supplemented with the measurement of a saturated line, $\Delta\lambda_c$ can be obtained and then atmospheric pressure P can be derived at δ and T known.

Absorption line measurements need high spectral resolution which sometimes is rather difficult to achieve. For absorption band interpretation, the curve of growth can be either measured experimentally or calculated by some model band. The average absorption of the band having lines of equal intensity spaced at an equal distance, d, and being pressure broadened (the so-called Elsasser model) is given by

$$R = 1 - e^{-\tau} = 1 - \frac{1}{2\pi} \int_{-\pi}^{\pi} \exp\left(-\frac{\beta x \operatorname{sh}\beta}{\operatorname{ch}\beta - \cos z}\right) dz$$

with $\beta = 2\pi\Delta\lambda_c/d$, $x = \sigma N/2\pi\Delta\lambda_c$. For $x > 1.25$ and $\beta < 0.3$ the equation is simplified to

$$R = \phi\left(\sqrt{\frac{2\pi\sigma N \Delta\lambda_c}{d^2}}\right)$$

where $\phi(x) = \left(\frac{2}{\pi}\right)^{1/2} \int_0^x e^{-t^2/2} dt$ is the probability integral. $R = 1 - e^{-\beta x}$ for $\beta > 3$ at any x.

CO_2 abundance and atmospheric pressure as deduced from spectroscopic observations. We will not dwell upon the development of these measurements and their interpretation (see Moroz 1967). Emphasis will be placed on the most important experiments prior to space flights to Mars.

Kaplan et al. (1964) measured the Martian near infrared spectrum with a resolution of $\lambda/\Delta\lambda \simeq 10^5$ featuring rotational lines of the very weak 870 nm band. Comparison with laboratory CO_2 spectra gave for the CO_2 column density on Mars a value of $(1.5 \pm 0.5) \times 10^{23}\,\mathrm{cm}^{-2}$ at $T = 230\,\mathrm{K}$. Owen (1964) compared this band with the same telluric band in solar spectra at great zenith angles. The intensities of these bands were found to be equal at an air mass of approx. 30, which corresponds to $2 \times 10^{23}\,\mathrm{cm}^{-2}$ of CO_2 on Mars.

To deduce atmospheric pressure Owen and Kuiper (1964) carried out the measurements of the 1.57 and 1.6 μm band widths at different pressures and CO_2 abundances (Fig. 3). Comparison of the data received with observations (air mass $M \simeq 3$ taken into account) gave a mean pressure of 4 mbar for pure CO_2 atmosphere and 7 mbar for 10% CO_2 + 90% N_2 mixture. The atmospheric pressure at the planetary surface is twice as large, or $8 < P < 15$ mbar.

Radio occultation measurements of atmospheric pressure. While passing through a medium with a refractive index different from unity, the electromagnetic waves experience a phase shift proportional to both path length and $m-1$. The refractive index, m, depends on gas density, chemical composition, and temperature, the dependence on density being the most pronounced. Ground-based measurements of the phase shift of a radio signal of a spacecraft entering behind a planetary limb make it possible to derive neutral atmospheric density, temperature, and pressure versus altitude, assuming that the atmospheric composition is known. Similarly, one can measure ionospheric electron number density, n_e, considering that in ionospheric plasma, m is below 1 and $(1-m)$ is proportional to n_e. This method was first used by

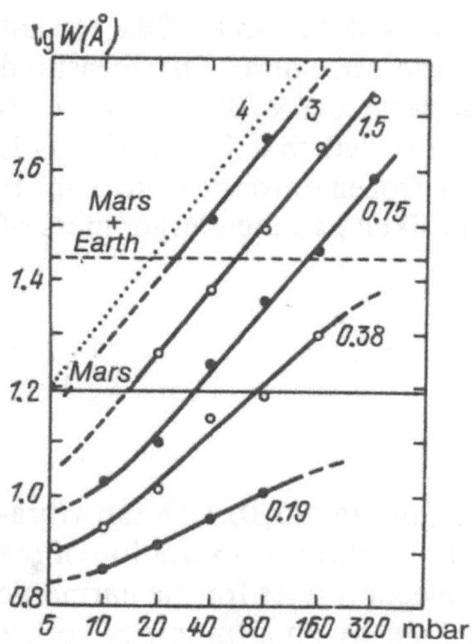

Fig. 3. Equivalent widths of CO_2 bands at 1.57 and 1.6 μm vs atmospheric pressure for different column amounts of CO_2 (in $10^{23}\,\mathrm{cm}^{-2}$). That for $4 \times 10^{23}\,\mathrm{cm}^{-2}$ is a result of extrapolation. *Horizontal lines* are the CO_2 band widths in the Martian spectrum with and without correction for telluric absorption (Owen and Kuiper 1964)

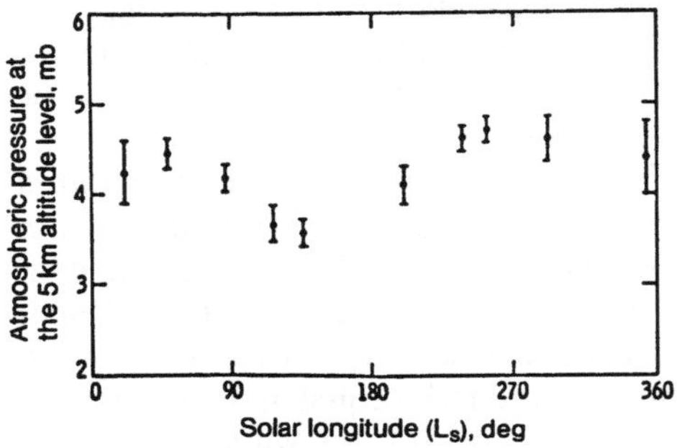

Fig. 4. Pressure changes observed at the 5-km altitude level during the Martian year. The time of year is marked along the abscissa axis in terms of aerocentric solar longitude (Lindal et al. 1979)

Mariner 4 (Kliore et al. 1967). Since then it was utilized by almost all space probes sent to Mars and gave a mean atmospheric pressure on Mars equal to 6 mbar. Correlation of this method with spectroscopic CO_2 observations shows that the atmosphere of Mars consists almost entirely of CO_2. These results were later confirmed by direct measurements on board Mars 6 (Avduevsky et al. 1975, Kerzhnovich 1977) and Viking 1, 2 (Seiff and Kirk 1977).

Seasonal variations of the atmospheric pressure were revealed by long-term radio-occultation measurements and Viking 1 and 2 landers (see Lindal et al. 1979). Solar aerocentric longitude, L_s, equal to 0° and 180° at vernal and autumnal equinoxes, respectively, defines the Martian seasons. The observations show two minima at $L_s = 150°$ and 350° and two maxima at 270° and 50° (Fig. 4). The maximum at 270° and the minimum at 150° are more pronounced, and their ratio equals 1.35. The explanation of this fact is the recycling of CO_2 between the polar caps and the atmosphere caused by the usual seasonal variations of solar heating as well as the eccentricity of the orbit. The perihelium corresponds to $L_s = 350°$ and the solar flux at that moment exceeds by a factor of 1.45 that at the aphelium. The measured pressure variations show the exchange of at least 0.8×10^{13} tons of CO_2 between the polar caps and the atmosphere, provided that first the CO_2 is released completely from one cap and thereafter frozen onto the other cap. If both processes act simultaneously (which is very likely) a much larger mass of CO_2 is actually exchanged.

1.2 CO and O_2 Mixing Ratios

Carbonmonoxide has (1,0) 4.66 μm, (2,0) 2.35 μm, and (3,0) 1.58 μm vibration-rotation bands. The numbers in brackets show vibration excitation of the upper (v') and lower (v'') levels, respectively. Selection rules for the harmonic oscillator allow transitions only for $\Delta v = 1$. Due to unharmonicity, other bands are present too, but their intensities are much lower and decrease with Δv.

A notable advance has been made in the ground-based investigations of planets due to the progress of the near infrared ($\lambda < 4$ μm) high resolution spectroscopy. The excellent spectra of Venus, Mars, Jupiter, and Saturn were measured by Connes et al. (1969). Martian spectra exhibit some lines of CO (2,0) and (3,0) bands (Kaplan et al. 1969) with stronger CO_2 lines overlapping CO lines of the (3,0) band. The P6 line equivalent width of 0.0088 cm^{-1} (in the wavenumber $1/\lambda$ scale) is the most readily defined. The same line of the (2,0) band is much stronger ($W = 0.049\ cm^{-1}$). From these data with the known σ one can obtain $f_{CO} = [CO]/[CO_2] = (8 \pm 3) \times 10^{-4}$ and $P = 5.3 \pm 1.5$ mbar.

Oxygen. Vibration-rotation bands of homonuclear diatomic molecules are strongly forbidden because such transitions would not change the dipole moment. So the O_2 spectroscopic detection may use electronic transitions only. The only allowed transitions from the ground state (only this state produces absorption) produce a Schumann-Runge band system at 175 – 200 nm and cannot be observed due to strong telluric (O_2 and O_3) and Martian atmospheric CO_2 absorption. The Herzberg band system is weak and located near the maximum of the ozone absorption. Thus, the measurements of planetary atmospheric absorption are possible only in the atmospheric ($b^1\Sigma_g^- - X^3\Sigma_g^-$) 762 nm and the infrared atmospheric ($a^1\Delta_g - X^3\Sigma_g^+$) 1.27 μm band systems. Both bands are forbidden; the 762 nm band being stronger by two orders of magnitude with a radiative lifetime of 12 s. A search for the very weak Martian lines, which are Doppler shifted with respect to the strong telluric O_2 lines, should be made to detect O_2. This requires high spectral resolution ($\lambda/\Delta\lambda \gtrsim 10^5$) and sensitivity of an instrument. Such observations made by Barker (1972) and Carleton and Traub (1972) gave the O_2 mixing ratio as $f_{O_2} = [O_2]/[CO_2] \simeq 1.3 \times 10^{-3}$. Recent observations result in $f_{O_2} = 1.1 \times 10^{-3}$ (Trauger and Lunine 1983).

Mass spectrometric measurements on board Viking landers (Owen et al. 1977) did not provide new information on CO and O_2 amounts. Both gases appear from CO_2 inside the mass spectrometric ion source. CO and N_2 are equal in molecular mass and a simultaneous measurement of these species is possible only when they are present in comparable quantities. To measure N_2 a chemical absorption of CO was used. The O_2 mixing ratio was evaluated as $(1-4) \times 10^{-3}$.

Microwave spectroscopy of the CO *2.6 mm line* (Clancy et al. 1983). This line is formed as a result of a rotational transition $I = 1 \rightarrow 0$. Its ground-based measurements gave the mean $f_{CO} = (2 \pm 1) \times 10^{-3}$, which is three times as high as a value derived from infrared spectroscopy. To explain their results Good and Schloerb (1981) suggest f_{CO} variations with a time scale of the order of several years. Thirteen years elapsed between the observations made by Kaplan et al. (1969) and by Good and Schloerb (1981). However, with all microwave measurements analyzed, Clancy et al. (1983) concluded that the microwave spectra of CO are not the best tool for measuring variations of CO abundance on Mars.

1.3 Ozone

The simplest and most sensitive method of ozone detection and measurement is optical spectroscopy in the Hartley continuum

$$O_3 + h\nu(\lambda < 300\,\text{nm}) \rightarrow O_2(a^1\Delta_g) + O(^1D)\,.$$

The absorption cross-section is very large and has a maximum of $10^{-17}\,\text{cm}^{-2}$ at 255 nm.

Assuming single scattering approximation, which in most cases is valid for the Martian atmosphere, the brightness coefficient of the atmosphere-surface system is

$$R = \frac{\pi I}{I_0} = \frac{\gamma(\varphi)}{4}\,\omega\,\frac{\mu_0}{\mu_0+\mu}\,[1-e^{-\tau M}] + A\mu_0(\mu\mu_0)^{\beta-1}e^{-\tau M}\,.$$

The first term describes the scattering of the atmosphere having the optical thickness, τ, single scattering albedo, ω, and a phase function, $\gamma(\varphi)$, μ_0 and μ being cosines of slope and optical axis angles and $M = 1/\mu_0 + 1/\mu$ being the air mass. The second term takes into account the surface reflection and depends on the albedo, A, and the parameter, β.

Let us consider the case when the contribution of the surface reflection is much less than that of the atmospheric scattering. It may be applied to UV observations outside polar caps. Let us introduce an additional restriction, $\tau < 1$, which is valid in the absence of dust storms. If the variations of atmospheric optical properties from the center to the edge of the absorption band are attributable only to ozone absorption, τ_0, then exponent expansion into three terms gives a center to edge ratio of the brightness coefficients as $1-(M\tau_0)/(2-M\tau)$. Here, τ is the atmospheric optical thickness without ozone taken into account. At the planetary disk where M equals ~ 3 the ozone sensitivity is $8 \times 10^{15}\,\text{cm}^{-2}$ for the albedo depression of $\sim 10\%$. This amount corresponds to 3 μm thickness of the gas layer under normal conditions. Ozone amounts are usually measured in micrometers (μm) and we will follow this. The above sensitivity gives the ozone mixing ratio, f_{O_3}, equal to 5×10^{-8}. The air mass is ~ 50 for limb observations, and the sensitivity is about an order of magnitude better.

In polar caps the atmospheric term is much smaller than the surface one and the center to edge of the absorption band ratio is $e^{-M\tau_0} \simeq 1 - M\tau_0$. The sensitivity is, thus, twice as good as that in the planetary disk at low and middle latitudes.

Mariner 7 and 9 observations. Ozone on Mars was discovered at the southern polar cap by means of the Mariner 7 UV spectrometer (Barth and Hord

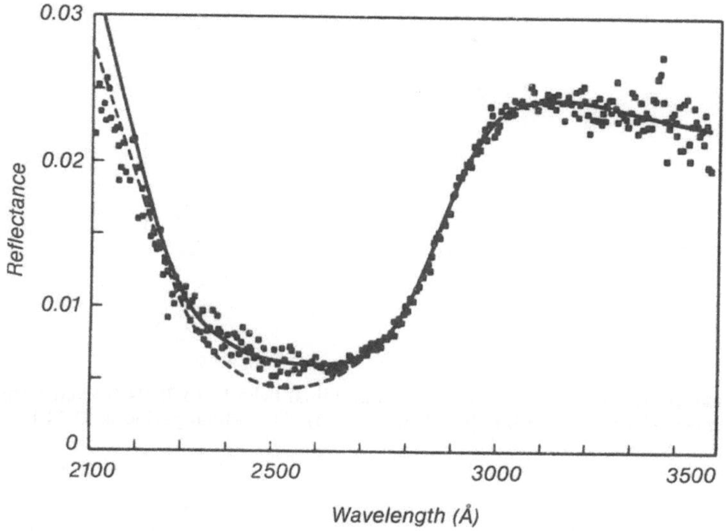

Fig. 5. A Martian spectrum measured at subpolar latitude (*dots*). *Solid line* is the best fit to these data using the ozone scale height as a parameter, *dashed line* is that for the homogeneous model (Wehrbein et al. 1979)

1971). Much more detailed measurements were made by the Mariner 9 orbiter. A monotonic brightness increase with decreasing wavelength in the range of 250 – 360 nm was observed at low and middle latitudes. This effect is due to atmospheric gas and aerosol scattering, no traces of ozone were found here. These spectra were used as comparison spectra rather than solar spectra to eliminate calibration errors and errors resulting from the difference in spectral resolution of solar and Martian spectra.

According to the measurements (Lane et al. 1973, Barth et al. 1973), the ozone amount exceeds the detection threshold only in the polar caps and polar hoods. The latter cover with clouds, fogs, and hazes much of the winter hemisphere at latitudes $\varphi > 45°$. Figure 5 shows a typical spectrum in this region. The polar hood begins to form in fall and disappears in spring. The ozone amounts undergo great seasonal, latitudinal, and local variations here.

Figures 6 and 7 give the seasonal variations of maximum ozone amounts measured by Mariner 9 in the polar caps and in a latitude belt of $\varphi = 50° - 75°$. Ozone appears only towards the end of summer at $\varphi = 60°$. At that time the polar hood haze is not detected by TV cameras, though a significant increase of the UV albedo at 305 nm is observed. One can observe hazes and clouds in TV images late in fall, with the ozone amounts being as high as 30 μm. The highest amounts (up to 60 μm) are observed in winter and are very variable from place to place. In spring, ozone abundance is smaller than in winter, and the polar hood gradually disappears.

Ozone variations in the polar caps are similar to those in the polar hoods (appearance at the end of summer, winter maximum, and disappearance at

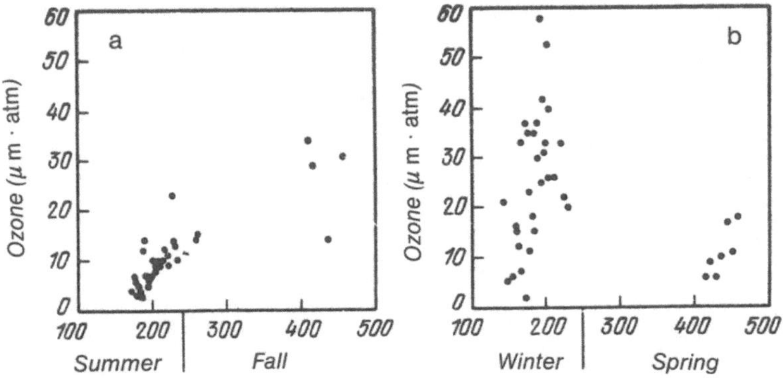

Fig. 6a, b. Maximal ozone amounts measured in the latitudinal belts 50 to 75°N (**a**) and S (**b**). *Abscissa* gives number of Mariner 9 orbits (Barth et al. 1973). The orbital period is 12.34 h

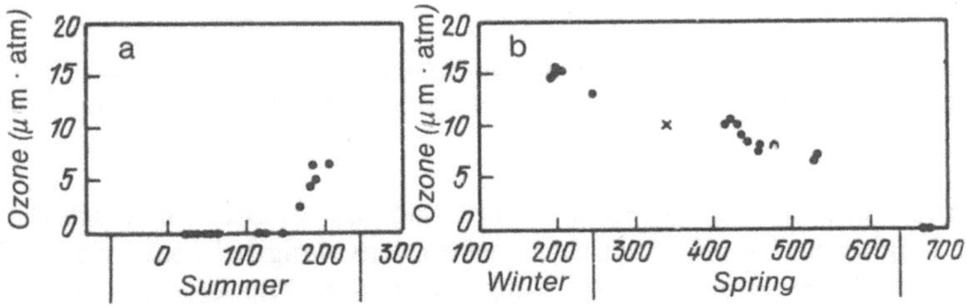

Fig. 7a, b. Ozone at polar caps (see Fig. 6)

the beginning of summer). However, these variations seem to be more regular and less local. The greatest variations were observed in the polar hoods, rather than in the polar caps.

A more detailed analysis of observations was done by Wehrbein et al. (1979) to derive both the ozone amount and H_0/H_s, the ratio of the ozone scale height to that of the scattering atmosphere. The atmosphere was assumed to be a conservative scattering medium with the scale height, H_s, optical thickness $\tau_\lambda = \tau(\lambda_0/\lambda)^k$, and Rayleigh phase function. The Lambertian surface ($\beta = 1$) with the albedo A was assumed. Ozone was given by the optical thickness, τ_0, at 255 nm and the scale height, H_0. Thus, the planetary reflection is defined by five parameters: τ, k, A, τ_0, H_0/H_s. In our opinion some of these assumptions are controversial. This refers first of all to the exponential description of the dependence of the scattering coefficient on height. In the polar caps and hoods aerosol formation at water condensation is of great importance. This aerosol should not be mixed uniformly in the atmosphere and forms a layer at some altitudes. Under the assumptions made, the ozone layer should form at the altitudes of the aerosol layer, which is not always the case. But this approach seems to be the best in the absence of data

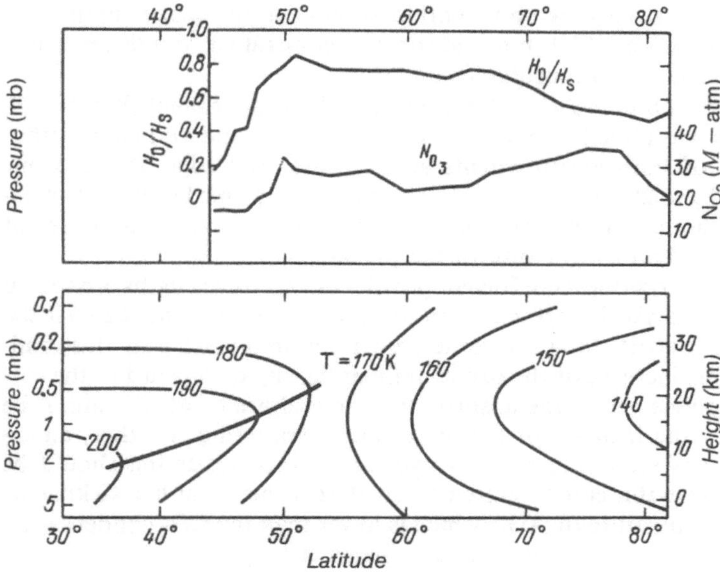

Fig. 8. Parameters of ozone distributions for orbits 202 in the northern hemisphere and the atmospheric temperature structure from infrared spectroscopic data measured at the same orbit. *Heavy line* indicates top of inversion layer (Wehrbein et al. 1979)

on the aerosol altitude distribution. It provides better agreement between calculated and measured spectra as compared to the case of a uniformly mixed atmosphere (Fig. 5).

The 103 spectra measured several days before the vernal equinox in the northern hemisphere were processed by this method. The spectra show almost equal ratios $H_0/H_s \simeq 0.6$ at the northern and southern polar regions ($\varphi > 61°$). Comparison of the ozone parameters and those of the atmospheric thermal structure for one orbit is shown in Fig. 8. At that time, late in winter, a strong temperature inversion occurs at middle latitudes. The atmospheric temperature has a maximum at 10−20 km and decreases with decreasing altitude, so ozone must be found mostly near the cold surface with more dry air than in warm, higher layers. At latitudes $\varphi > 55°$, a temperature minimum replaces the maximum, which is accompanied by simultaneous increase in the ozone scale height.

At high latitudes the temperature is so low that the ozone amount is defined by molecular hydrogen rather than water vapor and ceases to be temperature dependent there. The small scale height of ozone at middle latitudes is due to the fact that odd hydrogen (see Chap. 2.7) is not uniformly mixed in the atmosphere and its mixing ratio along with temperature increases with height. Hence, a mixing time is equal to or exceeds the photochemical lifetime of odd hydrogen (~3 days at middle latitudes).

Comparison of ozone data on two orbits shows that the ozone distribution is not a simple function of latitude and surface temperature, but depends in a complex way on several atmospheric conditions.

Measurements on board the Mars 5 orbiter. This experiment (Krasnopolsky et al. 1979) was aimed at obtaining ozone amounts and vertical distribution by means of limb measurements at the center and edge of the ozone absorption (260 and 280 nm). This task seemed to be feasible because, as mentioned above, the detection limit for limb observations was by an order of magnitude better than for the disk of the planet.

The measured brightness coefficient profiles at two wavelengths are shown in Fig. 9. Their analysis is performed on the assumption that the R260/R280 ratio minimum is attributable to ozone absorption in a layer near this minimum. Monotonic increase of the ozone number density downward to the surface should produce very large absorption at low altitudes which cannot be compensated by the increase of the atmospheric scattering. Further, an assumption that gas scattering prevails over an aerosol scattering should be ruled out because in this case an extremely small scale height at $h > 40$ km corresponds to a temperature of 90 K, which is lower than the CO_2 condensation temperature. Thus, aerosol scattering dominates at $h \gtrsim 40$ km.

The coincidence of the 260 and 280 nm brightness coefficients at $h > 40$ km denotes the similarity of optical properties of aerosol particles at these wavelengths. At $h \lesssim 20$ km where ozone absorption occurs this coincidence shows the single scattering albedo, ω, to be higher at 260 nm than at 280 nm. The narrow region of absorption indicates a sharp change in the aerosol properties at altitudes of 20 to 30 km. Calculations agree reasonably well with the observations only for two aerosol layers of different nature (see Fig. 10 and Table 1). The measurement latitude is 15°S, local time is 09.30 h.

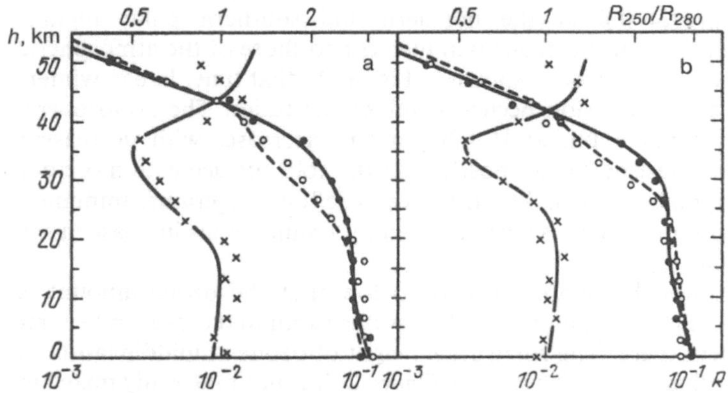

Fig. 9a, b. Photometric profiles of atmospheric brightness coefficients measured by two Mars 5 limb crossings. *Circles* and *black points* refer to 260 and 280 nm, respectively, *crosses* are their ratio, *lines* represent calculations (Krasnopolsky et al. 1979)

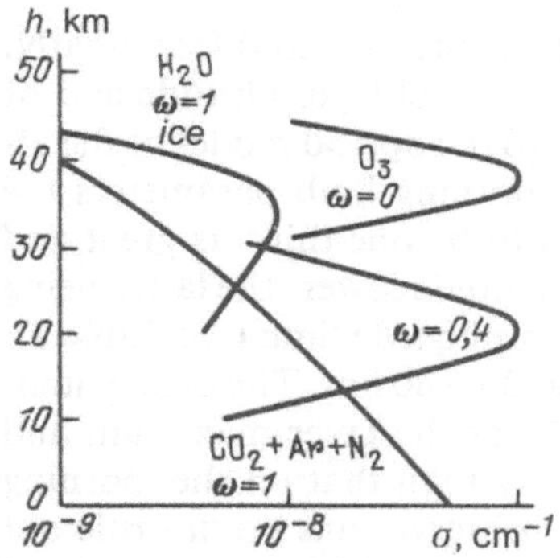

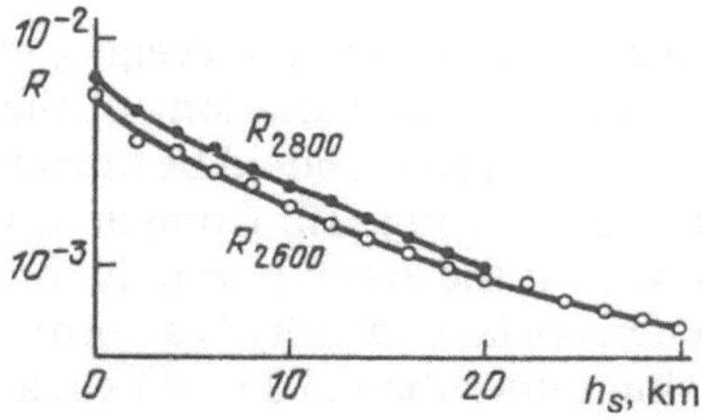

Fig. 11. Atmospheric brightness coefficient, R, vs shadow height, h_s, near evening terminator. *Dots* are measurements at two wavelengths, *lines* are calculations (Krasnopolsky et al. 1979)

Fig. 10. Optical model of the Martian morning side atmosphere at low latitudes for $\lambda = 260$ nm (Krasnopolsky et al. 1979) σ is the extinction coefficient

Table 1. Optical model atmosphere of Mars (Krasnopolsky et al. 1979)

Parameter[a]	Morning side limb, 09.30 h			Evening terminator, 18.00 h
	25. 2. 1974	27. 2. 1974	Mean values	
Ozone layer				
$[O_3]_{max}$, cm^{-3}	10^{10}	10^{10}	$(10\pm1.5)\times10^9$	$(3\pm1.5)\times10^9$
h_0, km	37	39	35 – 40	–
l, km	8	7	7.5 ± 2	–
Upper aerosol layer				
A, cm^{-1}	1.3×10^{-8}	1.3×10^{-8}	$(1.3\pm0.2)\times10^{-8}$	$(1.3\pm0.5)\times10^{-9}$
h_0, km	37	39	35 – 40	60 – 80
H_1, km	2.5	2.2	2.3 ± 0.2	–
H_2, km	15	15	15 ± 5	–
Lower aerosol layer				
A, cm^{-1}	2×10^{-7}	2×10^{-7}	$(2^{+2}_{-1})\times10^{-7}$	–
h_0, km	18	25	22 ± 4	–
l, km	10	10	10 ± 3	–
ω_{280}	0.25	0.25	0.25 ± 0.05	–
ω_{260}	0.4	0.4	$0.4^{+0.2}_{-0.1}$	–

[a] $O_{3\,max}$ is the ozone number density at the layer minimum, h_0 is the altitude of the layer maximum, l is the layer thickness, A is the extinction coefficient at the layer maximum, H_1 and H_2 are the upward and downward slopes of the upper aerosol layer, ω_{280} and ω_{260} are the single scattering albedo for the wavelengths shown

Measurements near the evening terminator allow the atmospheric structure to be defined as well. Figure 11 shows the dependence of the brightness coefficient on a shadow height, $h_s = R\beta^2/2$, where R is the planetary radius and β is the solar depression angle. Determining atmospheric structure from the terminator measurements is more ambiguous and sensitive to experimental errors as compared to limb measurements. Properties of the atmosphere near the morning side limb and the evening terminator turn out to differ greatly. To fit the observations, variations of the upper aerosol layer altitude and of the densities of all layers should be introduced into the optical model of the atmosphere shown in Fig. 10. Compared to the morning limb parameters the ozone amount at the evening terminator proves to be one-third as great and the upper aerosol layer density is an order of magnitude lower, the latter being at 50 – 70 km rather than at 35 – 40 km at the morningside limb (see Table 1).

The low latitude ozone layer is thus located at 35 – 40 km. The ozone number density in the evening is $(3 \pm 1.5) \times 10^9\,cm^{-3}$ at the layer maximum and closer to the day side equilibrium number density than that on the morning side limb, which is higher by a factor of 3. The observations do not rule out comparable ozone amounts in the lower troposphere which are undetected due to high atmospheric optical thickness on the planetary limb. The measured ozone number density is in accordance with an upper limit of $8 \times 10^{15}\,cm^{-2}$ deduced from the low latitude observations by Mariner 9.

O_2 *1.27 μm airglow*. High resolution near infrared spectroscopic measurements (Noxon et al. 1976, Traub et al. 1979) make it possible to detect emission of rotational lines of an $O_2\ (a^1\Delta_g - X^3\Sigma_g^+)$ band in the Martian spectrum. These lines were Doppler shifted from the telluric absorption lines of the same band in accordance with the relative velocity of Mars (see Fig. 12). The measurements were made by Connes with a 5 m telescope on Mount Palomar in 1973 and 1975. The data obtained in 1975 refer to the northern, southern ($\varphi > 45°$), and equatorial ($\pm 20°$) regions in winter in the northern hemisphere of Mars.

Excitation of the 1.27 μm airglow occurs at ozone photolysis

$$O_3 + h\nu \rightarrow O_2(a^1\Delta_g) + O(^1D), \quad I = 4 \times 10^{-3}\,s^{-1}.$$

The volume emission rate is

$$i = \frac{I[O_3]}{1 + \tau k n}$$

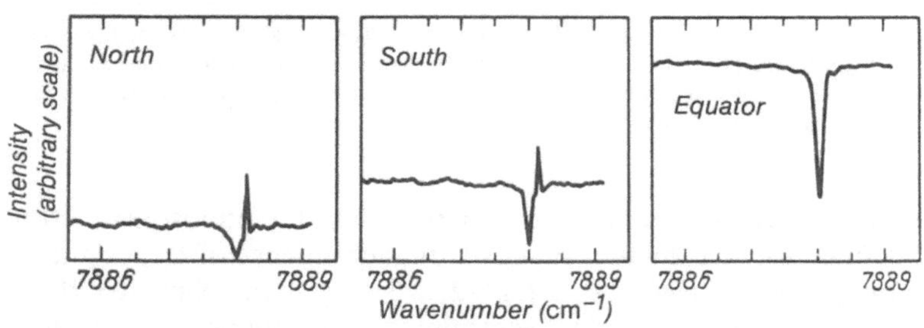

Fig. 12. Parts of high resolution spectra near a line of an O_2 1.27 μm band. Spectra consist of continuum radiation reflected by the surface of Mars, a broad telluric absorption line, and Martian Doppler shifted emission line (Traub et al. 1979)

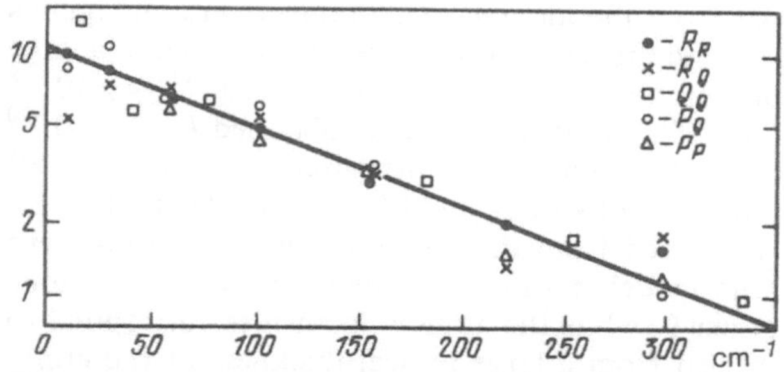

Fig. 13. Relative populations of upper rotational levels of the O_2 1.27 μm band vs their energy in cm^{-1}. Relative populations are proportional to the line intensity to line strength ratio; *different symbols* refer to different branches of the band. The average slope of the curve results in a rotational temperature of 200 K (Noxon et al. 1976)

Table 2. 1.27 μm airglow intensities and rotational temperatures on Mars (Traub et al. 1979)

Region	Band equivalent width (A)	Continuum brightness (MR/A)	Airglow intensity (MR)	Rotational temperature	Temperatures from Kong and McElroy (1977a)
North	2.42 ± 0.48	10.9	26 ± 5	152 ± 8	147
South	1.05 ± 0.19	10.6	11 ± 2	179 ± 10	170
Equator	0.21 ± 0.06	12.9	3 ± 1	211 ± 21	195

where $\tau = 1$ h is the radiative lifetime, n and k are CO_2 number density and quenching rate coefficient, respectively. The chemical formula within brackets denotes the number density of the substance indicated.

Rotational temperatures and airglow intensities were derived from the spectra. In each spectrum, equivalent widths of about 30 lines were measured to be typically $(0.5-1)\times 10^{-2}$ nm. Their equivalent width to the line strength ratios plotted versus the energies of upper rotational states form a straight line with a slope proportional to rotational temperature (Fig. 13). As the lifetime of $O_2(a^1\Delta_g)$ is large, this temperature is equal to the real atmospheric temperature averaged over the airglow altitude profile. The band intensity was taken as the sum of 39 lines with a correction for weak lines difficult to measure. The derived equivalent widths of the band were converted into band intensities by multiplying by continuum radiation reflected from the Martian surface. The results are shown in Table 2. One usually defines airglow intensities in Rayleighs[1], Kilorayleighs, and Megarayleighs (R, kR, MR).

[1] The Rayleigh unit (R) is a measure of the omnidirectional emission rate in a column of unit cross-section along the ling of sight, with 1 R ≡ 10^6 photons $cm^{-2}\,s^{-1}$

Ozone amounts derived from the airglow measurements are ambiguous due to the necessity of assuming the ozone altitude profile and due to a large uncertainty in the CO_2 quenching rate coefficient, $k = 1.5 \times 10^{-20} - 4 \times 10^{-18}$ $cm^3 s^{-1}$, see Noxon et al. (1976). Traub et al. (1979) obtained $k = 3 \times 10^{-20}$ $cm^3 s^{-1}$ by comparison of the 1.27 μm airglow measurements with a global ozone distribution model calculated by Kong and McElroy 1977a (see below). They explain the higher k derived from the measurements at the northern regions by the fact that the model is valid for maximum values of ozone amounts observed by Mariner 9, while the airglow intensities correspond to mean values. It may stem also from a large optical thickness of the atmospheric ozone in the winter hemisphere which decreases ozone photolysis and the O_2 $(a^1\Delta_g)$ formation rate.

The ozone altitude profile measured by Mars 5 at the evening terminator gives an airglow intensity of (5 ± 2.5) MR for $k = 3 \times 10^{-20}$ $cm^3 s^{-1}$ which is in accordance with the airglow observations within the experimental errors.

1.4 Water Vapor

Ground-based observations. After searching for water vapor over a long period of time, it was detected in the Martian atmosphere from the very weak absorption lines of the 820 nm band which were Doppler shifted from the telluric lines at the time of observation (Kaplan et al. 1964). It took a resolution power of $\sim 10^5$ to measure such lines. The 820 nm band does not belong among the strongest H_2O bands. However, the latter's measurements give no advantage due to an increase in telluric absorption.

Since then, ground-based measurements of water vapor have been repeatedly carried out. One of them was made by Barker (1976) with a 2.7 m telescope of the McDonald observatory with an echelle spectrometer having a resolution power of 2.7×10^5. With the instrument slit parallel to morning and evening terminators or to the equator during different seasons, Barker measured diurnal, latitudinal, and seasonal variations of H_2O amounts. The unit of water vapor column abundance is expressed in micrometer of the precipitable water layer and equals 3.3×10^{18} cm^{-2} of H_2O molecules. As an example, mean H_2O abundance at a different local time in fall $(170° < L_s < 269°)$ and vernal equinox $(332° < L_s < 5°)$ is shown in Fig. 14. A solar aerocentric longitude, L_s, equal to 0° and 180° at vernal and autumnal equinoxes, respectively, defines a season on Mars thus taking the place of months and dates for the Earth. Among other H_2O ground-based spectroscopic data, the H_2O rotational temperature should be noted to be 225 K (Tull 1970) and to exceed a mean atmospheric temperature of 204 K (Young 1971), thus indicating the excess amounts of water vapor in the warm low lying atmosphere.

Together with ozone, water vapor is a very variable atmospheric constituent. Yet, the ozone variability is of photochemical nature, while the H_2O

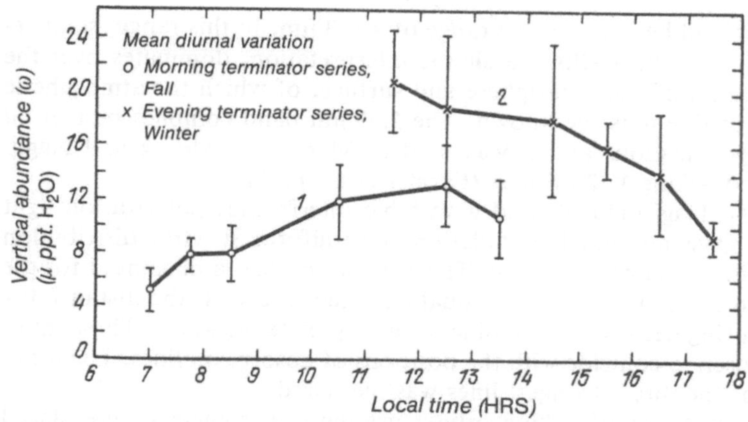

Fig. 14. Diurnal variations of water vapor amounts measured by ground-based spectroscopy. Local time represents positions of the instrument slit parallel to the morning and evening terminators. Slit sizes are (1 ×0.2) *R*, *R* is the planetary radius (Barker 1976)

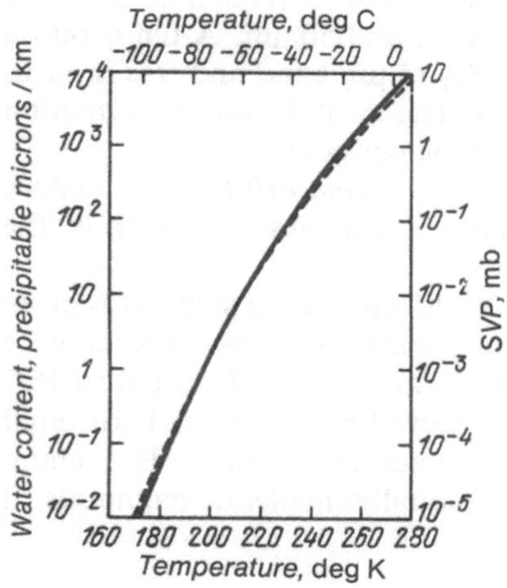

Fig. 15. Saturated water vapor density (*dashed line*) and pressure (*solid line*) vs temperature

variability results from a very strong dependence of H_2O saturated vapor pressure on temperature. With temperature variations from 145 to 270 K typical on Mars, H_2O saturated vapor pressure varies by a factor of 10^8 (Fig. 15). To now H_2O amounts in excess of 100 μm have not been observed on Mars with a detection threshold of 1 μm for orbiter observations. Thus, the H_2O amounts observed correspond to the condition of saturation at 190–220 K.

Orbiter measurements. Water vapor was investigated on board all Martian orbiters. Orbiters have the advantage of avoiding telluric H_2O absorption and of receiving much more detailed H_2O spatial distribution and variations. The absorption band to be used (Moroz 1978) should be strong, free of CO_2

bands, and should be in a spectral range of $\lambda < 3$ µm. In this range, solar reflected radiation, which allows a simple interpretation, dominates over the thermal radiation of the atmosphere and surface, of which the atmospheric temperature profile must be known. The 1.38 µm band complies best of all with these specifications and it was used at Mars 3, 5 (Moroz and Nagip 1975a, b), and Viking 1, 2 orbiters (Farmer et al. 1977).

Moroz and Nagip (1975) used a variable interference polarization light filter which gave no signal modulation for uniform spectral distribution within the filter band pass of 5 nm. The modulation had a maximum for the case when absorption occurred in equally spaced intervals, the distance between them being defined by a double refracting plate thickness. These intervals were chosen to coincide with the positions of absorption lines. The mean absorption of the three strongest lines was measured.

Five PbS detectors of Viking orbiter grating spectrometers were placed within and between the same three lines. A scanning system provided measurements in a band of 45 km along the orbit projections with a spatial resolution of 3 km.

An infrared Michelson interferometer on Mariner 9 (Hanel et al. 1972, Conrath et al. 1973) recorded spectra between 5 and 50 µm. A temperature profile was deduced from the shape of the CO_2 15 µm band and was in turn, used for H_2O 25 – 50 µm band interpretation. The most detailed information was obtained by Viking 1,2. Their results will be considered below.

The results of H_2O *measurements on board Viking orbiters.* To deduce water vapor amounts from the measurements, the curves of growth of the lines were calculated (Fig. 16).

The latitude distribution of water vapor at different L_s in summer is given in Fig. 17a. To reveal the variations, the results of measurements were averaged over time intervals corresponding to variations in L_s equal to 15°. Maximum abundances of ~100 µm were observed near the northern polar cap soon after the summer solstice. Late in summer the maximum H_2O abundance moves toward the equator, becoming gradually smaller in magnitude. It

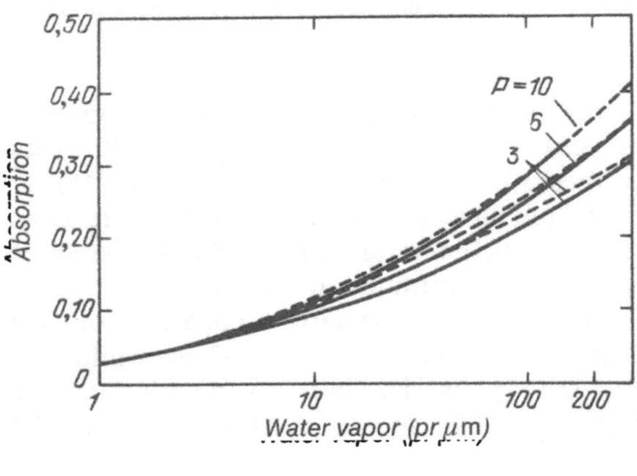

Fig. 16. Absorption at the center of 7232.2 cm^{-1} line vs water vapor amount for different atmospheric pressures (10, 6, 3 mbar) and temperature of 200 K (*solid lines*) and 250 K (*dashed lines*) (Farmer et al. 1977)

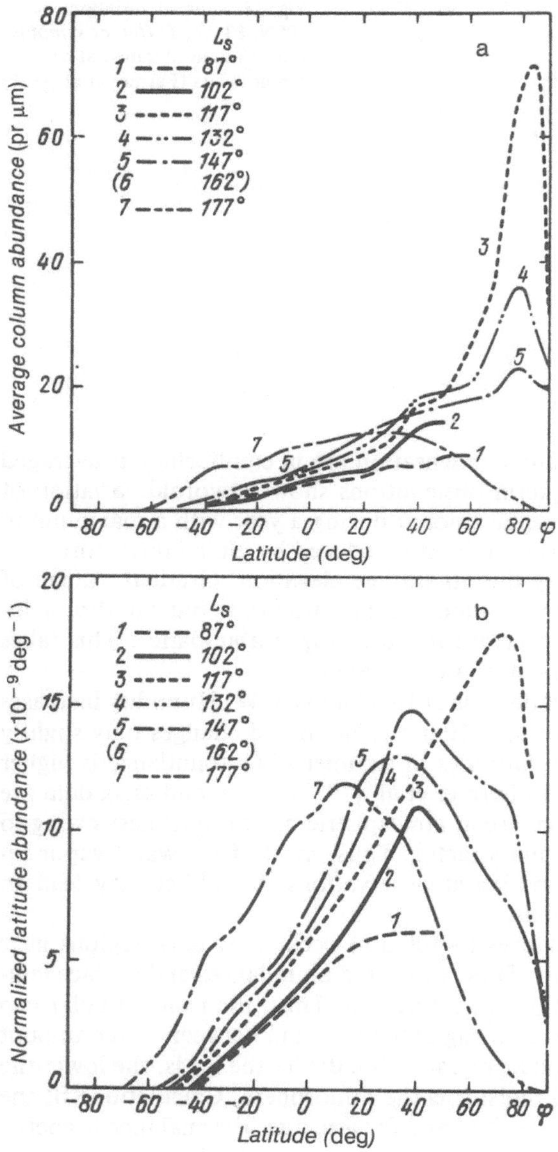

Fig. 17. **a** Latitude distribution of water vapor for various seasons; **b** seasonal redistribution of the water vapor shown in terms of the latitude-normalized abundances vs latitude (Farmer et al. 1977)

is believed that the latitudinal distribution is symmetric near the equator 2 weeks after the autumnal equinox. The curves in Fig. 17a can readily yield data on the total water amount inside various latitude belts (Fig. 17b). The observations during the first summer indicate that the total amount of water vapor in the atmosphere, defined as areas under the curves of Fig. 17b, is independent of L_s and equal to 1.3 km^3 of water ice. This means that at least

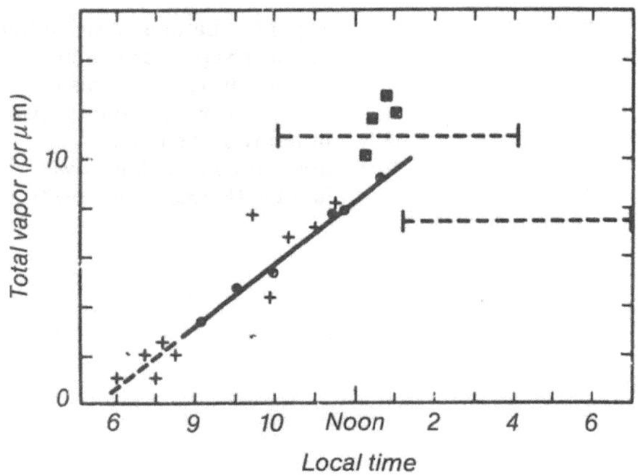

Fig. 18. Diurnal variations at 10°N, 83°W. *Different symbols* refer to data obtained at different orbits (Farmer et al. 1977)

during summer sublimation and condensation are in equilibrium as averaged over the entire planet. Long-term observations show a twofold variation of the total atmospheric water vapor amount during a year with a maximum in summer. In winter and fall, this amount is reduced by global dust storms.

Local variations are partly due to surface elevation. Gradual change of surface elevation does not entail water vapor variations, but an abrupt increase in elevation causes a decrease in water vapor abundance. This takes place, for instance, near mountains and volcanoes.

Figure 18 shows diurnal variations at 10°N and 83°W. Humidity increases linearly from ~1 μm at dawn to ~10 μm at noon and changes only slightly till dusk. At middle northern latitudes in summer H_2O abundance is higher and less variable. According to Farmer et al. (1977), dawn and dusk data are underestimated due to the increase in atmospheric optical thickness owing to condensation and a large air mass factor. Thus, most of the water vapor remains undetectable. In ground-based observations this effect may lead to overestimated diurnal variations.

Maximum column abundances (~100 μm) occur over dark regions near the residual summer polar cap. These measured abundances and surface temperatures indicate the conditions of saturation. Thus, the residual polar cap consists of water ice. There is a strong anticorrelation between water amount and surface albedo in some polar regions. The dirtier the ice is, the lower the albedo, and the warmer and moister is the atmosphere. Calculations fit the observations for an ice albedo of 0.34 and for a surface thermal inertia coefficient equal to that of water ice. Water vapor amounts in the polar areas are strongly influenced by the previous few years' dust storm activity which defines surface ice albedo. If conditions in the southern polar cap are the same as in the north and if under these conditions water vapor saturation is reached, the H_2O amounts should equal ~500 μm due to higher temperatures of the southern hemisphere in summer. But maximum amounts in southern polar regions turned out to be an order of magnitude lower than in northern ones, thus indicating that the southern polar cap does not consist of water ice.

Water vapor observations on board Viking orbiters were reproduced by a model (Davies 1981) based on surface temperature measurements. This model explains latitudinal and seasonal variations of water vapor by condensation in the polar regions. There are no permanent ice reservoirs at the middle and low latitudes. Maximum water ice amounts constitute 50 μm in the northern polar cap and a half of this value in the southern one, i.e., CO_2 is much more abundant than H_2O in the polar caps. It is interesting that the seasonal asymmetry of humidity produces water vapor meridional transport which is much more intense and accompanied by dust storms in the southern summer.

Numerous orbiter measurements of water vapor above the Viking lander site with different angular conditions have revealed for one occasion the vertical distribution of water vapor (Davies 1979). The observations give dependences of water vapor amounts on air mass for phase angles, i.e., angles between optical axis and direction to the Sun, of ~130° and ~80°. These dependences are in good agreement with calculations for an atmospheric optical thickness of $\tau \simeq 0.5$, an anisotropy factor of scattering of ~0.25 (the ratio of forward to isotropic scattering), a water vapor amount of 11 μm, and a homogeneous mixture of water vapor and dust in the lower scale height. This method yields water vapor vertical profiles relative to those of dust and thus cannot give absolute values of H_2O number densities. So despite many measurements of water vapor column abundances, there is hardly any vertical abundance profile available.

1.5 Composition of the Upper Atmosphere as Determined from Airglow Spectroscopy

Studies of the structure and composition of the Martian upper atmosphere by means of dayglow spectroscopy in a range of 110–400 nm (Fig. 19) were carried out on board Mariner 6, 7, 9 space probes (Barth et al. 1971, 1972, Stewart et al. 1972). Hydrogen 121.6 nm and oxygen 130 nm airglow was measured also by Mars 3 and 5 orbiters (Dementyeva et al. 1972, Kurt et al. 1974, Bertaux et al. 1975). An attempt to measure nightglow spectra in the visible range was undertaken by Krasnopolsky and Krysko (1976) with the Mars 5 orbiter.

Carbon dioxide. The analysis of excitation processes of UV dayglow emissions indicates their being produced almost completely by photodissociation, photoionization, and electron impact of CO_2 (Barth et al. 1971). This means that CO_2 is a predominant component of the upper atmosphere up to ~250 km where the spectra shown in Fig. 19 were measured.

The CO_2 number density profile was deduced by Krasnopolsky (1975) from CO_2^+ bands measured with Mariner 6 and 7 UV spectrometers. Vibra-

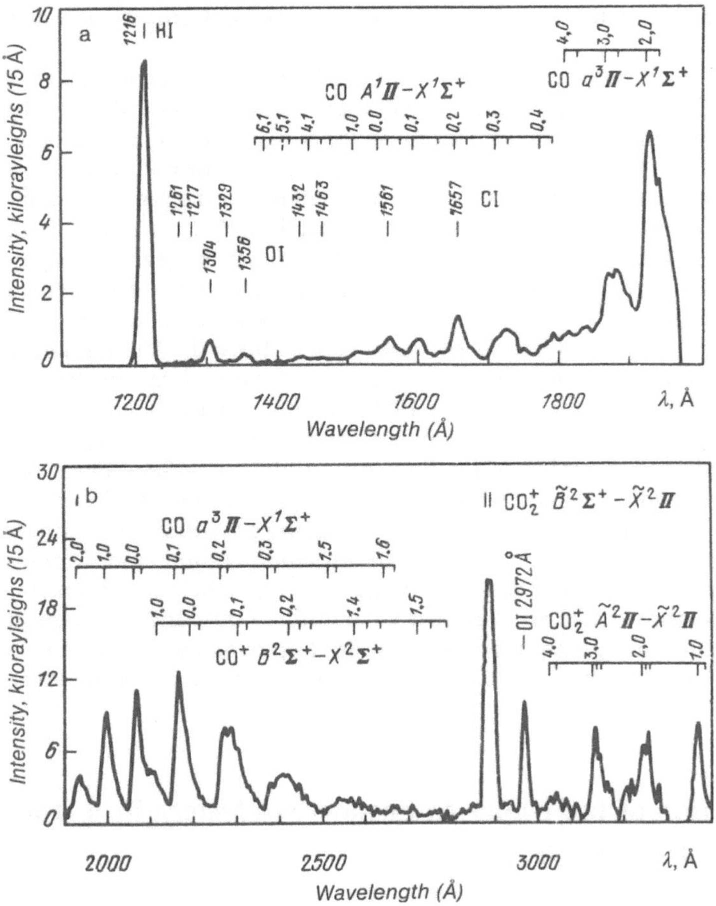

Fig. 19a, b. Martian dayglow spectra averaged on 120 limb crossings by Mariner 9. Positions of H, O, C lines and CO and CO_2 bands are shown (Barth et al. 1972)

tional distribution of the bands differs for the two main excitation processes: CO_2 photoionization and fluorescent scattering of CO_2^+. This makes it possible to separate contributions of these processes and to derive number densities of both components. By a similar technique CO_2 number densities were obtained from electron number densities, n_e, deduced from radio-occultation measurements (Fjeldbo et al. 1970) on the same probes. In this case the photoionization rate is given as $q = \alpha n_e^2$. The CO_2 number densities calculated with both the airglow and the radio-occultation data are in a good agreement (Fig. 20).

Atomic oxygen amounts can be derived from 130 nm airglow measurements and from the CO_2^+ number densities known. Analysis of Mariner

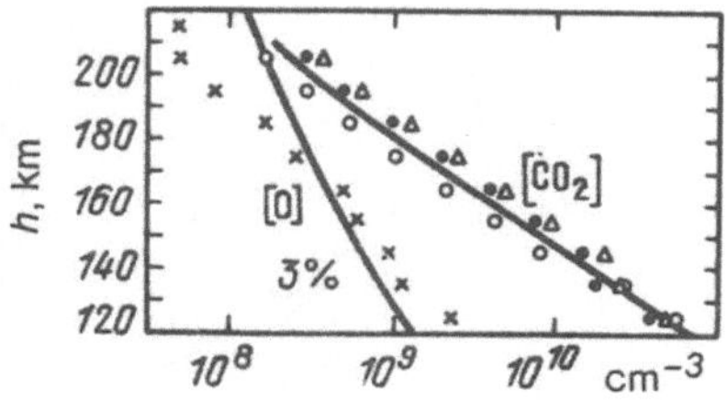

Fig. 20. CO_2 and O number densities vs height. CO_2 number densities were derived from the CO_2^+ dayglow (*points* and *triangles*) and from electron number densities (*circles*). Atomic oxygen number densities (*crosses*) were obtained from the CO_2^+ bands. *Lines* give mean profiles (Krasnopolsky 1975)

130 nm data was made by Strickland et al. (1972, 1973). At altitudes $h > 250$ km resonant scattering of atomic oxygen prevails over other numerous excitation mechanisms. Calculations utilize an atmospheric model for 100 to 700 km which is defined by the thermospheric temperature T_∞ and the atomic oxygen mixing ratio, $f_O = [O]/[CO_2]$ at an ionospheric maximum at 135 km. The absorption of CO_2 at 130 nm is neglected and superseded by the assumption of a black boundary at 100 km, where $\tau_{CO_2} \simeq 1$.

Next, radiative transfer calculations with the approximation of a plane-parallel optically thick atmosphere are carried out for the chosen atmospheric parameters to give the volume emission rate $i(h)$. After that, integrating along the optical axis, with the spheric structure and atmospheric absorption taken into account, yields the airglow intensity at limb crossing. The airglow profile obtained is multiplied by a third parameter, c, taking into consideration calibration errors and uncertainty in the measured solar 130 nm line flux. Least squares fitting to the measured airglow profile in Fig. 21 is shown in Fig. 22. The best fit gives $f_O \simeq 1\%$ and $c \simeq 4$ for $T_\infty = 300$ K according to Mariner 9 measurements. This analysis indicates large local variations of f_O from 0.1 to 10%.

Mars 3 orbiter filter measurements (Dementyeva et al. 1972) had three observation modes. The first utilized a LiF window with a sensitivity range of 105 – 134 nm. The second used a CaF_2 window with a range of 123 – 134 nm, and the third (the closed window) was intended for measuring cosmic ray background. Due to a large viewing field, limb measurements were poor and calculations were made to reproduce disk and limb measurements with the size of the viewing field taken into account. Assuming the presence of only 121.6 and 130 mm lines these calculations gave $f_O = (1.5 - 6)\%$ at 135 km.

Formation and destruction of CO_2^+ ions occur according to the following reactions:

$$
\begin{aligned}
CO_2 + h\nu &\rightarrow CO_2^+ + e^- \\
CO_2^+ + O &\rightarrow O_2^+ + CO \\
& O^+ + CO_2 \\
O^+ + CO_2 &\rightarrow O_2^+ + CO \\
CO_2^+ + e &\rightarrow CO + O \\
O_2^+ + e &\rightarrow O + O .
\end{aligned}
$$

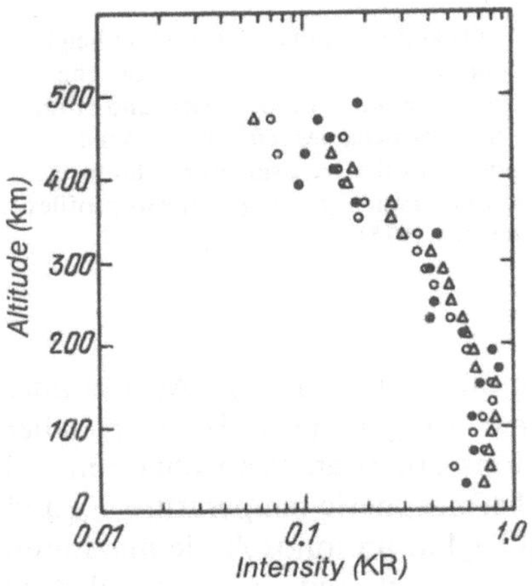

▲ **Fig. 21.** Atomic oxygen 130 nm dayglow measured at different limb crossing by Mariner 9 (Strickland et al. 1973)

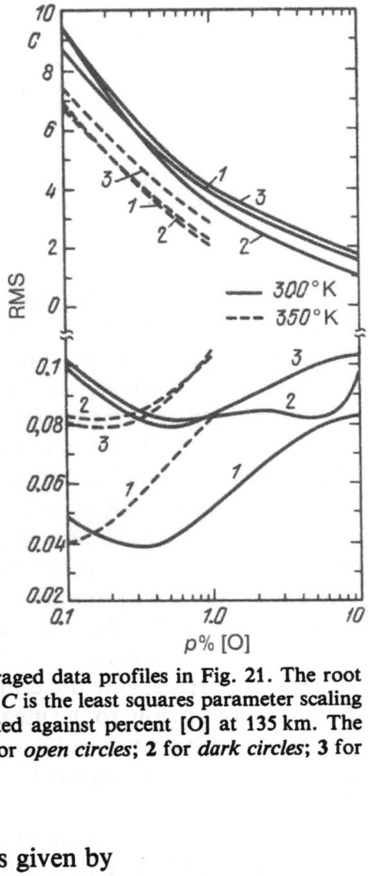

Fig. 22. Results of least squares fitting to the three averaged data profiles in Fig. 21. The root mean square (rms) error indicates the quality of fit, and *C* is the least squares parameter scaling the theoretical intensity. The *rms* error and *C* are plotted against percent [O] at 135 km. The *numbers of curves* correspond to symbols in Fig. 21: **1** for *open circles*; **2** for *dark circles*; **3** for *triangles* (Strickland et al. 1973)

Then the atomic oxygen number density is given by

$$[O] = \frac{\alpha_{O_2^+} n_e}{k} \left(\frac{n_e}{[CO_2^+]} - 1 \right)$$

where $\alpha_{O_2^+} = 1.6 \times 10^{-7} (300/T_e)^{1/2}$ cm^3 s^{-1} is the dissociative recombination coefficient of O_2^+, $k = 2.6 \times 10^{-10}$ cm^3 s^{-1} is the sum of the rate coefficients of both reactions between CO_2^+ and O. The derived CO_2^+ number densities reveal an atomic oxygen mixing ratio of 3% at 135 km (Fig. 20). Comparing the measured and calculated intensities of the CO_2^+ bands, Stewart (1972) found $f_O = 2\%$.

Thus, one may assume that the mean atomic oxygen mixing ratio is equal to (1 – 3)% at the height of an ionospheric maximum at 135 km. Local variations of f_O are in a range of (0.1 – 10)%.

Carbon monoxide. The relative intensities of the fourth positive bands of CO ($A^1\Pi - X^1\Sigma^+$) with $v'' = 0$ in the Martian dayglow (Fig. 19) proved to be much lower than those of the laboratory spectra. It is attributable to CO self-absorption, and may be used for the CO number density calculations.

In the Martian spectra the self-absorption of the (1,0) 151 nm band constitutes 60% (transmission of 40%) at 150 km. It depends on how much rotational distribution of the "hot" emission band coincides with that of the "cold" absorption band. The value of self-absorption obtained permitted Thomas (1971) to find $f_{CO} = [CO]/[CO_2] \simeq 3 \times 10^{-3}$ at 135 km from the theoretical consideration of both band structures.

The emission band is hot because it is produced mainly by CO molecules which are formed at the photoelectron impact dissociation of CO_2. Their kinetic, vibrational, and rotational temperatures are high and very uncertain. To alleviate the problem, the self-absorption of the CO emissions produced by 30 eV electron impacts with CO_2 was measured (Mumma et al. 1975). To provide 60% attenuation of the hot (1,0) band, the line-of-sight CO column density must be 6×10^{15} cm^{-2} (Fig. 23). Hence, $[CO] = 6 \times 10^{15} (\pi/2\, RH)^{-1/2} = 1.5 \times 10^8$ cm^{-3} at 150 km according to Mumma et al. (1975), where R is the planetary radius and H is the CO scale height.

More precise calculations (Krasnopolsky 1981a) should take into account line-of-sight absorption and emission of CO, absorption of solar radiation by CO_2, and fluorescent scattering by cold CO. Some of these corrections change the result by a factor of 2, but being opposite in their effects, the final value of $[CO] = 1.2 \times 10^8$ cm^{-3} differs only slightly from that of Mumma et al. (1975). This value corresponds to $f_{CO} = 1\%$ at the ionospheric maximum.

Atomic hydrogen in the Martian upper atmosphere and exosphere can be determined by means of H 121.6 nm airglow measurements. These measurements were made on board Mariner 6, 7, 9 and Mars 3, 5 space probes.

Airglow altitude profiles at the planetary limb reveal an exospheric temperature T_∞ and a hydrogen number density at a critical level, where an atomic and molecular mean free path equals an atmospheric scale height. The latter means that atoms and molecules with velocities in excess of a parabolic one may leave the atmosphere without collisions. An altitude profile of an escaping

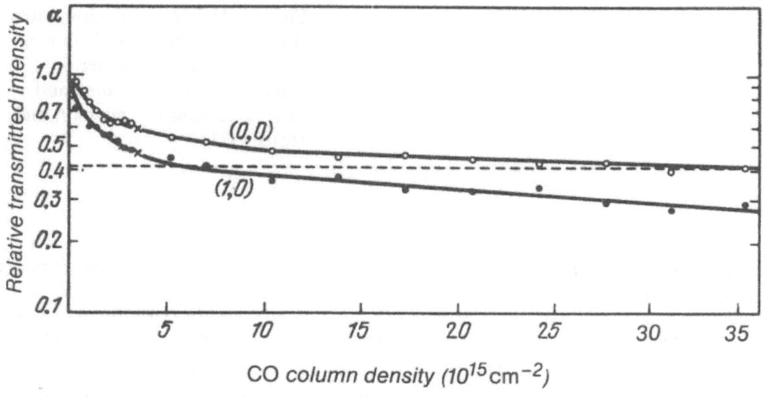

Fig. 23. Transmittance of hot CO (0,0) 155 nm and (1,0) 151 nm bands in cold carbon monoxide. These bands were excited by 30 eV electron impacts on CO_2 molecules (Mumma et al. 1975)

component above the critical level is determined unambiguously by T_∞ and n_c (Chamberlain 1963). The physical concepts underlying the altitude profile calculations are exospheric isothermicity, decrease of gravity acceleration with height, a flux of an escaping component, and the existence of three kinds of particle orbits (ballistic orbits with velocities at the critical level lower than the parabolic one, satellite orbits which do not intersect the critical level, and escaping orbits). For given T_∞ and n_c, hydrogen number density distribution was calculated to derive the altitude profile of H 121.6 nm airglow by means of the radiative transfer technique. Root mean square deviations of the calculated profiles from the measured one were at their lowest for $T_\infty = 350$ K, $n_c = 3 \times 10^4$, and 2.5×10^4 cm^{-3} for Mariner 6 and 7 measurements, respectively (Anderson and Hord 1971, Fig. 24). Hydrogen number densities versus height are given in Table 3. Vertical optical thickness of hydrogen equals unity at a critical level of 250 km.

Hydrogen distribution below the critical level is determined from planetary disk measurements. Anderson (1974) considered seven [H] model distributions over the range of 250 to 80 km (Fig. 25), where model 1 represents diffusive equilibrium without escape and model 7 assumes total hydrogen production below 80 km. Since maximum hydrogen production occurs near the ionospheric peak, all models except model 1 coincide above this peak. Radiative transfer calculations for Mariner 6 observations are shown in Fig. 26.

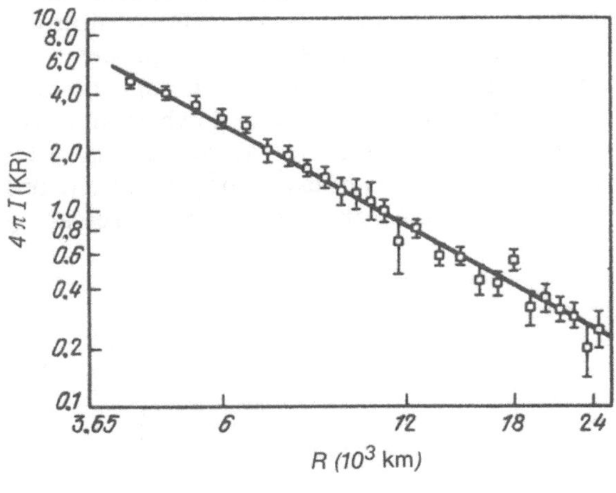

Fig. 24. H 122 nm airglow intensity vs planetocentric distance, *R*, measured by Mariner 6 (*squares with error bars*) and calculated (*line*) (Anderson and Hord 1971)

Table 3. Atomic hydrogen number densities deduced from Mariner 6 and 7 airglow measurements (Anderson and Hord 1971)

h, km	250	500	750	1000	2500
[H], 10^4 cm^{-3}	3.0	2.1	1.6	1.3	0.42
h, km	5000	7500	10000	12500	15000
[H], 10^4 cm^{-3}	0.13	0.058	0.032	0.020	0.014

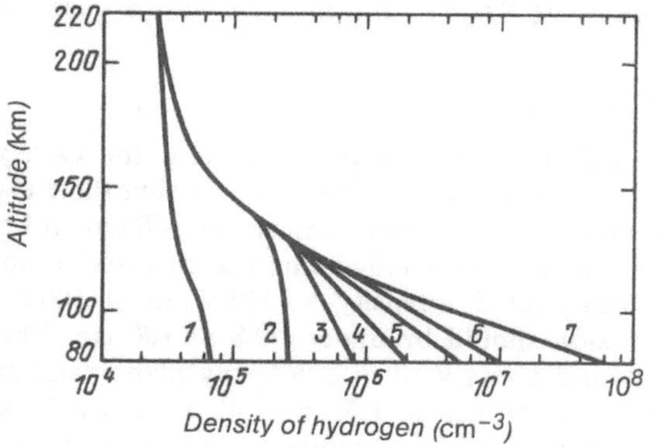

Fig. 25. Model distributions of atomic hydrogen at heights of 80 – 220 km (Anderson 1974)

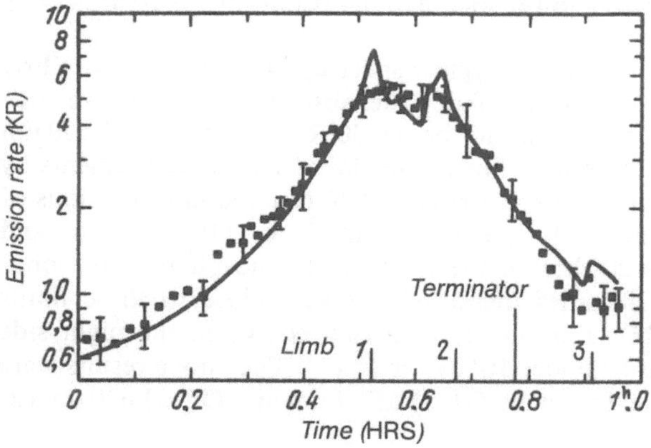

Fig. 26. 122 nm intensity variations caused by change of the line of sight position during Mariner 6 flyby. *Line* represents calculations with model 4 and $n_c = 2.5 \times 10^4\ cm^{-3}$, $T_\infty = 350$ K (Anderson 1974)

Model 4 with $\tau = 2.2^{+1.2}_{-0.6}$ and model 6 with $\tau = 5^{+5}_{-1}$ are the best to fit the Mariner 6, 7, and Mariner 9 data, respectively.

Similar interpretation of Mars 3 measurements (Dementyeva et al. 1972) gives $n_c = (1.85 \pm 0.5) \times 10^4\ cm^{-3}$ and $T_\infty = 315 \pm 30$ K (Dostovalov and Chuvahin 1973). As in the case of Mariner 6 and 7, the optical thickness at the critical level is unity. In the range of 250 – 100 km, $\tau = 2.3$, i.e., twice as small as that measured by Mariner 9.

Mariner 6 and 7 observations were carried out in 1969 during an increased level of solar activity ($F_{10.7\ cm} \simeq 180$); $F_{10.7\ cm} \simeq 120$ for Mariner 9 and Mars 3 measurements in 1971 – 1972 and $F_{10.7\ cm} \simeq 80$ during Mars 5 operations in 1974. Solar activity is measured in $10^{-22}\ W\ m^{-2}\ Hz^{-1}$ of 10.7 cm solar radiation intensity. According to Mars 5 data (Bertaux et al. 1975) $n_c = 2.3 \times 10^4\ cm^{-3}$, $T_\infty = 325 \pm 30$ K, $\tau = 1$ at 200 km and $\tau = 3$ from 87 to 200 km.

An estimation of 121.6 nm radiation at the night side limb of Mars was obtained by Krasnopolsky and Krysko (1976) from the absence of 557.7 nm nightglow. Its slant intensity did not exceed 50 *R* at 80 km. The CO_2 absorption of 121.6 nm radiation excites $O(^1S)$ 557.7 nm airglow with a yield of 15%:

$$CO_2 + h\nu(121.6\,\text{nm}) \rightarrow CO + O(^1S)\,.$$

The total absorption cross-section is $\sigma = 8\times10^{-20}\,\text{cm}^2$ and the $O(^1S)$ quenching rate coefficient by CO_2 is $k = 2.5\times10^{-14}\,\text{cm}^3\,\text{s}^{-1}$. Using the CO_2 model distribution, one can deduce the 121.6 nm intensity at 100 km to be under 50 *R* (there are some errors in the paper cited and here a corrected result is presented). Interplanetary background intensity at 121.6 nm is 500 *R*, hence, the atmospheric transmission should be under 10% at 100 km. The transmission of an isotropically and conservatively scattering plane-parallel system with a black low boundary is defined as $1.3/(\tau+1.4)$, thus, $\tau \gtrsim 9$. A comparison with a value of $\tau = 4$ on the day side obtained on the same satellite (see Bertaux et al. 1975) indicates high diurnal variations of hydrogen. Yet a Mariner data extrapolation to a lower solar activity results in $\tau \gtrsim 10$ under day side conditions as well.

Upper limits. From the Mariner dayglow spectra, Dalgarno and McElroy (1970) obtained $f_{N_2} < 5\%$ at 135 km from the absence of the N_2 Vegard-Kaplan band system having an intensity below 50 *R*. The 557.7 and 630 nm oxygen nightglow (an upper limit of line intensities of 2 *R*) measurements by the Mars 5 orbiter (Krasnopolsky and Krysko 1976) resulted in restrictions of $[O] < 10^{11}\,\text{cm}^{-3}$ at 80 km and $n_e \lesssim 7\times10^3\,\text{cm}^{-3}$ at 110–130 km and $n_e \lesssim 2\times10^3\,\text{cm}^{-3}$ at 150–200 km in the night side ionosphere. The ionospheric electron number densities measured are very close to these limits (Savich and Samovol 1976). Further electron flux energy in the night side ionosphere is calculated to be under $10^9\,\text{eV cm}^{-2}\,\text{s}^{-1}$. The above results were obtained with the use of the following $O(^1S)$ 557.7 nm and $O(^1D)$ 630 nm excitation mechanisms:

$$O + O + M \rightarrow O_2^* + M$$
$$O_2^* + O \rightarrow O_2 + O(^1S)$$
$$O_2^+ + e \rightarrow O(^1D,\ ^1S) + O(^3P)$$
$$CO_2 + e \rightarrow CO + O(^1S) + e\,.$$

It follows from the absence of hydroxyl OH bands and from the ozone excitation process:

$$H + O_3 \rightarrow OH^* + O_2$$

that $[O_3]\times[H] < 10^{15}\,\text{cm}^{-6}$ at 60 km.

The upper limits of the meteoritic atom ion abundances are deduced from the absence of corresponding emissions in the night and twilight airglow (Krasnopolsky 1978). It is of interest that sodium abundance is smaller than that in the terrestrial atmosphere by an order of magnitude. The Na 590 nm nightglow on Mars is more than two orders of magnitude weaker than that on the Earth.

1.6 Mass Spectrometric Measurements of Atmospheric Composition

A sample of an atmospheric gas can be ionized, for instance, by an electron beam. The behavior of ions in electric and magnetic fields depends on their mass to the electric charge ratio. This allows different types of electrostatic and magnetic analyzers to be developed for measuring the mass spectra of an atmospheric gas. From these spectra, the chemical composition of the atmosphere can be obtained. Mass spectrometry is a powerful and universal tool for investigations of the isotopic and chemical composition of planetary atmospheres. Mass spectrometric measurements were carried out in the upper atmosphere of Mars by Viking 1 and 2 entry probes at 115 – 200 km (Nier and McElroy 1977) and near the planetary surface (Owen et al. 1977).

Upper atmosphere. Figure 27 shows a typical mass spectrum at an altitude of 140 km. The 44, 22, and 12 mass peaks provide the CO_2 number density, inasmuch as the first and the second peaks are the result of one- and twofold ionization, respectively, and the third corresponds to dissociative ionization with C^+ formation. Other component contributions to these peaks should be ruled out because of the coincidence of the CO_2 number densities derived from these peaks by the results of laboratory calibration (Fig. 28). Similarly, argon-40 gives the 40 and 20 mass peaks.

The 28 peak is attributed to CO and N_2, and dissociative ionization of N_2 produces the 14 peak. Thus, CO and N_2 may be separated and the CO contribution to the 28 peak constitutes about two-thirds (Fig. 29). The 30 mass peak can include NO and $^{12}C^{18}O$. The C and O isotopic ratios may be found from the 44, 45, and 46 peaks attributable to CO_2. Then, the NO and CO contributions to the 30 peak are almost equal (Fig. 29). The 32 mass peak is caused by CO_2 dissociative ionization and by molecular oxygen. The derived altitude profiles of CO_2, N_2, A_2, CO, O_2, and NO number densities are shown in

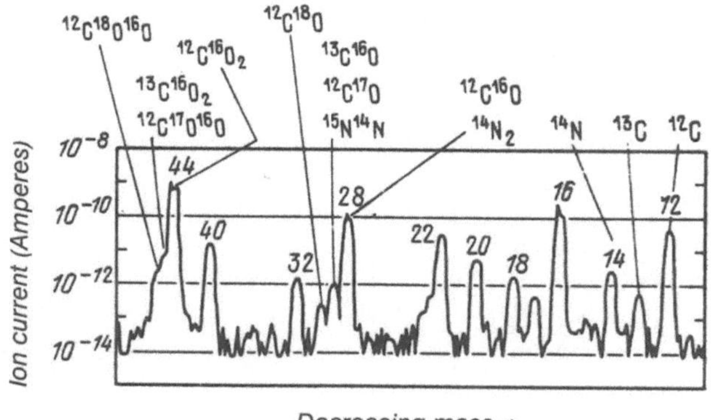

Fig. 27. Mass spectrum measured by the Viking 1 entry probe mass spectrometer at 140 km (Nier and McElroy 1977)

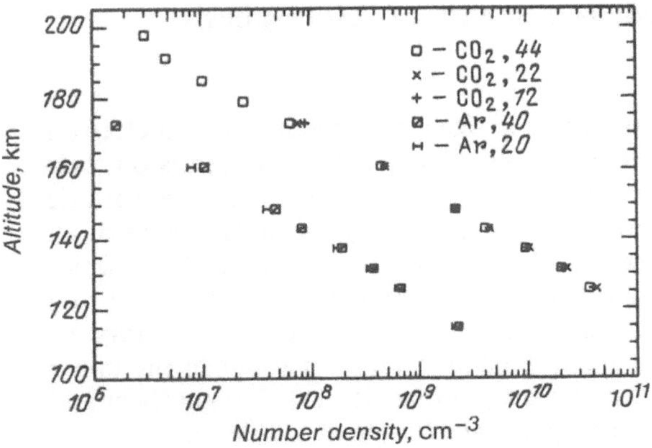

▲
Fig. 28

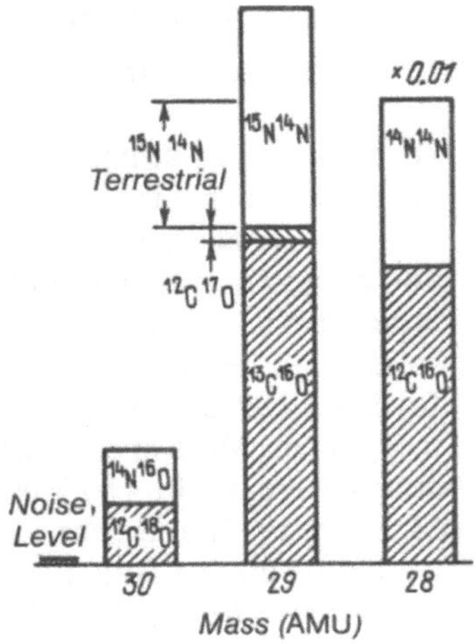

Fig. 29 ▶

Fig. 28. CO_2 and Ar number densities derived from mass peaks at 44, 22, and 12 for CO_2 and 40 and 20 for Ar (Nier and McElroy 1977)

Fig. 29. Contributions of different components to the peaks at 28, 29, and 30 in Fig. 27 (Nier and McElroy 1977)

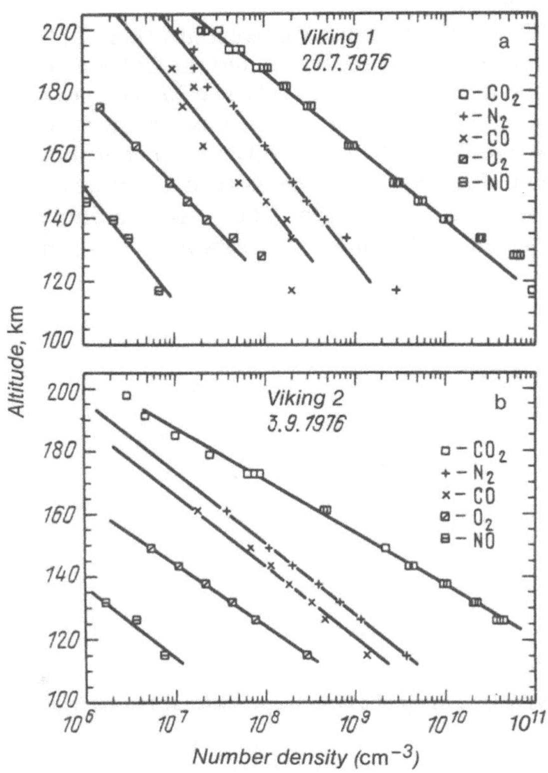

Fig. 30 a, b. Height profiles of atmospheric constituent number densities (Nier and McElroy 1977)

Table 4. Martian lower atmospheric composition based on spectroscopic and mass spectrometric measurements (Owen et al. 1977)

Gas	Mixing ratio	Gas	Mixing ratio
CO_2	95.32%	H_2O[a]	0.03%
N_2	2.7%	Ne	2.5 ppm
Ar	1.6%	Kr	0.3 ppm
O_2	0.13%	Xe	0.08 ppm
CO	0.07%	O_3[a]	0.03 ppm

[a] variable

Table 5. Isotopic ratios of atmospheric constituents on the Earth and Mars (Owen et al. 1977)

Isotopic ratio	Earth	Mars
$^{12}C/^{13}C$	89	90
$^{18}O/^{18}O$	499	500
$^{14}N/^{15}N$	277	165
$^{40}Ar/^{36}Ar$	292	3000
$^{129}Xe/^{132}Xe$	0.97	2.5

Fig. 30. The measurement of atomic oxygen presents great difficulties because of wall recombination and CO_2 dissociative ionization. In fact, it has not been measured by the Viking mass spectrometers.

Composition of the lower atmosphere was studied by mass spectrometers on Viking 1 and 2 landers (Owen et al. 1977). As a result, nitrogen, oxygen, and noble gases were detected (Table 4). Mixing ratios of atmospheric constituents are usually measured in percent, parts per million (ppm), and parts per billion (ppb). The detection and measurement of xenon demonstrate the excellent possibilities of the mass spectrometric technique because noble gases do not decompose inside the instrument ion source.

Nitrogen and argon mixing ratios measured by different instruments during the entry into the atmosphere and after landing coincide. The accuracy of O_2 measurements near the surface was not very high. The CO measurements failed due to the great amounts of CO_2 decomposed to CO by the ion source and due to the CO_2 and CO absorption by Ag_2O and LiOH used for enrichment of atmospheric samples with other constituents.

The Viking 1 and 2 mass spectrometric data in Table 4 apply to low latitudes at $L_s \simeq 100°$ when the CO_2 amount in the atmosphere equals ~0.9 of the mean one. One should expect variations of the nitrogen and noble gas mixing ratios by a factor of 1.3 at low and middle latitudes due to the seasonal effect of the CO_2 amount. In polar regions condensation and evaporation of CO_2 result in much enhanced increase and decrease, respectively, of these mixing ratios.

Isotopic ratios of elements in the Martian and Earth's atmospheres are shown in Table 5. For carbon and oxygen, deviations are within the limits of experimental errors. Nitrogen, argon, and xenon ratios differ considerably from those on the Earth. These differences are connected with the processes of the planets' formation and evolution and are of great importance for solving the problem of the solar system's origin.

1.7 Ionospheric Composition

Three techniques were utilized in Martian ionospheric studies: radio-occultations, optical measurements, and direct measurements by entry probes.

Radio-occultation. This method, described in Chap. 1.1, was used for the first time on Mariner 4 (Kliore et al. 1967) and later on almost all space probes. All these measurements provide altitude profiles of electron number densities, n_e, at different local time and solar activity. A great body of information was obtained by Mariner 9 (Kliore et al. 1972, Fig. 31). The first and only measurements of the night-side ionosphere were made by Mars 4 (Savich and Samovol 1976, Fig. 32).

Optical measurements by the Mariner UV spectrometers make it possible to obtain CO_2^+ ion number densities from the CO_2^+ bands in dayglow spectra. Stewart (1972) calculated the CO_2^+ band system intensities for both pure CO_2^+ ionospheric composition and for that of O_2^+. In both cases, the contribution

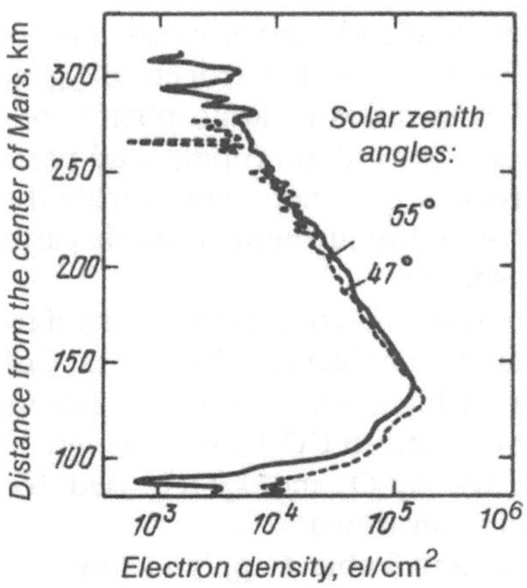

Fig. 31

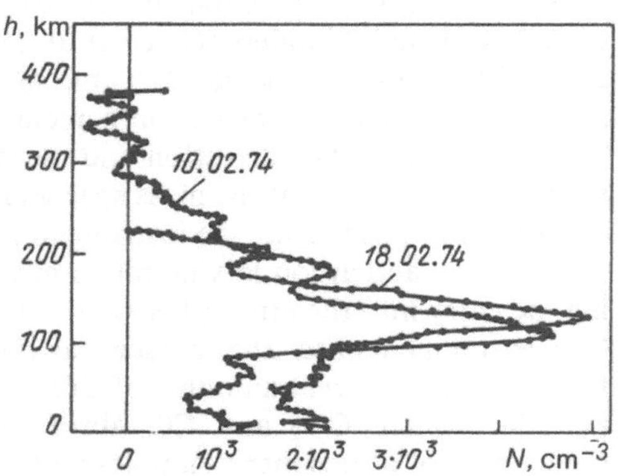

Fig. 32

Fig. 31. Typical electron number density profiles measured by Mariner 9 in the Martian ionosphere (Kliore et al. 1972)

Fig. 32. Electron number density profiles in the nighttime ionosphere measured by Mars 4 (Savich and Samovol 1976)

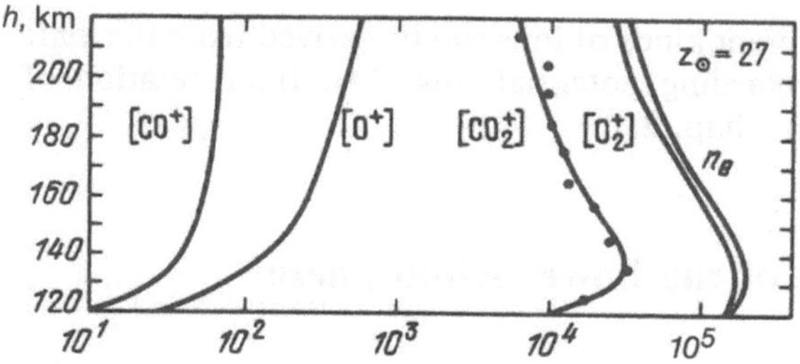

Fig. 33. Martian model ionosphere at $z_0 = 27°$. CO_2^+ number densities were derived from the CO_2^+ dayglow, n_e from the radio-occultation measurements, CO^+ and O^+ are results of calculations (Krasnopolsky 1975)

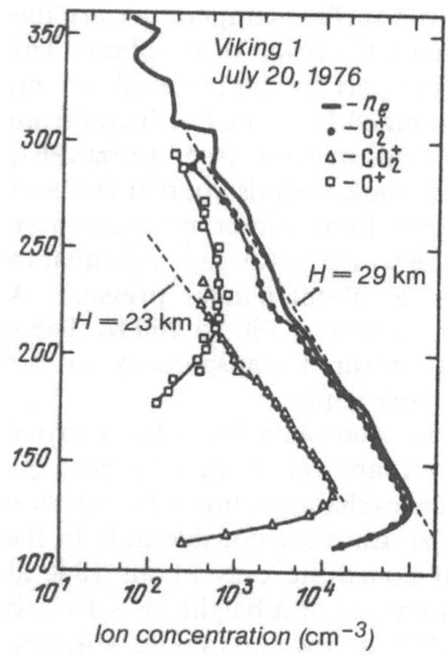

Fig. 34. Martian ionospheric composition as measured by the Viking 1 entry probe (Hanson et al. 1977)

of CO_2 photoionization excitation is rather high. The best fit with measurements is for $[O_2^+]/[CO_2^+] \simeq 2$.

Krasnopolsky (1975) used the same experimental data, but alongside the CO_2^+ band system intensities, took into account their vibrational distribution. This allowed CO_2 photoionization and CO_2^+ fluorescent scattering to be separated and to obtain the $[CO_2^+]$ altitude profile (Fig. 33). Comparison with the n_e profile measured by the radio-occultation technique on the same space probes gave $[O_2^+]/[CO_2^+] \simeq 6$ with $[O_2^+] + [CO_2^+] = n_e$ predicted by ionospheric photochemistry.

Ionospheric composition as measured by the Viking entry probes. The first in situ measurements of ionic composition were made by retarding potential analyzers on the Viking entry probes. In such mass spectrometers, all kinds of ions have equal initial relative velocities determined by spacecraft velocity and plasma temperature. An electrostatic field retards ions and a measured current is formed by ions with an energy in excess of the retarding potential. The energies of the ions are proportional to their masses, so the number densities and ionic temperatures of different kinds of ions can be derived from the ionic current dependence on the retarding potential (Fig. 34). Interpretation of these data will be discussed in Chap. 2.1.

1.8 Temperature Profile of the Lower Atmosphere

Surface temperature. Because a planetary surface is in thermal contact with the lower atmosphere, the measurements of the surface temperature are important in order to study the thermal regime of the lower atmosphere. The Martian atmosphere is transparent in a major part of the infrared region. Therefore, the surface temperatures can be obtained from the Earth and from Mars orbiters by means of infrared radiation measurements (see Moroz 1967, 1978). Maximum temperatures of low latitude dark regions constitute about 320 K at noon at the moment of a planetary perihelium (winter in the northern hemisphere). Minimum temperatures are in the winter polar cap and equal to the CO_2 condensation temperature of 148 K at about 6 mbar pressure. A typical range of diurnal variations near the equator is 180 to 260 K. These large variations are connected with a high infrared transparency of the atmosphere caused by very low amounts of water vapor.

The Martian surface is heated directly by the solar radiation, while thermal conditions of the boundary atmospheric layer are determined by heat exchange with the surface. A typical time for this exchange is much longer than a day. Therefore, diurnal variations of temperature are considerable in the lowest atmospheric layer ~1 km thick, but constitute only about 10 K at higher altitudes. Thus, the atmospheric temperature at a height of ~1 km is equal to a mean surface temperature and is 220 K at low and middle latitudes.

This mean surface temperature can be derived from a simple equation

$$\tfrac{1}{4}(1-A)F_0=\sigma T^4$$

where $A = 0.24$ is the integral albedo of the planet, $F_0 = 6\times10^5\,\mathrm{erg\,cm^{-2}\,s^{-1}}$ is the solar radiation on the Martian orbit. $\frac{1}{4}$ is the ratio of the disk to sphere areas. This equation results in the global average temperature of 210 K. The atmosphere is transparent to the bulk of both solar and thermal surface radiations, except solar UV radiation with $\lambda < 200$ nm, which constitutes only 0.7×10^{-4} of the total solar radiation and surface radiation inside the band of CO_2 at 15 μm which is near the spectral maximum of the Martian thermal radiation. The atmosphere absorbs and radiates at this band about 15% of the total thermal radiation of the surface. Radiative transfer at this band determines thermal structure of the lower and middle atmosphere, and dust opacity in the visible range is an additional source of heating here. The mean rate of atmospheric heating with approximatively equal contributions of both the sources is defined as a ratio of the absorbed energy flux to the product of atmospheric mass and specific heat and equals about $3\times10^{-4}\,\mathrm{K\,s^{-1}}$ or

30 K day^{-1}. That is why diurnal temperature variation in the lower and middle atmosphere is about 10 K. Absorption of solar UV radiation occurs in the upper atmosphere and results in formation of the thermosphere with a high heating rate. With a heating efficiency, ε, defined as the fraction of solar UV energy converted to thermal energy, equal to 0.2 at ionospheric altitudes (Fox and Dalgarno 1979a), the atmospheric heating rate constitutes $\sim 10^{-2}$ K s^{-1} in optically thin conditions.

Temperature profile derived from a CO_2 *band of 15 μm.* Spectroscopy of this band can be used to obtain the atmospheric temperature profile. Intensity of radiation at any wavelength is determined by the atmospheric temperature at a level of $\tau \simeq 1$. Moreover, if at this level the temperature gradient is negative, the corresponding spectral feature is observed as an absorption feature. The positive temperature gradient corresponds to emission. Both the absorption coefficient of the 15 μm band and the level of $\tau \sim 1$ in the atmosphere are dependent on wavelength, thus making the measurements of the altitude profile possible. This typical problem of remote sensing is solved by an integral equation with a detailed band structure taken into account.

The spectra of the 15 μm band were measured by an infrared spectrometer on board the Mariner 9 orbiter (Hanel et al. 1972). Figure 35 shows the spectra typical for low latitudes and the polar caps. In the first case, the band is observed in absorption, thus indicating a monotonic temperature decrease with height. The same band is observed in emission with some absorption in the center at the polar caps. This means that temperature increases up to a certain altitude and then decreases. A temperature profile of low latitude atmosphere soon after a dust storm derived from the 15 μm band measurements is shown in Fig. 36. These measurements give the temperature profile up to 0.1 – 0.05 mbar level ($\sim$40 km on Mars).

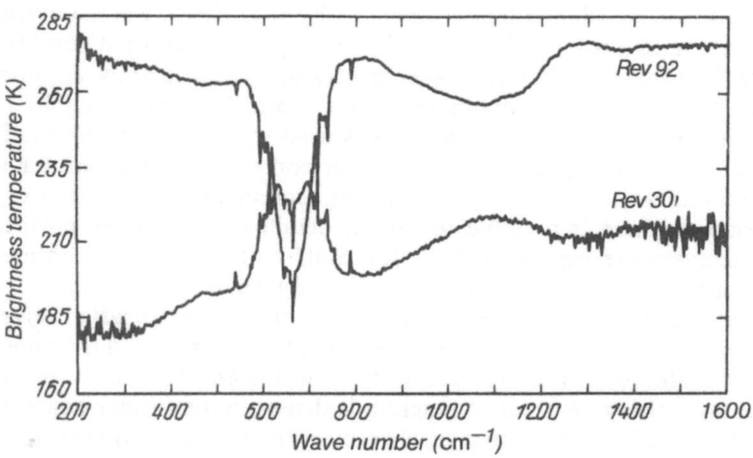

Fig. 35. Infrared spectra of Mars at middle latitude (rev 92) and at the southern polar cap (rev 30) (Hanel et al. 1972)

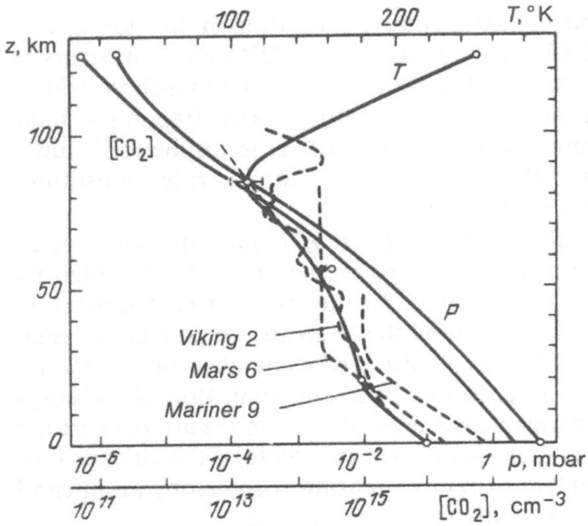

Fig. 36. Temperature, pressure, and CO_2 number density vs height (Krasnopolsky 1979a, Krasnopolsky et al. 1980). Temperature profiles measured directly by the Mars 6 (Kerzhanovich 1977) and Viking 2 (Seiff and Kirk 1977) landing probes and derived from the Mariner 9 infrared spectrometric observations (Hanel et al. 1972) are also shown

Radio-occultation measurements provide temperature profiles as derivatives of the logarithm of atmospheric pressure with respect to height. Many temperature profiles up to 30 km were obtained by Mariner 9 (Kliore et al. 1972).

Direct measurements. One usually applies three techniques for the measurement of temperature and related parameters (atmospheric density and pressure): thermometric, manometric, and densitometric measurements during the parachute phase of landing and the measurements of probe deceleration by accelerometers and Doppler shift of radiowaves. These observations were first made on the Mars 6 lander (Avduevsky et al. 1975, Kerzhanovich 1977) and the temperature profile obtained was approximated by a constant temperature T_0 down to z_0 and a constant temperature gradient below z_0 (three parameters). The best fit to observations is for the profile given in Fig. 36.

A more sophisticated instrumentation was used by Viking landers (Seiff and Kirk 1977). Besides an accelerometer, thermometer, and manometer it had two radars for altitude and Doppler shift measurements. A high accuracy of measurements made it possible to obtain detailed temperature profiles along with temperature variations (Fig. 36). Seiff and Kirk (1977) consider these variations as being produced by inertia gravity waves.

Temperature profile deduced from properties of the condensation layer. The Mars 5 limb measurements (Krasnopolsky et al. 1979) revealed the upper aerosol layer at altitudes of 20 – 40 km in the morning and 50 – 70 km in the evening; the layer being an order of magnitude denser in the former case (see Chap. 1.3). High variability of the layer characteristics and their good correlation with an approximate evaluation of atmospheric temperatures have led to the conclusion that this layer consists of water ice particles.

Saturated water vapor number densities in a temperature range of 150 – 190 K can be written as

$$n_s = 2.2 \times 10^{12} e^{(T-180)/\alpha} \mathrm{cm}^{-3}, \quad \alpha = 5.2\,\mathrm{K}.$$

Total flux is zero for the two phase system equilibrium, i.e.,

$$\frac{dm}{dz} + \frac{dn_s}{dz} + \frac{m+n_s}{H} + m\frac{V}{K} = 0$$

where m is the number of water molecules in solid phase per cubic cm, V is the Stokes velocity of ice particles. It gives

$$m = n_{0s} \frac{\gamma/\alpha - 1/H}{V/K - \gamma/\alpha + 1/H} [e^{-\gamma z/\alpha} - e^{-z(V/K+1/H)}]$$

with V/K and the temperature gradient, γ, assumed to be constant.

The integration of this formula yields the total amount of the condensate in the layer, while its differentiation gives m_{max} and z_{max}. For a more accurate description of the layer, a temperature profile should be given by $T_0 - \gamma z - \delta z^2$. The best agreement with observations is for $\gamma = 0.45\,\mathrm{K\,km^{-1}}$; $\delta \simeq 10^{-2}\,\mathrm{K\,km^{-2}}$, $V/K = (0.07 \pm 0.03)\,\mathrm{km^{-1}}$. The total amount of the condensate is $7 \times 10^{-6}\,\mathrm{g\,cm^{-2}}$ with its maximum density of $5 \times 10^{-12}\,\mathrm{g\,cm^{-3}}$.

If the atmospheric water vapor amount is near a mean value of 15 μm and the bottom of the layer is at 20 km, then $T_0 = 180$ K at 20 km meets the condensation conditions. So the temperature profile obtained can be given by

$$T(z) = 180° - 0.45(z-20) - 0.01(z-20)^2 = 185 - 0.05z - 0.01z^2$$

at 20 – 40 km with z in km.

If in the evening the altitude of the layer is ~60 km, then the condensation conditions give $T = 160$ K. The atmosphere being warmer by ~10 K in the afternoon one obtains $T \simeq 150$ K at 60 km in the morning side atmosphere, which agrees with the derived temperature profile.

The Mars 5 measurements also gave an atmospheric scattering scale height of 6 ± 0.5 km at 80 – 90 km (Krasnopolsky et al. 1979). If it equals an atmospheric scale height, then $T = 110 \pm 10$ K. So the temperature profile relation obtained is valid in the range of 20 – 90 km. It correlates with other temperature measurements. To complete this profile one can make use of the mean surface temperature at low latitudes ($T = 220$ K) and $T = 250$ K at 125 km from Mariner 6 and 7 dayglow measurements (see below). This curve conforms rather well to atmospheric pressures of 6 mbar near the surface and 1.7×10^{-6} mbar at 125 km (see Fig. 20).

1.9 Temperature of the Upper Atmosphere

Radio-occultation measurements. The first evaluations of the upper atmospheric temperature were based on ionospheric electron number density profiles measured by the radio-occultation technique. It can be shown that the atmosphere is optically thin for solar UV radiation at 150–220 km, i.e., the photoionization rate is

$$q = In = In_0 e^{-z/H} .$$

In this altitude range, the condition of photochemical equilibrium may be used

$$q = \alpha n_e^2 \quad \text{or} \quad n_e = \left(\frac{In_0}{\alpha}\right)^{1/2} e^{-z/2H} .$$

So the ionospheric plasma scale height, H_e, is twice as high as that of the neutral atmosphere. Hence, the temperature of the neutral atmosphere may be derived with its molecular mass being known. Measurements give $H_e = 29$ km for $F_{10.7} = 80$ (Mariner 4, Kliore et al. 1967), $H_e = 46$ km for $F_{10.7} = 180$ (Mariner 6 and 7, Fjeldbo et al. 1970), $H_e = 39$ km for $F_{10.7} = 120$ (Mariner 9, Fig. 31). $F_{10.7}$ defines here a level of solar activity. Mariner 9 measurements reveals no systematic variations with solar zenith angle, local time, and solar activity despite a large range of H_e variations (25–48 km). This is due to the effects of very intense gravity waves formed by the global dust storm at the time of observation (Klore et al. 1972).

However, comparison of the mean H_e at 150–200 km measured by Mariner 4, 6, 7, 9 and Viking radio-occultations with solar 10.7 cm flux adjusted to the Mars orbit with its eccentricity taken into account reveals a strong correlation (Lindal et al. 1979).

The conversion of H_e into temperatures assuming that the atmospheric molecular mass equals that of CO_2 gives overestimated values ($T = 285$, 450, and 380 K, respectively). This is accounted for by the atomic oxygen amounts lowering the mean atmospheric molecular mass. A detailed analysis of radio-occultation data of Mars 2, 3, 5 to obtain the properties of the neutral atmosphere was carried out by Michailov and Ivanov-Kholodny (1975). They derived $T_\infty = 380$ K, $f_O = 1\%$ and CO_2 number densities. Unfortunately, they assumed equal electronic and neutral temperatures which have led to the overestimation of the CO_2 number density by a factor of 2–3 at $h > 150$ km. In the interpretation of Mariner 6 and 7 data, Krasnopolsky (1975) utilized the calculated electronic temperatures and obtained atmospheric temperatures and CO_2 number densities to correlate with dayglow measurements (see below).

Optical measurements. Numerous determinations of temperature are based on airglow measurements. Exospheric temperature $T_\infty = 300–350$ K was derived from the vertical distributions of H 121.6 nm airglow observed by the Mariner and Mars space probes (see Chap. 1.5). Temperature in the range

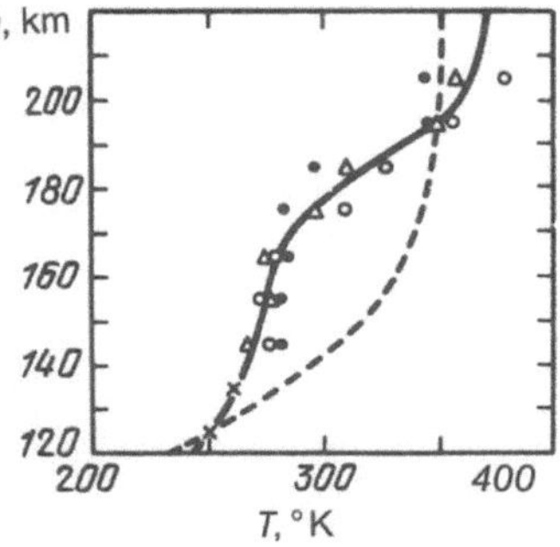

Fig. 37. Altitude profiles of upper atmospheric temperature deduced from the measurements of electron number densities (*circles*) and of the CO_2^+ airglow (*points* and *triangles*) (Krasnopolsky 1975). *Dashed line* is that of the model atmosphere by Barth et al. (1972)

of 150–200 km can be obtained from the scale height of dayglow emissions. Cameron bands of CO give the most precise values ($T_\infty = 325$ K, see Stewart et al. 1972).

Direct measurements on board the Mars 5 orbiter (Bertaux et al. 1975) utilized the Doppler line broadening. The H 121.6 nm intensities with a cold and hot filament inside a cell with molecular hydrogen were compared. In the last case, partial thermodissociation of H_2 occurs and H atoms thus formed absorb airglow within their line width. The difference of intensities depends evidently on the airglow line width, thus giving $T = 340 \pm 30$ K.

Krasnopolsky (1975) obtained a temperature profile from the CO_2^+ bands and n_e measured by Mariner 6 and 7 (Fig. 37). This profile differs greatly from a typical thermospheric profile having isothermicity above 160 km (Stewart 1972). This may be explained by heating caused by inertia gravity and acoustical waves dissipating in the atmosphere. On the other hand, these waves stimulate atmospheric mixing and cool the thermosphere by the increased heat sink to the mesosphere. The temperature profile in Fig. 37 yields an energy flux of gravity waves of 4×10^{-2} erg cm^{-2} s^{-1}, heating efficiency of solar radiation of $\varepsilon \simeq 0.1$, eddy diffusion coefficient $K = 2 \times 10^8$ cm^{-2} s^{-1} at 160 km, where $[CO_2] = 5 \times 10^9$ cm^{-3}.

Mass spectrometric measurements on board Viking entry probes (Nier and McElroy 1977). The measured CO_2 number density profiles (Fig. 38) result in the temperature profiles shown in Fig. 38. The error bars reflect the spread in temperatures derived from 44, 22, and 12 mass peaks. The thermospheric temperatures have proven to be very low (~180 and ~130 K for Viking 1 and 2, respectively). The complicated vertical structure of the temperature profiles is similar to that in the lower atmosphere and must probably reflect the influence of inertia gravity waves.

The thermospheric temperature distribution is characterized not only by considerable deviations from isothermicity, but also by large local variations. It should be pointed out that the temperature profiles in Fig. 37 differ from those in Fig. 38 in that they reflect the mean characteristics of the atmosphere because they have been derived by averaging four limb profiles of the CO_2 bands and two profiles of $n_e(h)$ measured in different locations.

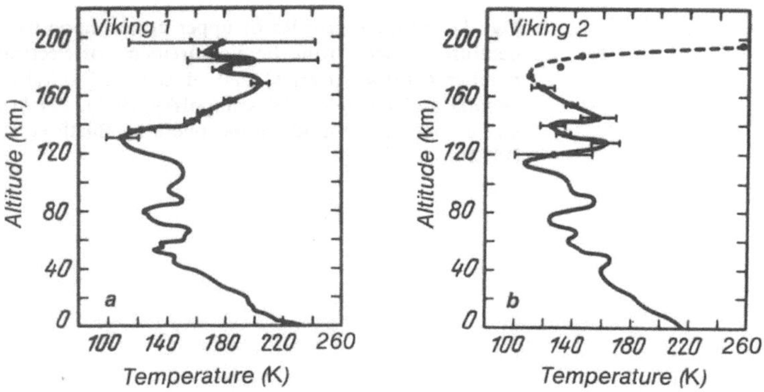

Fig. 38a, b. Temperature profiles of the Martian atmosphere measured by Viking 1 and 2 entry probes. Data above 120 km were obtained by mass spectrometry (Nier and McElroy 1977) and below 120 km by accelerometers, altimeters, and manometers (Seiff and Kirk 1977)

Izakov (1978) argued that the Viking mass spectrometric measurements can be described by height independent temperature and the eddy diffusion coefficient.

1.10 Eddy Diffusion Coefficient

The eddy diffusion coefficient is an important characteristic of the atmosphere because the relationship between molecular diffusion, eddy diffusion, and photochemical times defines the vertical distribution of atmospheric constituents. Similar to the molecular diffusion coefficient, this coefficient is introduced in the continuity equation with the individual scale height H_i being replaced by the average scale height H_a. It equals the product of the mean size of the atmospheric eddies and gas velocities inside them in the same way as the molecular diffusion coefficient is a product of molecular mean free path and velocity.

Atmospheric mixing by eddies enhances atmospheric thermal conductivity which may be considered as a heat source or sink given by

$$q_m = \frac{\partial}{\partial z}\left[\rho c_P K\left(\frac{\partial T}{\partial z} + \frac{g}{c_P}\right)\right]$$

where ρ is the atmospheric density, c_P is the specific heat at constant pressure, g is the gravity acceleration, and $g/c_P = \Gamma \simeq 5\ \text{K km}^{-1}$ is the adiabatic lapse rate. Dissipation of turbulent energy heats the atmosphere, and this source is described by

$$q_d = \frac{K\rho g}{TR_{f_0}}\left(\frac{\partial T}{\partial z} + \frac{g}{c_P}\right)$$

where R_{f_0} is the Richardson number which is about 0.2 in the middle atmosphere of Mars. Their heating rates (see Chap. 1.8) are

$$t_m = \frac{\partial}{\partial z}\left[K\left(\frac{\partial T}{\partial z}+\Gamma\right)\right] - \frac{K}{H}\left(\frac{\partial T}{\partial z}+\Gamma\right) \quad \text{and} \quad t_d = \frac{K\Gamma}{TR_{f_0}}\left(\frac{\partial T}{\partial z}+\Gamma\right),$$

respectively. A reasonable assumption is t_m, t_d to be smaller or of the order of atmospheric heating rate by absorption at the 15 μm band and dust absorption in the visible range, i.e., 3×10^{-4} degree s^{-1} (Chap. 1.8). Therefore, with $\partial T/\partial z \ll \Gamma$ in the lower and middle atmosphere one obtains

$$\frac{K\Gamma}{H}, \frac{K\Gamma^2}{TR_{f_0}} \lesssim 3\times10^{-4}\,\text{degree s}^{-1} \quad \text{or} \quad K \lesssim 0.6\times10^{7}\,\text{cm}^2\,\text{s}^{-1}.$$

The eddy thermal conductivity acts as a heat sink in most of the atmosphere.

Due to absorption of solar UV radiation, the heating rate is much higher in the upper atmosphere (10^{-2} degree s^{-1}, see Chap. 1.8), so the above relations give $K \lesssim 2\times10^{8}\,cm^2\,s^{-1}$. However, this approach appears to be valid in the bulk atmosphere and not in the upper atmosphere, because the energy of gravity waves generated in the lower atmosphere may exceed the local UV heating in the upper atmosphere.

Processes causing atmospheric turbulent motion are, e.g., meteorological phenomena, heat exchange between the atmosphere and planetary surface, and atmospheric tides. These processes produce inertia gravity waves which move upwards and grow in amplitude in the absence of dissipation as $\varrho^{-1/2}$, where ϱ is the atmospheric density. This may proceed until wave-induced temperature gradients reach a superadiabatic value, thus forming convective motions. Since theory cannot provide reliable data on the eddy diffusion coefficient, K, it may be deduced from the inhomogeneous distribution of some measured atmospheric constituents.

A search for water vapor condensation in the upper aerosol layer detected by the Mars 5 orbiter (Krasnopolsky et al. 1979, see Chap. 1.8) gave a ratio of the Stokes velocity of ice particles to K equal to (0.07 ± 0.03) km. With the total amount and density of a condensate being known, one can fit the calculated brightness of the layer to the measured one by the proper choice of the mean radii of particles. This radius is equal to 1.5 μm with an optical thickness of the aerosol layer, $\tau=0.07$, correlating with the Mariner 6 observations. Then, the Stokes velocity is $V=1.5\,cm\,s^{-1}$, thus $K=(2\pm1)\times10^{6}$ $cm^2\,s^{-1}$ at heights of about 30 km, being derived from the V to K ratio.

Observations from TV cameras of aerosol layers at planetary limbs from the Mariner 9 orbiter (Anderson and Leovy 1978) can give K values, too. With an assumed particle radii of 1 μm they provide $K=2\times10^{6}\,cm^2\,s^{-1}$ and $7\times10^{6}\,cm^2\,s^{-1}$ at 35 and 45 km, respectively, after the end of a dust storm. During the dust storm K is equal to $(2-14)\times10^{7}\,cm^2\,s^{-1}$ at 50 – 60 km.

Analyzing diurnal variations of water vapor amounts Krasnopolsky and Parshev (1977) (see also Chap. 2.6) deduced $K = 3 \times 10^6\,\mathrm{cm^2\,s^{-1}}$ in the lower troposphere. Wehrbein et al. (1979) ascribed a very small ozone scale height at middle latitudes in winter to temperature and odd hydrogen number densities increasing with height in the lower troposphere. This is valid for odd hydrogen not being mixed well, i.e., its mixing time exceeds the photochemical lifetime equal to 3 days. Hence, $K \lesssim 4 \times 10^6\,\mathrm{cm^2\,s^{-1}}$.

Nier and McElroy (1977) obtained $K(z)$ by correlating the altitude profiles of constituents with different molecular masses measured by the Viking landers. If the vertical distribution of a component is determined by transport processes (eddy and molecular diffusion), its scale height is given by

$$H = (K + D_i)\left(\frac{K}{H_a} + \frac{D_i}{H_i}\right)^{-1}$$

where D_i is the molecular diffusion coefficient, H_a and H_i are the average scale height and that of the i component, respectively. With other values known from the Viking measurements, this relation gives K. Four components (N_2, Ar, CO, O_2) were used to obtain K with their lower atmospheric mixing ratio measurements by optical and mass spectrometry taken into account. CO and O_2 were corrected for chemical sources and sinks. The $K(z)$ obtained are listed in Table 6. Figure 39 shows the CO number densities measured and calculated for $K(z)$ from Table 6 and $K(z)$ fixed at $5 \times 10^7\,\mathrm{cm^2\,s^{-1}}$. These data are not consistent with the idea of Izakov (1978) that the measured number densities are described by the constant temperature and the eddy diffusion coefficient. Values of K for altitudes above 140 km are subject to large errors. The turbopause is located at 120 – 125 km with $[CO_2] = 5 \times 10^{10}\,\mathrm{cm^{-2}}$.

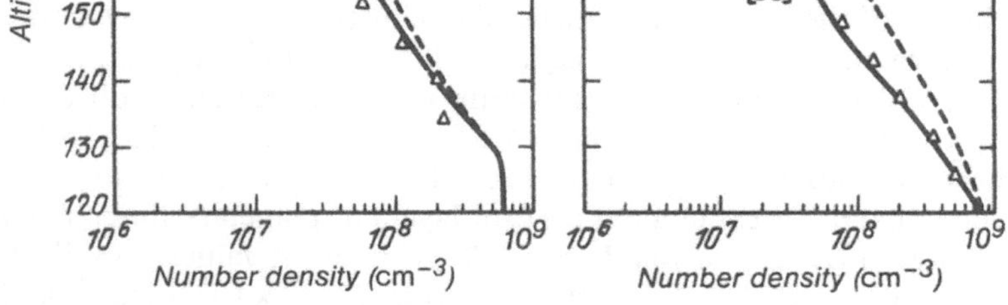

Fig. 39a, b. CO number densities vs height measured by Viking mass spectrometers (*triangles*) and calculated with the eddy mixing coefficient taken from Table 6 (*solid lines*) and equal to $5 \times 10^7\,\mathrm{cm^2\,s^{-1}}$ independent of height (*dashed lines*) (Nier and McElroy 1977)

Table 6. Eddy diffusion coefficients based on the Viking entry probe mass spectrometric measurements (Nier and McElroy 1977)

z, km		100	110	120	130	140	150	160	170
K, $cm^2\,s^{-1}$	$V1$	5+7[a]	5+7	5+7	6.2+7	1.3+8	2.3+8	5.9+8	1.2+9
	$V2$	2.1+7	4.6+7	9.8+7	2.1+8	4.4+8	9.3+8	2.0+9	4.2+9

[a] 5+7 means 5×10^7

The value of $K=2\times10^8\,cm^2\,s^{-1}$ at 160 km ($[CO_2]=5\times10^9\,cm^{-2}$) derived by Krasnopolsky (1975) from dayglow measurements agrees well with the Viking data ($K=(1.3-4.4)\times10^8\,cm^2\,s^{-1}$ at 140 km where $[CO_2]=5\times10^9\,cm^{-3}$).

Von Zahn et al. (1980) argue that $K(z)$ given in Table 6, as well as measured in the Venusian thermosphere at 130–170 km (Chap. 3.13) and in the Earth's atmosphere from 30 to 85 km, can be represented by $K=1.4\times10^{13}\,n^{-1/2}\,cm^2\,s^{-1}$. Individual properties of the planets are revealed via the range of number density, n, which is valid for this expression. This range is $10^{12}-10^9\,cm^{-3}$, i.e., 100–170 km, on Mars.

The dependence of $K=An^{-1/2}$ reflects the increase of amplitudes of gravity waves with height discussed above. However, the coincidence of A for the three planets is very surprising and has not been explained.

2 Photochemistry of the Martian Atmosphere

2.1 Ionosphere

A problem and ways to solve it. Consider a problem of the calculation of ionospheric composition with the neutral upper atmospheric composition being known. The ionization rate q_i of component i is given by

$$q_i = \int_0^\infty I_\lambda n_i \sigma_{i\lambda} \alpha_{i\lambda} \exp\left(-\sum_k \sigma_{k\lambda} n_k H_k \operatorname{s} z_0\right) d_\lambda .$$

The terms are:

I_λ the intensity of solar UV radiation in quanta $\mathrm{cm}^{-2}\,\mathrm{s}^{-1} \cdot \mathrm{nm}$
$\sigma_{i\lambda}$ the absorption cross-section in cm^2
$\alpha_{i\lambda}$ the quantum yield of ionization (I_λ, $\sigma_{i\lambda}$, $\alpha_{i\lambda}$ are given by Torr et al. 1979)
H_i the scale height of component i
z_0 the solar zenith angle.

Dissociative ionization producing minor ionospheric components should be taken into account.

Photoionization forms photoelectrons with energies which can be sufficient for secondary ionization. Approximate calculation of the multiple ionization may be done on the assumption that the primary photoelectron energy is equal to that of the photon minus 16 eV lost in the photoionization process and that every ionization process consumes 32 eV of the photoelectron energy. Therefore, the ionization quantum yield exceeds unity for photons of $\lambda \lesssim 40$ nm and equals (Krasnopolsky 1973)

$$\alpha_{i\lambda} = 1 + \frac{\sigma_{ie} \sum_k n_k \dfrac{\sigma_{k\lambda}}{\sigma_{i\lambda}}}{\sum_k \sigma_{ke} n_k} \left(\frac{39}{\lambda\,(\mathrm{nm})} - 0.5\right)$$

where σ_{ie} is the electron impact ionization cross-section. It may be assumed to be proportional to the number of O, N, C atoms in a molecule, thus eliminating absolute values of σ_{ie} from the calculations.

The previous consideration of photoelectron ionization alongside a more accurate technique taking into account all kinds of electron energy loss processes (ionization, dissociation, and excitation of atmospheric gases) presupposes that photoelectrons appear and vanish at the same altitude. This approach is applicable even at heights where the electron mean path is comparable with the atmospheric scale height.

For the calculation of photoelectronic processes the energy loss functions or stopping cross-sections must be known. They are given by

$$L_i(E) = \frac{dE}{n_i dx} = \sum_k \sigma_{ike}(E)\, \Delta E_{ik}$$

where σ_{ike} is the cross-section of a k process of a photoelectron and molecule i. $L_i(E)$ was calculated by Dalgarno et al. (1969), Sawada et al. (1972), and Fox and Dalgarno (1979a). The number of acts of the j process along the electron path from the point of appearance a to the point of stop b is

$$\int_a^b \sigma_{ij}(E)\, n_i dx = n_i \int_0^E \frac{\sigma_{ij}(\varepsilon)\, d\varepsilon}{\sum_k n_k L_k(\varepsilon)} \qquad \text{since} \qquad dx = \frac{dE}{\sum_k n_k L_k(E)} .$$

The number of electrons with energy E appearing in 1 $cm^3\, s^{-1}$ is

$$P(E) = \int_E^\infty p(\varepsilon)\, d\varepsilon$$

where $p(\varepsilon)$ is the photoelectron production rate per unit energy or the differential spectrum of photoelectron generation and $P(E)$ is the integral spectrum. The assumption of a continuous energy loss is made, i.e., every electron with an energy in excess of E will slow down to energy E. This assumption becomes inadequate at low energies and gives an overestimated value of the j rate (by a factor of 2) for a threshold energy of a few electron volts.

Photoelectron generation spectra should be distinguished from those of photoelectron fluxes which can be measured in the atmosphere and are given by

$$\phi(E) = \frac{P(E)}{\sum_k n_k L_k(E)} .$$

Calculations of photoelectron fluxes are described in detail by Fox and Dalgarno (1979b) and Mantas and Hanson (1979).

Therefore, the rate of process j with a component i involved (for example, its ionization) is

$$q_{ij} = n_i \int_0^\infty \frac{P(\varepsilon)\, \sigma_{ij}(\varepsilon)\, d\varepsilon}{\sum_k n_k L_k(\varepsilon)} .$$

Calculations of the integral production rate spectra do not need high accuracy and may be done by integration of differential spectra given by

$$p(E) = I(E + 16\ \text{eV}) \sum_i n_i \sigma_i(E + 16\ \text{eV})\, \alpha_i(E + 16\ \text{eV}) \exp\left(-\sum_j n_j \sigma_j H_j \,\text{s}\, z_0\right)$$

on the assumption that photon energy deposition to ionization constituted 16 eV. This expression is similar to that for the photoionization rate and differs only in that cross-sections and intensities depend on energy rather than on wavelength.

Ions formed enter chemical reactions and recombine. A term describing the ion X_i sink in the continuity equation is $[X_i] \sum_l k_l n_l = l_i$ where k_l is the reaction rate coefficient and n_l is the reactant number density. k_l depends usually on neutral, ionic, or electronic temperatures (T, T_n, T_e). Sometimes it depends strongly on the vibrational or electronic excitation of reactants. A relationship between ionic and neutral temperatures is determined by the number of collisions of ions and neutral molecules during lifetimes of the former and by a relationship between their molecular masses. The most effective energy exchange takes place at equal masses of collisional partners. For a large number of collisions ionic temperatures coincide with neutral ones. As a rule, this is valid up to the height where neutral atmosphere number density is $\sim 10^9\,\mathrm{cm}^{-3}$. Further, with decreasing number densities, the ionic temperature begins to exceed the neutral one. Masses of electrons and neutral molecules differ much more than those of ions and neutral molecules; that is why a great number of collisions are needed to equalize T_e and T. Energy exchange in this case is proportional to the mass ratio m/M, the number of collisions during the lifetime depends on particle velocity being proportional to $m^{-1/2}$. Therefore, with $T_i \simeq T$ valid for $n \gtrsim 10^9\,\mathrm{cm}^{-3}$, the equality of T_e and T needs $n \gtrsim 10^9 \sqrt{M/m} \simeq 3 \times 10^{11}\,\mathrm{cm}^{-3}$. We shall not consider T_i and T_e calculations (see Shimizu and Ashihara 1972).

In most of the ionosphere (up to ~ 220 km for Mars) the photochemical lifetime $\tau_C = l_i^{-1}$ is much smaller than the ambipolar diffusion time $\tau_D = \dfrac{H_i^2}{D_i}$ where $H_i = \dfrac{k(T_i + T_e)}{m_i g}$ is the ion scale height and D_i is the ambipolar diffusion coefficient. In this case the approximation of photochemical equilibrium is valid, and the calculations of ionospheric composition need a system of algebraic equations given by

$$q_i = [X_i] \sum_l k_l n_l$$

to be solved. One must take into account secondary ions resulting from ionic reactions along with primary ones produced by photoionization and electron impact ionization. This system may be reduced to a polynomial equation in terms of electron number density, n_e, by a substitution of X_i in the plasma neutrality equation, $n_e = \sum_i [X_i]$, and solved by Newton's method.

At $\tau_D < \tau_C$ ionic diffusion prevails, and the distributions of ion number densities are characterized by a scale height which is twice as large as that of neutral molecules of equal mass at a mean temperature of electrons and ions. To evaluate τ_D one must know the diffusion coefficients, D_i, being of the order of $(10^{16} T_i)/n\ \mathrm{cm}^2\,\mathrm{s}^{-1}$ for ions ($D \simeq (10^{17} T^{0.75})/n$ for neutral species).

In the intermediate region, where $\tau_D \sim \tau_C$, one should solve a system of continuity equations (see Introduction) with D_i dependent on altitude.

The Martian upper atmosphere is optically thin in the altitude range where the neutral number density varies from 10^8 to $3\times10^9\,\mathrm{cm}^{-3}$ and the exponential term in the photoionization relation may be neglected. In this range the approximation of photochemical equilibrium is valid, so

$$q_i = n_i \int_0^\infty I_\lambda \sigma_{i\lambda} \alpha_{i\lambda} d\lambda = n_i J_i$$

and the total ionization rate equals that of recombination:

$$q(h) = \sum_i J_i n_{0i} e^{-(h/H_i)} \simeq J n_0 e^{-(h/H)} = \alpha n_e^2(h)\,.$$

Here, J, α, H are the mean values of the ionization and recombination coefficients and scale height, respectively. Hence,

$$n_e(h) = \left(\frac{J n_0}{\alpha}\right)^{1/2} e^{-(h/2H)}$$

the ionospheric scale height is twice as large as that of the neutral atmosphere. Therefore, below a region of $\tau_D \sim \tau_C$ an ionospheric scale height describes the temperature and molecular mass of neutral molecules and, above this region, those of ions. The level of $\tau_D \sim \tau_C$ almost coincides with that of $[CO_2] \simeq [O]$ so the increase of the ionospheric scale height occurs here due to both the change in ionospheric conditions described and the sharp increase of the neutral atmosphere scale height.

The *ionospheric composition on Mars* was calculated by Stewart et al. (1972) and Krasnopolsky (1975) on the basis of dayglow and radio-occultation measurements from Mariner 6 and 7 and by Hanson et al. (1977) and Fox and Dalgarno (1979b) on the basis of neutral and ionospheric composition measurements by the Viking entry probes (Nier and McElroy 1977, Hanson et al. 1977). One can find some data on absorption and ionization cross-sections, rate coefficients, and solar radiation in reviews by Schunk and Nagy (1980) and Torr et al. (1979).

Figure 40 shows a neutral model atmosphere based on the Viking data. The homopause is assumed to be at 120 km where $K = 5\times10^7\,\mathrm{cm^2\,s^{-1}}$. Below the homopause gas mixing ratios are assumed to be constant with the number densities distributed according to the individual scale heights above it. Thus, the number density profiles adopted are inadequate near the homopause, though being quite applicable for ionospheric calculations.

The atomic oxygen number densities were not measured by the Viking mass spectrometers and the curve in Fig. 40 is chosen to fit $[CO_2^+]$ measurements by the relation

$$[O] = \frac{k_4}{k_1 + k_2}\,\frac{[O_2^+]}{[CO_2^+]}\,n_e\,.$$

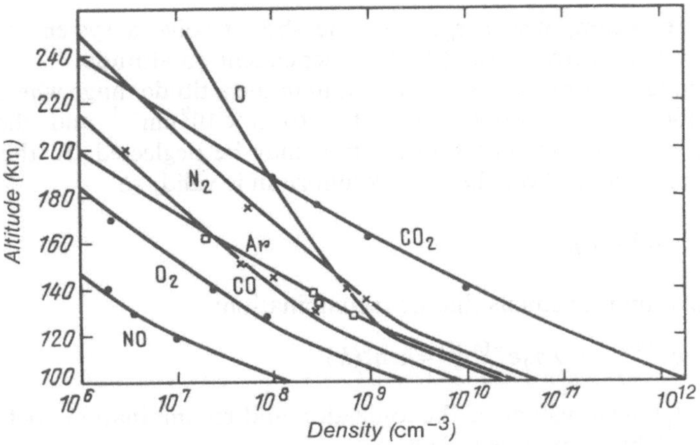

Fig. 40. Neutral composition of the day side upper atmosphere based in measurements of Viking 1 and adopted by Fox and Dalgarno (1979b) for ionospheric calculations

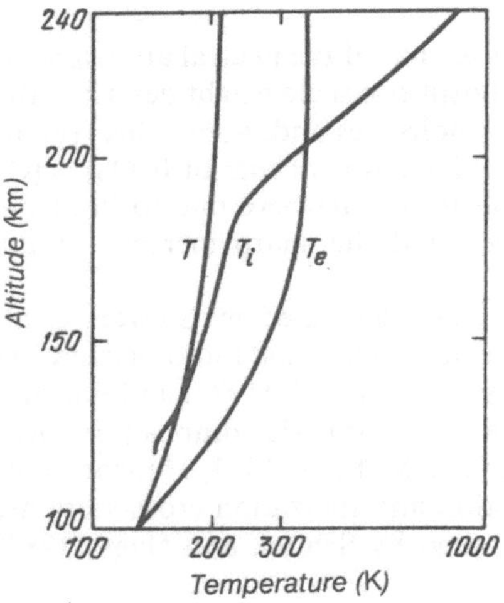

Fig. 41

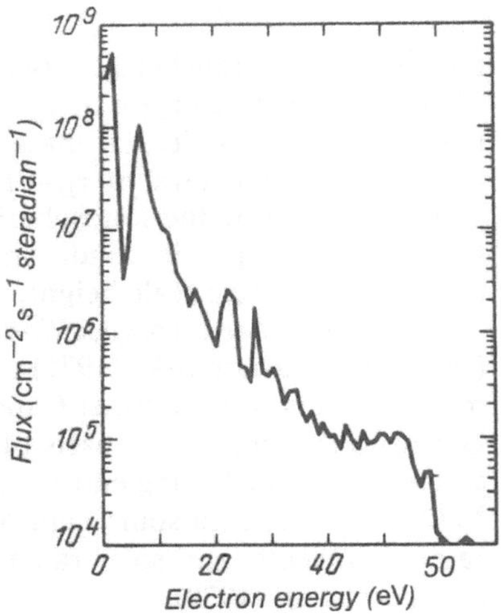

Fig. 42

Fig. 41. Day side neutral, ion, and electron temperatures adopted by Fox and Dalgarno (1979b). T_i was taken from data by Hanson et al. (1977)

Fig. 42. Photoelectron flux at 130 km calculated by Fox and Dalgarno (1979b)

The ionospheric composition measurements result in $[O_2^+]/[CO_2^+] \simeq 8\%$ and $[O]/[CO_2] \simeq 3\%$ at 135 km in accordance with the estimates obtained by Krasnopolsky (1975) from the CO_2^+ band spectroscopy and those of Kurt et al. (1974) from atomic oxygen dayglow measurements.

Exospheric temperature (Fig. 41) is assumed to be 225 K to fit ion composition measurements (Hanson et al. 1977). This value is slightly in excess of that deduced from mass spectrometric measurements (see Fig. 38). The ion temperature profile adopted is in accordance with measurements as well. T_e was chosen to fit the observed O_2^+ number densities and appears to be too low ($T_e \ll T_i$).

The calculated photoelectron fluxes are shown in Fig. 42. The integrated production rates of the different electronic states of CO_2^+, N_2^+, O^+, and Ar^+ ions as well as CO^+, C^+, N^+, and some part of O^+ ions formed by dissociative ionization are given in Table 7. We have added some CO^+ production from the direct photoionization of CO. Chemical reactions in the Martian atmosphere are listed in Table 8. The ion number densities calculated in the approximation of photochemical equilibrium are shown in Fig. 43 along with the measured ones. Above 200 km O^+ diffusion dominates, thus making the calculation invalid.

Figure 43 does not show CO^+ number densities. From Tables 7 and 8 one may obtain $[CO^+] \simeq [N_2^+]$, which is confirmed by calculations (see Fig. 33).

Recently, the ionospheric composition was recalculated by Fox and Dalgarno (1983) by means of a continuity equation for 12 ions and 5 neutral species. An improved photochemistry which was developed for the interpretation of the Venusian ionospheric composition and included metastable oxygen ions and nitrogen atoms (see Chap. 4.3, Table 32) was used. Figure 43

Table 7. Integrated ion production rates (Fox and Dalgarno 1979b)

Species	Photo-ionization	Electron impact ionization	Total
CO_2^+			
($X^2\Pi_g$)	3362[a]	2219	5581
($A^2\Pi_u$)	1415	397	1812
($B^2\Sigma_u^+$)	1548	237	1785
C^+	57	52	109
CO^+	250[b]	118[b]	368[b]
O^+	694	87	781
Ar^+	52	58	110
N_2^+	184	75	259
($X^2\Sigma_g^+$)	67	59	126
($A^2\Pi_u$)	83	12	95
($B^2\Sigma_u^+$)	17	4	21
N^+	31	11	42
O^+	192	105	297
(^4S)	73	36	109
(^2D)	64	45	109
(^2P)	39	24	63
($^4P^0$)	11	...	11
($^2P^0$)	6	...	6

[a] Values are given in units of $10^6\ cm^{-2}\ s^{-1}$
[b] See text

Table 8. Chemical reactions in the Martian ionosphere (Fox and Dalgarno 1979b)

Reaction no.	Reaction	Rate coefficient, $cm^3 s^{-1}$	Energy released, eV
(R1)	$CO_2^+ + O \rightarrow CO + O_2^+$	$k_1 = 1.6 \times 10^{-10}$	1.33
(R2)	$CO_2^+ + O \rightarrow CO_2 + O^+$	$k_2 = 1.0 \times 10^{-10}$	0.13
(R3)	$O^+ + CO_2 \rightarrow O_2^+ + CO$	$k_3 = 9.4 \times 10^{-10}$	1.20
(R4)	$O_2^+ + e \rightarrow O + O$	$k_4 = 1.9 \times 10^{-7}(300/T_e)^{0.5}$	4.76
(R5)	$CO_2^+ + e \rightarrow CO + O$	$k_5 = 3.8 \times 10^{-7}$	4.56
(R6)	$CO_2^+ + NO \rightarrow NO^+ + CO_2$	$k_6 = 1.2 \times 10^{-10}$	4.51
(R7)	$CO_2^+ + O_2 \rightarrow CO_2 + O_2^+$	$k_7 = 5 \times 10^{-11}$	1.71
(R8)	$N_2^+ + CO_2 \rightarrow N_2 + CO_2^+$	$k_8 = 7.7 \times 10^{-10}$	1.81
(R9)	$N_2^+ + O \rightarrow NO^+ + N(^2D)$	$k_9 = 1.4 \times 10^{-10}(300/T_n)^{0.4}$	0.94
(R10)	$N_2^+ + e \rightarrow N + N(^2D)$	$k_{10} = 3.5 \times 10^{-7}(300/T_e)^{0.5}$	3.44
(R11)	$N_2^+ + CO \rightarrow N_2 + CO^+$	$k_{11} = 7.4 \times 10^{-11}$	1.57
(R12)	$Ar^+ + CO_2 \rightarrow Ar + CO_2^+$	$k_{12} = 4.6 \times 10^{-10}$	1.97
(R13)	$Ar^+ + N_2 \rightarrow Ar + N_2^+$	$k_{13} = 2.2 \times 10^{-11}$	0.18
(R14)	$N^+ + CO_2 \rightarrow N + CO_2^+$	$k_{14} = 1.1 \times 10^{-9}$	0.78
(R15)	$NO^+ + e \rightarrow N(^2D) + O$	$k_{15} = 2.3 \times 10^{-7}(300/T_e)^{0.5}$	0.38
(R16)	$O_2^+ + NO \rightarrow NO^+ + O_2$	$k_{16} = 4.4 \times 10^{-10}$	2.71
(R17)	$C^+ + CO_2 \rightarrow CO^+ + CO$	$k_{17} = 1.1 \times 10^{-9}$	2.91
(R18)	$CO^+ + CO_2 \rightarrow CO_2^+ + CO$	$k_{18} = 1.1 \times 10^{-9}$	0.24

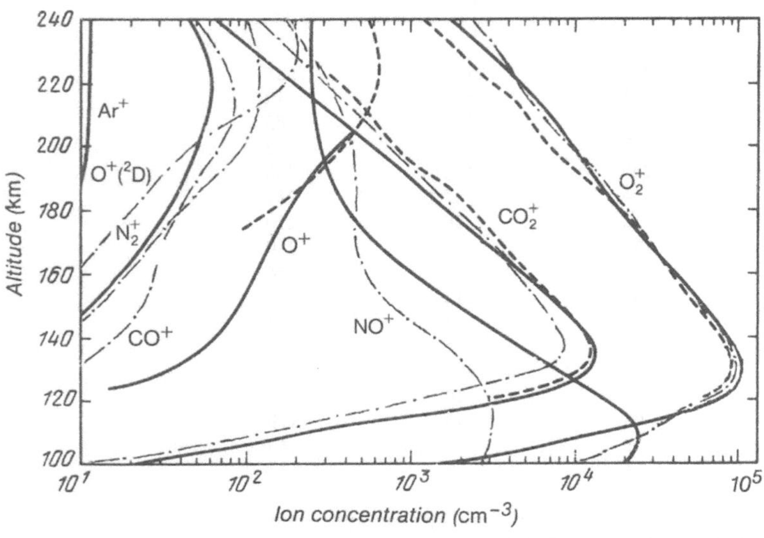

Fig. 43. Martian ionospheric composition calculated by Fox and Dalgarno (1979b) (*solid lines*), measured by Viking 1 (*dashed lines*), and calculated by Fox and Dalgarno (1983) (*dot-dashes*)

demonstrates the results of the calculations. A twofold difference in $[N_2^+]$ at 200 – 240 km is due to the reaction

$$O^+(^2D) + N_2 \rightarrow O + N_2^+ .$$

An order of magnitude difference in $[NO^+]$ at 100 – 120 km was not discussed by Fox and Dalgarno (1983). In our opinion it is due to very high neutral NO number densities measured by the Viking mass spectrometers which greatly exceed those from photochemical calculations (see Chap. 2.2).

2.2 Photochemistry of Nitrogen

Ionospheric processes markedly affect the chemical composition of the neutral atmosphere. This is evidenced, for example, by the formation of nitrogen compounds. The most abundant of them, NO, was detected by the Viking mass spectrometers (Fig. 30) with a mixing ratio of $[NO]/[CO_2] \simeq 7 \times 10^{-5}$ at 120 km. The photochemistry of nitrogen in the Martian atmosphere was considered by Yung et al. (1977) and McElroy et al. (1976a, 1977).

Atomic nitrogen is mainly formed by the electron impact dissociation of N_2:

$$e + N_2 \rightarrow e + N + N(^2D) .$$

Analysis of this process by Organ et al. (1975) has shown that half of all the atoms produced are electronically excited at a ^2D metastable state, giving 520 nm ($^4S - {}^2D$) airglow with a radiative lifetime of 10^5 s. Dissociate recombination of N_2^+ ions is an additional source of N atoms:

$$N_2^+ + e \rightarrow N + N(^2D)$$

though most of the N_2^+ ions are removed by a charge exchange reaction with CO_2. Production rates of both processes are shown in Fig. 44. However, this figure does not include the following reactions

$$N_2^+ + O \rightarrow NO^+ + N(^2D)$$
$$NO^+ + e \rightarrow N(^2D) + O .$$

The contribution of these reactions to atomic nitrogen formation somewhat exceeds that of N_2^+ recombination. Also not taken into account is the N_2 predissociation in a spectral range of 800 – 1000 Å:

$$N_2 + h\nu \rightarrow N + N(^2D)$$

the contribution of which is comparable with the photoelectron impact dissociation of N_2. Processes of atomic nitrogen formation as applied to the Earth's atmosphere were repeatedly considered in detail (see Oran et al. 1975, Strobel et al. 1976, Ogawa and Kondo 1977, Krasnopolsky 1979b) and are almost the same in the atmospheres of terrestrial planets. Their quantum yields

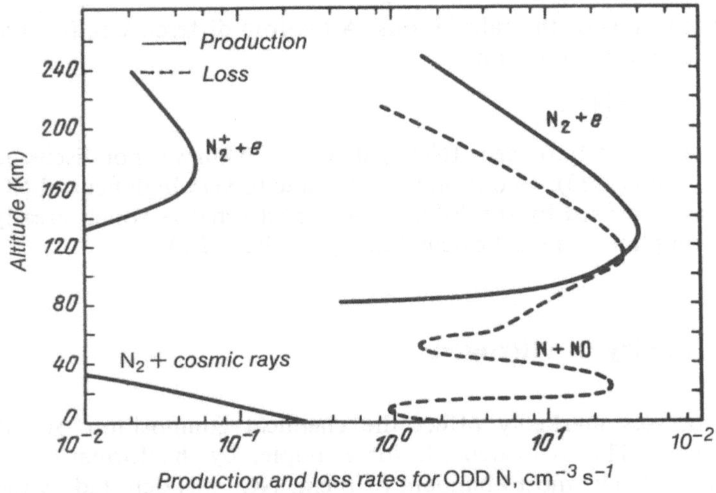

Fig. 44. Odd nitrogen production and loss rates (Yung et al. 1977)

of $N(^2D)$ are given from the papers cited. Almost all $N(^2D)$ atoms are converted into NO by

$$N(^2D)+CO_2 \rightarrow NO+CO$$

because physical quenching fails spin conservation. Nitric oxide reacts with odd oxygen O, O_3, and odd hydrogen H, OH, HO_2 to form a number of oxides NO_2, NO_3, N_2O_5, N_2O_4 and acids HNO_2 and HNO_3. As in the case of the Earth, nitrogen compounds vanish in the reaction

$$N+NO \rightarrow N_2+O$$

or on the planetary surface. A list of the reactions of nitrogen compounds and their rate coefficients are given in Table 9. Odd oxygen and hydrogen number densities were taken from the paper by Kong and McElroy (1977b). They considered two types of boundary conditions on the surface: a chemically

Table 9. Reactions and their rate coefficients for the nitrogen chemistry on Mars (Yung et al. 1977)[a]

$NO+h\nu \rightarrow N(^4S)+O(^3P)$	$J_1 = 1.724 \times 10^{-4}$
$NO_2+h\nu \rightarrow NO+O$	$J_2 = 1.767 \times 10^{-4}$
$NO_2+h\nu \rightarrow NO+O_2$	$J_3 = 4.31 \times 10^{-3}$
$N_2O_5+h\nu \rightarrow NO_2+NO_3$	$J_4 = 3.32 \times 10^{-5}$
$HNO_2+h\nu \rightarrow OH_2+NO$	$J_5 = 5.6 \times 10^{-4}$
$HNO_3+h\nu \rightarrow OH+NO_2$	$J_6 = 3.71 \times 10^{-5}$
$N_2O+h\nu \rightarrow N_2+O(^1D)$	$J_7 = 3.19 \times 10^{-7}$
$N(^4S)+NO \rightarrow N_2+O$	$k_0 = 2.2 \times 10^{-11}$
$N(^4S)+OH \rightarrow NO+H$	$k_1 = 6.8 \times 10^{-11}$

[a] Photolysis rates are in s^{-1} for optically thin atmosphere.

Table 9 (continued)

$N(^4S) + O_2 \rightarrow NO + O$	$k_2 = 1.5 \times 10^{-11} \exp(-3580/T)$
$N(^4S) + NO_2 \rightarrow N_2O + O$	$k_3 = 7.9 \times 10^{-12}$
$N(^4S) + NO_2 \rightarrow NO + NO$	$k_4 = 6.1 \times 10^{-12}$
$N(^4S) + O_3 \rightarrow NO + O_2$	$k_5 = 2.0 \times 10^{-12} T^{1/2} \exp(-1200/T)$
$NO + O_3 \rightarrow NO_2 + O_2$	$k_6 = 1.5 \times 10^{-12} \exp(-1330/T)$
$NO + O \rightarrow NO_2 + h\nu$	$k_7 = 6.4 \times 10^{-17}$
$NO + O + M \rightarrow NO_2 + M$	$k_8 = 6.0 \times 10^{-33} \exp(965/T)$
$NO + OH + M \rightarrow HNO_2 + M$	$k_9 = \dfrac{1.8 \times 10^{-11} \exp(806/T)}{2.3 \times 10^{20} + M}$
$NO + HO_2 \rightarrow OH + NO_2$	$k_{10} = 2.0 \times 10^{-11} \exp(-1200/T)$
$NO_2 + O_3 \rightarrow NO_3 + O_2$	$k_{11} = 1.3 \times 10^{-13} \exp(-2475/T)$
$NO_2 + O \rightarrow NO + O_2$	$k_{12} = 9.1 \times 10^{-12}$
$NO_2 + OH + M \rightarrow HNO_3 + M$	$k_{13} = \dfrac{3.5 \times 10^{-13} \exp(880/T)}{2.6 \times 10^{18} + M}$
$NO_2 + O + M \rightarrow NO_2 + M$	$k_{14} = 5.0 \times 10^{-31}$
$HNO_2 + O \rightarrow OH + NO_2$	$k_{15} = 1.6 \times 10^{-11} \exp(-1860/T)$
$HNO_2 + OH \rightarrow H_2O + NO_2$	$k_{16} = 5.0 \times 10^{-13}$
$HNO_3 + OH \rightarrow H_2O + NO_3$	$k_{17} = 8.9 \times 10^{-14}$
$HNO_3 + O \rightarrow OH + NO_3$	$k_{18} = 1.0 \times 10^{-11} \exp(-1860/T)$
$HNO_3 + H \rightarrow H_2 + NO_3$	$k_{19} = 1.0 \times 10^{-12} \exp(-1180/T)$
$HNO_3 + H \rightarrow OH + HNO_2$	$k_{20} = 1.0 \times 10^{-11} \exp(-1180/T)$
$NO_2 + NO_3 \underset{r}{\overset{f}{\rightleftarrows}} N_2O_5^*$	$\overset{f}{k_{21}} = 1.0 \times 10^{-12}$
	$\overset{r}{k_{21}} = \dfrac{1.0 \times 10^3 + M}{2.6 \times 10^{19} + M} + 1.0 \times 10^7$
$N_2O_5 + M \underset{r}{\overset{f}{\rightleftarrows}} N_2O_5^* + M$	$\overset{f}{k_{22}} = 0.9 \times 10^{-5} \exp(-9700/T)$
	$\overset{r}{k_{22}} = 3.7 \times 10^{-11}$
$N(^4S) + CO_2 \rightarrow NO + CO$	$k_{23} < 1.0 \times 10^{-19}$
$N + O_2^+ \rightarrow NO^+ + O$	$k_{24} = 1.8 \times 10^{-10}$
$N + CO_2^+ \rightarrow NO^+ + CO$	$k_{25} = 1.0 \times 10^{-11}$
$NO + O_2^+ \rightarrow NO^+ + O_2$	$k_{26} = 6.3 \times 10^{-10}$
$NO + CO_2^+ \rightarrow NO^+ + CO_2$	$k_{27} = 1.2 \times 10^{-10}$
$NO^+ + e \rightarrow N(^2D) + O$	$k_{28} = 1.0 \times 10^{-7} \exp(T/1000)^{-1.5}$
$N(^2D) + CO_2 \rightarrow NO + CO$	$k_{29} = 1.8 \times 10^{-13}$
$N_2 + e \rightarrow 2N(^2D) + e$	k_{30}
$\rightarrow 2N(^4S) + e$	k_{31}
$N_2O + O(^1D) \rightarrow 2NO$	$k_{32} = 3.0 \times 10^{-10} \exp(430/T)$
$N_2^+ + e \rightarrow N(^2D) + N(^2D)$	k_{33}
$\rightarrow N(^2D) + N(^4S)$	k_{34}
$\rightarrow N(^4S) + N(^4S)$	k_{35}
$N_2^+ + O \rightarrow NO^+ + N(^2D)$, 90–100% $\rightarrow NO^+ + N(^4S)$, 0–10%	$k_{36} = 1.4 \times 10^{-10} \left(\dfrac{300}{T_1}\right)^{0.44}$
$N_2^+ + NO \rightarrow NO^+ + N_2$	$k_{37} = 3.3 \times 10^{-10}$
$N^+ + CO_2 \rightarrow CO_2^+ + N$	$k_{38} = 1.3 \times 10^{-9}$
$N^+ + CO \rightarrow CO^+ + N$	$k_{39} = 5.0 \times 10^{-10}$
$N^+ + O_2 \rightarrow O_2^+ + N$	$k_{40} = 3.0 \times 10^{-10}$
$N^+ + O_2 \rightarrow NO^+ + O$	$k_{41} = 3.0 \times 10^{-10}$
$N^+ + NO \rightarrow NO^+ + N$	$k_{42} = 8.0 \times 10^{-10}$
$O^+ + N_2 \rightarrow NO^+ + N(^4S)$	$k_{43} = 6.0 \times 10^{-13}$
$O^+ + NO \rightarrow NO^+ + O$	$k_{44} = 2.0 \times 10^{-11}$

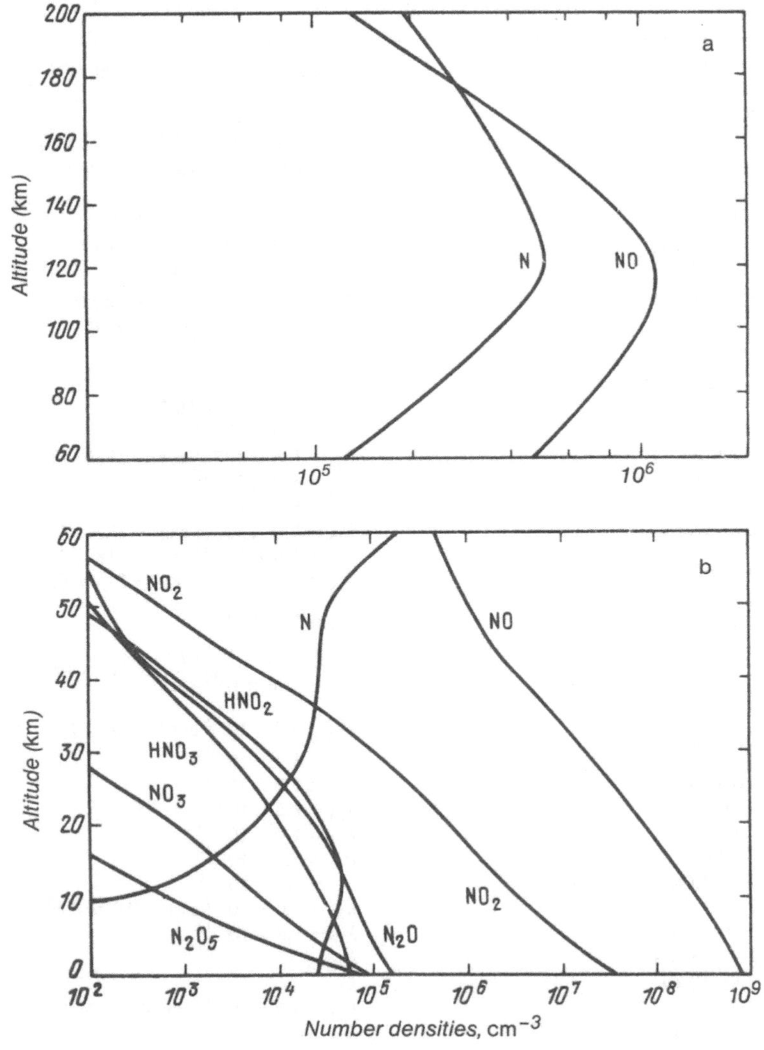

Fig. 45a, b. Number densities of nitrogen compounds (Yung et al. 1977)

passive surface with fluxes of components equal to zero, and a chemically active surface with fluxes of HNO_2 and HNO_3 equal to $-\gamma_i n_i v_i$ where γ_i is the rate coefficient of a corresponding heterophase reaction and v_i is the thermal velocity. The calculated number densities of the nitrogen compounds for the passive surface are shown in Fig. 45. The most abundant are NO, N, and NO_2. The results indicate clearly that a blue haze on Mars cannot consist of N_2O_4 (see Sagan et al. 1965) with a number density of only about $10\ cm^{-3}$.

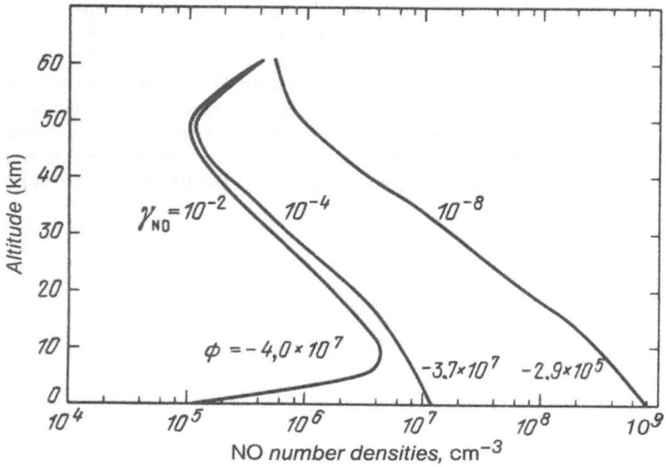

Fig. 46. Nitric oxide number densities calculated with various rate coefficients, γ_{NO}, of a heterogeneous sink at the surface. $\phi(0)$ is the flux of NO to the surface (Yung et al. 1977)

The main source of odd nitrogen compounds is atomic nitrogen produced by ionospheric reactions. These compounds appear in the middle and lower atmosphere as a result of downward transport from the ionosphere, and their concentrations depend strongly on the transport rate and surface sink (Fig. 46), the former being defined by the eddy diffusion coefficient. All calculations were performed for a very high K equal to $10^8\ cm^2\ s^{-1}$ throughout the atmosphere. A surface may be considered passive if the surface reaction rate coefficient is small, $\gamma \lesssim 10^{-8}$. With γ increasing up to 10^{-4} NO number densities are descreasing rather significantly below 60 km, and further increase of γ does not produce any effect above $\sim$10 km. For lower K, the solution become less sensitive to the chosen value of γ.

If the surface sink of nitrogen is rather high (γ exceeds 10^{-2}) and leads to the formation of stable components like $Ca(NO_3)_2$ and other nitrites and nitrates, the rate of this deposition under present conditions is about 6×10^5 N equivalent atoms $cm^{-2}\ s^{-1}$. Thus, the atmosphere could supply a surface layer 0.3 cm thick in 4.5×10^9 yr. In the past, the atmospheric abundance of N_2 could have been higher, so this value may be higher, too.

Calculated NO concentrations turned out to be smaller than the measured ones (Fig. 47). In our opinion, this can be partly accounted for by the lack of some processes of atomic nitrogen formation in the calculations (see above). McElroy et al. (1976a, 1977) obtained higher [NO] (curve b in Fig. 47) with electron impact dissociation cross-sections increased by a factor of 5 for $E<40$ eV. The NO number densities are still higher if the $N(^2D)$ yield is increased up to 75% for N_2 dissociation by electrons and cross-sections simultaneously increased by a factor of 3 for $E<40$ eV (curve c in Fig. 47). The

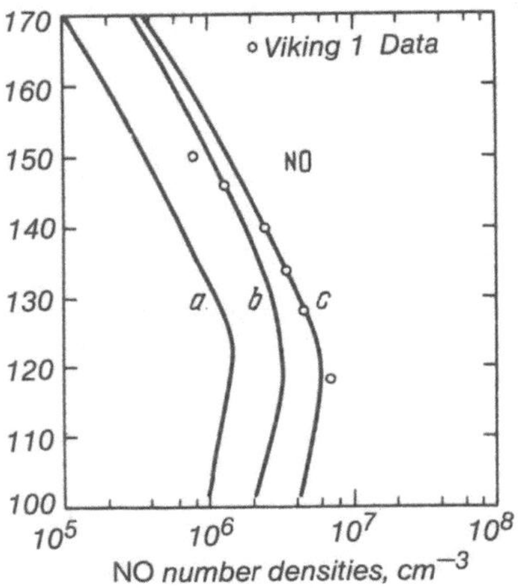

Fig. 47. Nitric oxide in the Martian upper atmosphere. *Circles* are measurements, *curve a* represents calculations by Yung et al. (1977), *curve b* was calculated with N_2 electron impact dissociation cross-sections increased by a factor of 5 for $E < 40$ eV, *curve c* is the same by a factor of 3 and with the $N(^2D)$ yield by ionospheric reactions equal to 75% (McElroy et al. 1976a)

curve correlates with the observations rather well, and a steeper decrease of NO at $h > 140$ km may be attributed to a colder atmosphere than adopted in the calculations.

However, new laboratory measurements confirm initial values of the cross-sections discussed (see Zipf and McLaughlin 1978). Krasnopolsky (1983b) considered the problem of N and NO for both Martian and Venusian upper atmospheres. His analysis confirmed mass spectrometric measurements of N in the Venusian atmosphere, while NO number densities measured in the Martian atmosphere and NO nightglow intensity in the Venusian atmosphere tend to be overestimated (see Chap. 4). On Mars [NO] $\simeq$ [N] $\simeq 1.5 \times 10^6$ cm^3 at 125 km in accordance with curve a in Fig. 47.

2.3 H_2 Dissociation and Escape of Atomic Hydrogen

Another important result of ionospheric processes is dissociation of molecular hydrogen occurring in the following reaction:

$$H_2 + CO_2^+ \rightarrow CO_2H^+ + H$$
$$CO_2H^+ + e \rightarrow CO_2 + H \, .$$

The rate coefficient of the first reaction is 1.4×10^{-9} cm^3 s^{-1} and the recombination coefficient of CO_2H^+ is assumed to be 3×10^{-7} cm^3 s^{-1}.

Molecular hydrogen forms in the lower atmosphere as a result of the interaction between products of H_2O and CO_2 photolysis. Upward transport supplies it to the ionosphere where some part of H_2 undergoes dissociation. This forms near the ionospheric maximum upward and downward fluxes of atomic hydrogen. The upward flux escapes completely together with some portion of H_2. The escaping flux, due to thermal energies of the constituents, is given by

$$\phi = n_c \sqrt{\frac{kT}{2\pi m}} \left(1 + \frac{\gamma M m}{kTR_c}\right) e^{-\frac{\gamma M m}{kTR_c}}$$

where n_c is the number density of an escaping component at a critical level in the atmosphere where the mean free path of molecules is equal to an atmospheric scale height (200 – 250 km on Mars for different solar activity), T is the temperature, m is the molecular mass, γ is the gravitational constant, M is the mass of a planet, and R_c is the distance from the center of a planet to its critical level. The power of the exponent is the ratio of the gravity energy of molecules to their thermal energy.

The problem of diffusion and escape of H_2 and H in the Martian thermosphere as applied to Mariner 6, 7, and 9 hydrogen airglow measurements was discussed by Anderson (1974). He adopted $[CO_2^+] = 0.3\ n_e$ with n_e derived from radio-occultation measurements. This relation agrees with the model ionosphere of Barth et al. (1972) and appears to be somewhat overestimated (see Chap. 2.1). The eddy diffusion coefficient was assumed to be $K = 5 \times 10^8$ $cm^2\,s^{-1}$ and independent of height, which is also too high even near the ionospheric maximum (see Chap. 1.10). At a lower boundary of 80 km, the H_2 mixing ratio was varied to fit the observations. According to Mariner 6 and 7 data, $n_c = 2.5 \times 10^4\ cm^{-3}$ and $T_c = 350$ K with an optical thickness of $\tau = 2.2^{+1.2}_{-0.6}$ at 80 km. Figure 48 illustrates a good agreement between the altitude profile of [H] deduced from 121.6 nm airglow measurements and the calculated one with $f_{H_2} = 2 \times 10^{-5}$ at 80 km. Figure 49 demonstrates the H_2 number densities and the atomic hydrogen production, $p(z)$, in the reactions of H_2 with CO_2^+ for the solution obtained. Atmospheric temperature and the thermal diffusion factor were assumed to be 350 K and 0.25 respectively, and the flux of atomic hydrogen escape was $\phi_H = 1.4 \times 10^8\ cm^{-2}\,s^{-1}$. The H_2 number density exceeds by an order of magnitude that of H at the critical level. Fluxes of atomic and molecular hydrogen versus height are given in Fig. 50. The H_2 flux from the lower atmosphere constitutes $3 \times 10^8\ cm^{-2}\,s^{-1}$, with most of the H_2 molecules undergoing dissociation in the ionosphere and only about 10% escaping. The H_2 dissociation gives rise to the escaping and downward fluxes of H.

As was mentioned above, the assumed K and $[CO_2^+]$ are somewhat overestimated. The smaller K is, the lower is the homopause and the higher are the H_2 ionospheric number densities at the same mixing ratio at the lower boundary. Therefore, some compensation occurs and the H_2 dissociation rate changes only slightly. Further, in the model obtained, the atomic hydrogen number density was varied at the critical level to fit the Mariner 9 observations which give $\tau = 5^{+5}_{-1}$ at 80 km. $[CO_2^+]$, K, f_{H_2}, and T were assumed to be the same. Indeed, as will be shown later the H_2 lifetime constitutes 200 yr, thus, f_{H_2} cannot markedly change in 2 yr. According to the measurements, $T_c = 325$ K for the condition of Mariner 9 observations rather than 350 K for the Mariner 6 and 7 and electron number density profiles are very much the same for both cases. The best agreement with Mariner 9 observations is for

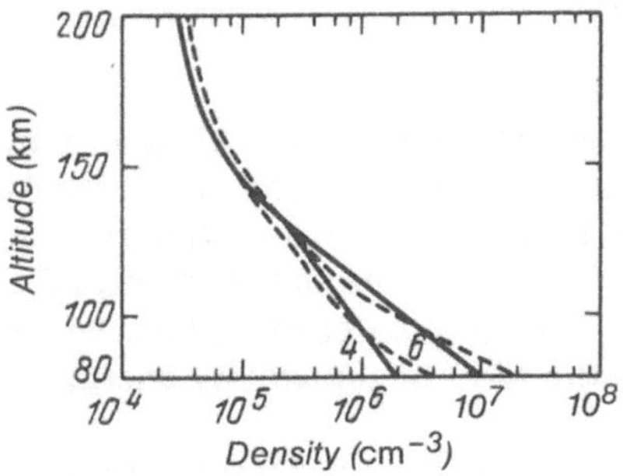

Fig. 48. Altitude profiles of atomic hydrogen number densities. *Solid lines* show Mariner 6, 7 and Mariner 9 data which can be described by models 4 and 6, respectively (see Fig. 25), *dashed lines* are the results of photochemical calculations with $f_{H_2} = 2 \times 10^{-5}$ (Anderson 1974)

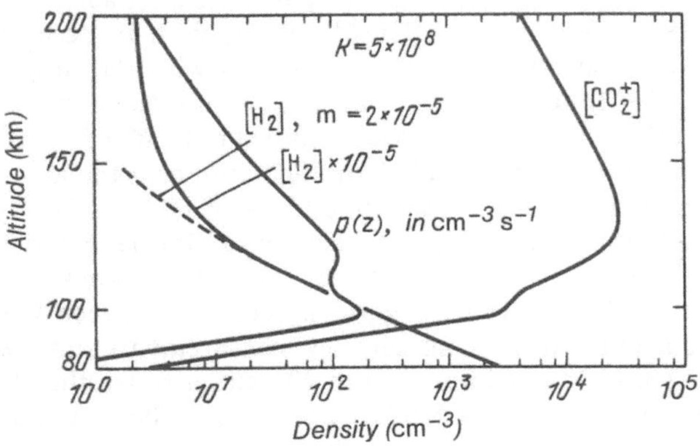

Fig. 49. Calculated H_2 number densities and production rate of H with $K = 5 \times 10^8\ cm^2\ s^{-1}$ and CO_2^+ shown (Anderson 1974)

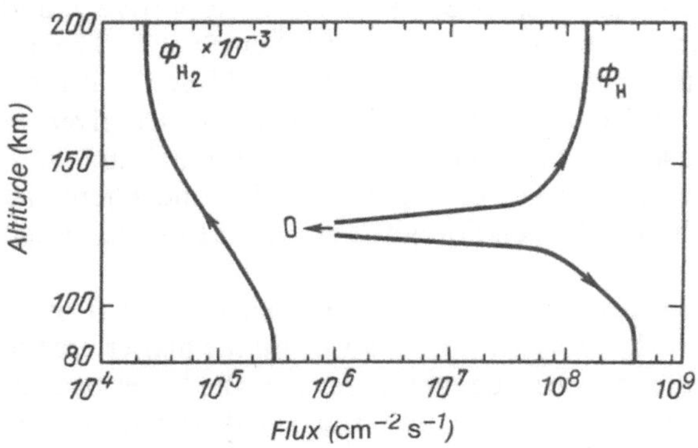

Fig. 50. H and H_2 fluxes in the upper atmosphere (Anderson 1974)

$n_c = 3 \times 10^4\ cm^{-3}$ and $\phi_H = 1.7 \times 10^8\ cm^{-2}\ s^{-1}$ (Fig. 48). The difference in [H] for both solutions is 20% at the critical level and five fold at 80 km. This difference is attributed to a downward flux decrease of about 20%. This decrease may result from a more stable lower atmosphere and correspondingly a smaller K below 80 km during Mariner 9 observations. Reduction of the sink of H below 80 km produces a similar effect.

Thus, the optical depth of the thermosphere and atomic hydrogen amount at 80 km are very sensitive to small variations of the flux of H. In our opinion, this high sensitivity makes the initial assumption of CO_2^+, K, and T being the same at the moments of Mariner 6 and 7 and Mariner 9 observations doubtful. For example, n_c varies by a factor of 1.4 with temperature varying from 350 to 325 K the escape flux to keep constant. However, the fact of such a strong variability in itself is certainly of interest.

2.4 Nonthermal Escape and Isotopic Composition of Oxygen and Nitrogen

Oxygen. With a mean rate of hydrogen escape of $\sim 10^8\ \mathrm{cm^{-2}\,s^{-1}}$ Mars should lose a total water amount equivalent to a layer 2 m thick during its age of 4.6 billion years. Molecular oxygen produced should exceed the measured abundance by a factor of 10^4. However, some processes of nonthermal escape of oxygen hinder its accumulation (McElroy 1972).

Parabolic velocity, which is necessary for the gravity field of Mars to be overcome, is equal to $4.7\ \mathrm{km\,s^{-1}}$ at the critical level. It corresponds to the energy of the oxygen atom of 1.97 eV. Oxygen atoms formed by recombination of O_2^+ and CO_2^+

$$O_2^+ + e \rightarrow O(^3P) + O(^1D)$$

$$CO_2^+ + e \rightarrow CO(X^1\Sigma) + O(^1D)$$

acquire a kinetic energy of 2.5 and 4 eV, respectively, and can leave the Martian atmosphere. The escape flux of atomic oxygen is given by

$$\phi_0 = 0.5\,\alpha n_{ec}^2 H_e$$

where n_{ec} is the electron number density at the critical level equal to that of O_2^+ ions, $\alpha = 1.6 \times 10^{-7} \left(\frac{300}{T_e}\right)^{1/2}\ \mathrm{cm^3\,s^{-1}}$ is the O_2^+ recombination rate coefficient, and H_e is the plasma scale height. The factor 0.5 is due to the following: (1) n_e^2 has a scale height of 0.5 H_e; (2) half of the oxygen atoms formed with upward velocities can escape; (3) each recombination act gives two oxygen atoms. Under the conditions of Mariner 6 and 7 observations $n_{ec} = 1.5 \times 10^4\ \mathrm{cm^{-3}}$, $H_e = 46$ km, $\alpha = 10^{-7}\ \mathrm{cm^3\,s^{-1}}$ at the critical level of 250 km, thus giving $\phi_0 = 0.5 \times 10^8\ \mathrm{cm^{-2}\,s^{-1}}$. Taking into account some contribution of photoelectron impact dissociation of CO_2 producing oxygen atoms with a rather high energy, the oxygen escape rate is half that of hydrogen, i.e., conforms to the stoichiometric ratio of H_2O. This relationship between the oxygen and hydrogen escape rates is not accidental, and a more detailed analysis made by McElroy and Donahue (1972) and Liu and Donahue

(1976) shows that a reduced oxygen escape rate (as a result of lower solar activity, for example) provides the increase of oxygen and ozone amounts in the lower atmosphere. With the constant H_2O photolytic rate the molecular hydrogen production should be at almost the same level too; however, part of the H_2 transported to the upper atmosphere should be reduced due to the destruction in the lower atmosphere in the following reactions

$$O_3 + h\nu \rightarrow O_2 + O(^1D)$$

$$O(^1D) + H_2 \rightarrow OH + H\,.$$

Thus, a decrease in the escape rate of O leads to the reduced escape rate of H, and the initial relation $2\phi_O = \phi_H$ will be maintained.

Nonthermal escape of atomic oxygen may increase the isotopic ratio $^{18}O/^{16}O$ in the atmosphere. Ground-based measurements of the absorption bands of $^{12}C^{16}O^{18}O$ and $^{13}C^{16}O_2$ have indicated strong similarity of the isotopic ratios of oxygen and carbon on Mars and on the Earth (see Moroz 1967, 1978). Much more precise measurements by the Viking landers (see Chap. 1.6, Table 5) confirm this result. The analysis of $^{18}O/^{16}O$ enrichment by nonthermal escape during the planetary lifetime has shown (Nier et al. 1976) that the upper limit for the observed enrichment of less than 3% implies a huge reservoir of oxygen in the form of CO_2 or H_2O on or under the surface, which is at least as large as 5×10^{25} atoms cm^{-2}, i.e., exceeds an equivalent atmospheric pressure of 2 bar.

Nitrogen (Nier et al. 1976, McElroy et al. 1976b, 1977). The isotopic abundance ratio $^{15}N/^{14}N$ on Mars is 1.75 times as high as that on the Earth with $f = {}^{15}N/^{14}N = 0.0064$ and 0.0037, respectively. The enrichment may arise as a result of preferential escape of ^{14}N through dissociative recombination of N_2^+ and electron impact dissociation of N_2. The former gives nitrogen atoms with velocities of 4.96 km s^{-1} and 4.7 km s^{-1} for ^{14}N and ^{15}N, respectively. To form N atoms with a sufficient energy to escape Mars, N_2 molecules should acquire an energy of excitation of at least 15.6 eV in the latter process. With an energy of excitation in excess of 19 eV autoionization is a dominant reaction path. According to calculations, about 16% of all dissociation events result in production of atoms with velocities sufficient for escape.

Given equal initial nitrogen isotopic ratios on the Earth and Mars, initial nitrogen abundance on Mars from the presently observed enrichment in ^{15}N can be deduced. A parameter $R = f_c/f$ is introduced into calculations to account for depletion in ^{15}N in the exosphere owing to diffusive separation where f_c is the isotopic ratio at the critical level. R depends on the eddy diffusion coefficient; the higher K is, the smaller is 1-R. The enrichment of the present atmosphere with respect to the initial one is given by $\varepsilon(t) = f(t)/f(0)$. With equal probabilities for both isotopes to escape from the exosphere one obtains

$$d^{15}N = f_c d^{14}N \quad \text{or} \quad \frac{d^{15}N}{^{15}N} = R\frac{d^{14}N}{^{14}N}\,.$$

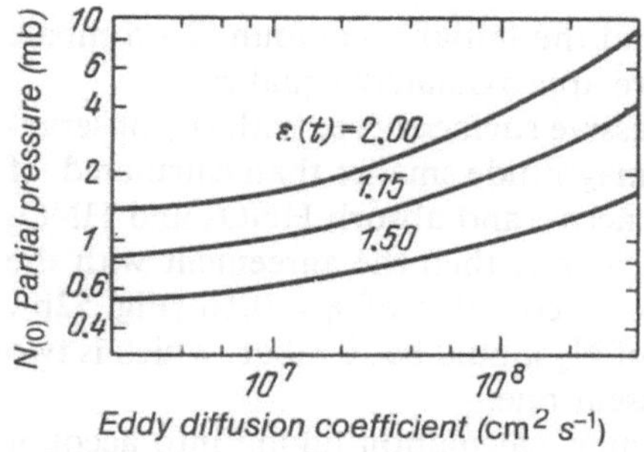

Fig. 51. Initial amounts of N_2 in the Martian atmosphere 4.6 b.y. ago required to supply enrichments of ^{15}N in the present atmosphere of the magnitude 50, 75, and 100%, as functions of the eddy diffusion coefficient in the upper atmosphere (Nier et al. 1976)

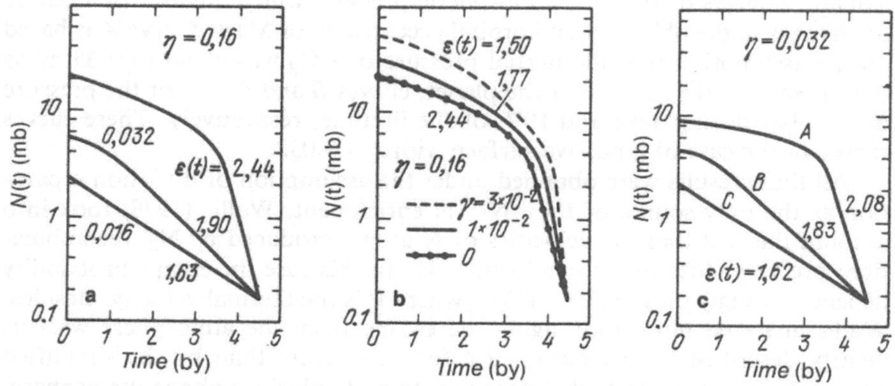

Fig. 52a–c. Abundance of N_2 as a function of time (in billion years) **a** for passive surface and different escape efficiencies, η, of electron impact dissociation of N_2; **b** for $\eta = 0.16$ and different rate coefficients, γ, of the heterogeneous sink of HNO_2 and HNO_3 at the surface; **c** for $\eta = 0.032$, passive surface, and climatic changes of CO_2 pressure: $A - p < 0.65$ mbar for one-third of a climatic cycle, B and $C - p = 65$ mbar for one-third and one-tenth of a climatic cycle, respectively. $K = 3 \times 10^8\ cm^2\ s^{-1}$ (McElroy et al. 1976b)

If R is assumed to be constant, then

$$^{15}N(t)/^{15}N(0) = [^{14}N(t)/^{14}N(0)]^R \quad \text{or} \quad \varepsilon(t) = [^{14}N(0)/^{14}N(t)]^{1-R}.$$

Values of nitrogen abundances on Mars 4.6 b.y. ago as functions of the eddy diffusion coefficient near the homopause are shown in Fig. 51. With $K = 10^8\ cm^2\ s^{-1}$ the initial abundance of nitrogen is 2 mbar, i.e., an order of magnitude higher than now.

The above calculation does not depend on the nature of escape processes and their efficiencies. It compares the initial and final isotopic ratios for the case of an adopted passive surface. The initial isotopic ratio on Mars is taken as equal to the current isotopic ratio on Earth.

The abundance of N_2 as a function of time is given in Fig. 52a for different escape efficiencies at electron dissociation of N_2. The surface is passive, i.e., does not absorb nitrogen compounds. To fit the observations the escape efficiency should be $\eta = 0.016 - 0.032$ and the initial N_2 amount $2-5$ mbar. With this η both processes of escape have approximately equal rates.

Therefore, the calculations for the passive surface agree with the observations for the values of η by an order of magnitude smaller than calculated. If the surface is assumed to be chemically active and absorb HNO_2 and HNO_3 with the heterogeneous reactivity coefficient γ, then the agreement with the observations can be obtained for a calculated value of $\eta = 0.16$ (Fig. 52b). However, in this case the initial amount of N_2 should be 30 mbar, which is two orders of magnitude higher than the present one.

Figure 52c shows the results of a similar calculation taking into account climatic changes of the CO_2 atmospheric pressure which might arise because of changes in the obliquity and orbital eccentricity of Mars. Curve *A* is based on the assumption that the partial pressure of CO_2 was equal to 0.65 mbar during 30% of the lifetime of the planet, curves *B* and *C* are for the pressure of 65 mbar during 30% and 10% of the lifetime, respectively. These curves represent the case of a passive surface with $\eta = 0.032$.

All these results were obtained under the assumption of diffusion separation as the only source of the $^{15}N/^{14}N$ enrichment. Wallis (1978) took into account the fact that the velocities of N atoms produced by N_2^+ recombinations are very close to a critical value, V_c. In this case the escape probability depends strongly on $y = (V^* - V_c)/V_i$ where V_i is the thermal velocity. Besides, Wallis analyzed more thoroughly the escape from the atmosphere with its density described by the barometric formula rather than by the simplified approach, which neglects the escape up to and collisions above the exobase. He obtained the escape probabilities of 42% and 24% for ^{14}N and ^{15}N, respectively, produced by N_2^+ recombination. According to the results obtained by Nier et al. (1976) and McElroy et al. (1976b), $R = 0.82$ with $K = 3 \times 10^8$ $cm^2 s^{-1}$, i.e., $^{15}N/^{14}N$ shows an 18% decrease in the exosphere as compared to the lower atmosphere. With the probabilities of the escape taken into account, the effective value of R is $0.82 \times (0.24)/(0.42) = 0.42$ for the recombination. For the ratio of escape rate for electron impact dissociation to that for N_2^+ recombination, β, the effective value of R is given by

$$R = (0.42 + 0.82\,\beta)/(1 + \beta)\,.$$

For the passive surface and $\beta = 0$ calculations give the enrichment $\varepsilon(t) = 1.51$ rather than the 1.75 measured. Both parameters β and the initial nitrogen amount can be obtained from the relations

$$\frac{^{14}N(t)}{^{14}N(0)} = 1.51^{-(1+\beta)} = 1.75^{-1/(1+R)}\,.$$

They result in $\beta = 4.3$, i.e., the contribution of the electron impact dissociation of N_2 into the escape of nitrogen atoms is four times more effective than N_2^+ recombination and the initial amount of N_2 is 1.4 mbar.

Table 9a. ^{14}N escape fluxes (Fox and Dalgarno 1983)

Process	Low solar flux	High solar flux	Average
$N_2 + h\nu \rightarrow N + N$	5.9×10^4	1.4×10^5	1.0×10^5
$N_2 + h\nu \rightarrow N^+ + N + e$	1.7×10^4	6.0×10^4	3.0×10^4
$N_2 + e \rightarrow N^+ + N + 2e$	1.4×10^4	4.8×10^4	3.1×10^4
$N_2^+ + e \rightarrow N + N$	8.4×10^4	5.0×10^5	2.9×10^5
$N_2^+ + O \rightarrow NO^+ + N$	4.7×10^4	1.3×10^5	8.9×10^4
$N_2 + O^{++} \rightarrow N^+ + N + O^+$	4.8×10^3	1.4×10^4	9.4×10^3
Total	2.3×10^5	8.9×10^5	5.6×10^5

In $cm^{-2} s^{-1}$

Further study of the problem was made by Fox and Dalgarno (1983), who used more detailed photochemistry (see Chap. 2.1) and took into account recent laboratory measurements which gave very low kinetic energy of nitrogen atoms produced by electron impact dissociation of N_2. Thus, this process can be eliminated. They found that only half the N_2^+ ions are not vibrationally excited, therefore, the isotopic fractionation for N_2^+ recombination is not as high as calculated by Wallis. Calculated escape fluxes are given in Table 9a. Integration over the lifetime of the planet resulted in an initial nitrogen amount of ~ 2 mbar and the isotopic enrichment of 2.5, much larger than the measured one. The enhanced solar fluxes and much larger CO_2 amounts in the past as well as the exchange with regolith materials may be responsible for this difference.

Thus, in the past the nitrogen amount on Mars exceeded the present values of 0.16 mbar and might have been in a range of 1 – 30 mbar. This conclusion is of great importance for the problem of the evolution of the solar system.

2.5 Dissociation of CO_2: Atomic Carbon in the Upper Atmosphere

Another result of the ionospheric processes is dissociation of CO_2. In fact each ionization of CO_2 and O produces CO and O in recombination:

$$CO_2 + h\nu \rightarrow CO_2^+ + e$$
$$CO_2^+ + O \rightarrow O_2^+ + CO$$
$$O_2^+ + e \rightarrow O + O$$

$$\text{net } CO_2 + h\nu \rightarrow O + CO$$

CO_2^+ recombination gives CO and O as well, and in the case of O^+ this process is preceded by charge exchange with CO_2. Therefore, a rather intense

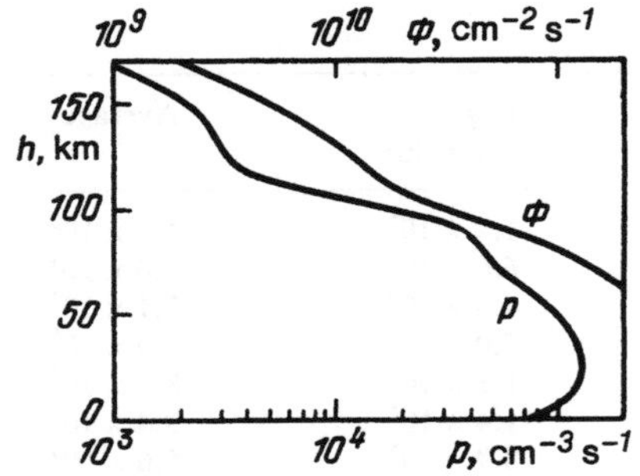

Fig. 53. CO_2 photolytic rate and downward flux of CO and O (Krasnopolsky 1981a)

formation of O and CO occurs at the ionospheric altitudes. The three body reactions to form O_2 and CO_2 are not effective here due to the low number densities of the components ($[CO_2] = 2 \times 10^{10}$ cm^{-3}, $f_O = (1-3)\%$, $f_{CO} = 1\%$ at the ionospheric maximum). At the end of the 1960's, low mixing ratios of O and CO in the Martian upper atmosphere were rather difficult to explain because on the Earth $[O]/[O_2] \simeq 2$ at 125 km where $[O_2] = 2 \times 10^{10}$ cm^{-3}. Now it is attributed to a considerably higher rate of eddy diffusion on Mars as compared to the Earth.

To calculate the CO_2 photolytic rate one needs absorption cross-sections, quantum yield of dissociation, and intensities of solar UV radiation. The cross-sections were taken from measurements of Lee et al. (1973) for $\lambda < 70$ nm, of Nakata et al. (1965) for a range of 70 – 170 nm, and of Shemansky (1972) at 170 – 210 nm. The quantum yield was assumed to be unity at all wavelengths except (1) at $\lambda < 40$ nm, where ionization by photoelectrons calculated on the basis of the data of Fox and Dalgarno (1979a) is possible; and (2) at a range of 195 – 210 nm with a considerable contribution of Rayleigh scattering in the extinction cross-sections. Intensities of solar UV radiation were utilized from Torr et al. (1979) and Mount et al. (1980). The CO_2 number densities used are those shown in Figs. 20 and 36. These curves correspond to observations by the Mariners, which gave the most accurate data on atomic oxygen altitude profiles. Calculations were made with a solar zenith angle of $z_0 = 60°$. To satisfy global average conditions the results obtained were reduced by a factor of 2 (Fig. 53). Processes of O and CO removal occur mostly in the lower atmosphere and may be neglected down to 50 – 60 km. This simplifies calculations of the number densities of these components at $h > 60$ km:

$$\phi_i(z) = -(K+D_i)\frac{dn_i}{dz} - n_i\left(\frac{K}{H} + \frac{D_i}{H_i}\right) = -\int_z^\infty p(x)\,dx\,.$$

Values of $\int_z^\infty p(x)\,dx$ are shown in Fig. 53, and $D_i = 10^{17}\,T^{0.75}/[CO_2]$ cm^2 s^{-1}.

For $z < 100$ km where $K \gg D$ one can integrate this equation:

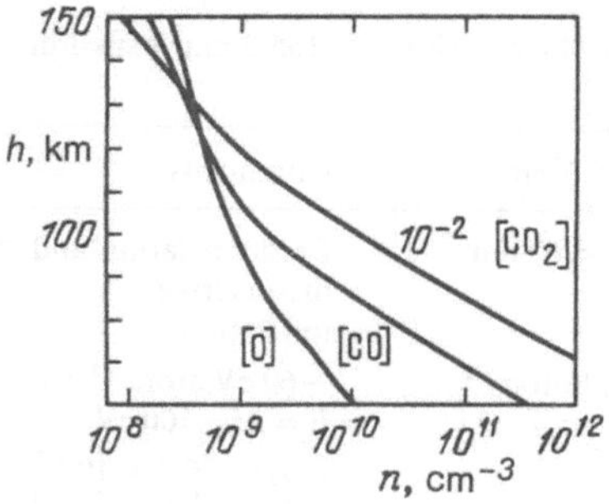

Fig. 54. CO and O number densities height for $K = 1.4 \times 10^{10} M^{-1/2}$ $cm^2 s^{-1}$ above 100 km and $K = 10^7$ $cm^2 s^{-1}$ below 100 km

$$\phi_i(z) = -K\left(\frac{dn_i}{dz} + \frac{n}{H}\right) = -K[CO_2]\frac{df_i}{dz} = -\int_z^\infty p(x)\,dx \quad \text{or}$$

$$f_i(z) = \int_{60\,\text{km}}^{z} \frac{\int_x^\infty p(y)\,dy\,dx}{K(x)[CO_2]} + f_i(60\,\text{km})\,.$$

As in the lower atmosphere, one may adopt $f_{CO} = 2 \times 10^{-3}$ at 60 km as a mean value of infrared and microwave spectroscopic data (Chap. 1.2). The atomic oxygen mixing ratio is unknown at 60 km; however, the results for $h \gtrsim 80$ km are rather insensitive to the value of $f_O(60\,\text{km})$, and thus, $[O] \simeq 3 \times 10^{10}$ cm^{-3} or $f_O(60\,\text{km}) = 7.5 \times 10^{-5}$ may be assumed.

The results of calculations with $K = 1.4 \times 10^{13} [CO_2]^{-1/2}$ $cm^2 s^{-1}$, being in best agreement with the Viking data (see von Zahn et al. 1980) above 100 km, and with $K = 10^7$ $cm^2 s^{-1}$ below 100 km, are shown in Fig. 54. The integral in the relation for f_i being smaller than $f_i(60\,\text{km})$ for CO up to ~100 km, the CO mixing ratio is nearly constant in this range. Diffusion equilibrium distribution approximates adequately O and CO number densities above 150 km. Number densities of CO at 90–150 km and of O below 150 km cannot be described by the barometric relation either with the average or individual scale heights, and the assumption of the average scale height below the homopause (115 km in our case) and of the individual scale height above it is not valid. This is due to the high production rates of carbon oxide and atomic oxygen.

Calculations give $f_{CO} = 1\%$ and $f_O \simeq f_{CO}$ at 135 km in accordance with the measurements (see Chap. 1.5). Therefore, the eddy diffusion coefficients derived from the Viking data in the upper atmosphere are applicable to the conditions of the Mariner 6 and 7 measurements despite a great difference in solar activity.

Atomic carbon in the upper atmosphere. The study of the photochemistry of atomic carbon was prompted by the identification of carbon lines in the Mars dayglow spectra (Fig. 19). The problem was discussed first by McElroy and McConnell (1971) and much later by Krasnopolsky (1982a). The processes of atomic carbon formation and excitation of 165.7 nm emission (the

Table 10. Processes of atomic carbon formation and their yields of 165.7 nm emission (Krasnopolsky 1982a)

No.	Reaction	Yield of 165.7 nm	Comments
1	$CO + h\nu(\lambda < 111.6\ \text{nm}) \rightarrow C + O$	0.4% for $\lambda < 67$ nm	Predissociation and dissociative ionization
2	$CO + e \rightarrow C + O + e$	~2000 eV photon^{-1}	~60 eV atom^{-1} for $E = 25 - 100$ eV
3	$CO^+ + e \rightarrow C + O$	0	$\alpha_{CO^+} = 6.5 \times 10^{-7}$ $(300/T_e)^{1/2}$ $cm^3\ s^{-1}$
4	$CO_2 + h\nu(\lambda < 75\ \text{nm}) \rightarrow C + O + O$	0.24% for $\lambda < 52$ nm	2% of the absorption cross-section
5	$CO_2 + e \rightarrow C + O + O + e$	~9000 eV photon^{-1}	~1000 eV atom^{-1}
6	$C + h\nu(\lambda = 165.7\ \text{nm}) \rightarrow C^*$	–	$g = 1.4 \times 10^{-5}\ s^{-1}$ on Mars

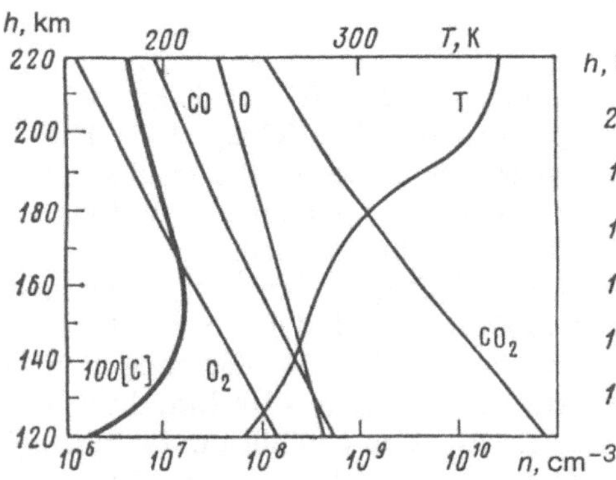

Fig. 55

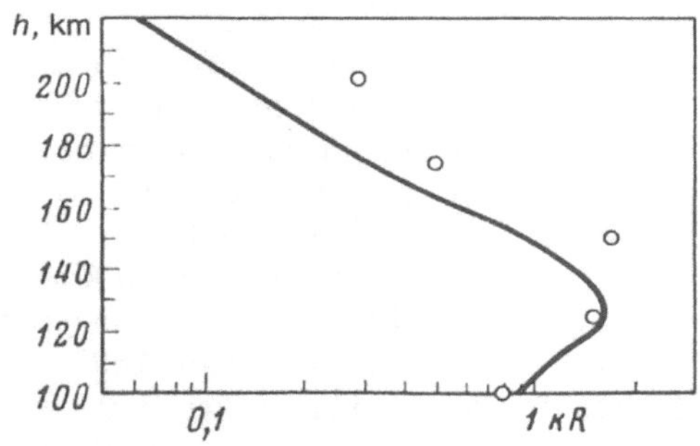

Fig. 56

Fig. 55. Atomic carbon number densities in the Martian atmosphere.calculated with altitude profiles of atmospheric components shown (Krasnopolsky 1982a)

Fig. 56. Atomic carbon 166 nm line intensities at the Martian limb measured by Mariner 6, 7 and calculated by Krasnopolsky (1982a)

strongest carbon line) are summarized in Table 10. The only important reaction of the removal of C atoms is

$$C + O_2 \rightarrow CO + O\,, \qquad k = 2.6 \times 10^{-11}\ cm^3\ s^{-1}\,.$$

Number densities of C atoms calculated for $F_{10.7} \simeq 170$ appropriate to Mariner 6 and 7 conditions are shown in Fig. 55. Atomic carbon is three orders of magnitude less abundant in the upper atmosphere than atomic oxygen. Resonance scattering of 165.7 nm radiation given a zenith intensity of $9\,R$ and only $45\,R$ at limb maximum, due to the effect of saturation. Other excitation mechanisms produce "hot" emission as soon as radiating C atoms have excess kinetic energies. Thus, their emission is not subjected to the effect of saturation. Calculated and measured limb intensities of 167.7 nm line are given in Fig. 56.

2.6 Diffusion and Photolysis of Water Vapor

Now we turn our attention to the photochemistry of the Martian atmosphere below 60 km where water vapor plays a significant part. The problem of water on Mars is closely connected with a search for biological species on the planet, with its geochemistry, origin, and evolution. Despite numerous measurements of water vapor amounts, very limited information about its vertical distribution is now available. The main results in this field are the Viking orbiter measurements indicating water vapor to be mixed uniformly with dust in the lowest scale height (see Chap. 1.4). Hence, the most natural assumption is a uniform vertical distribution of $\sim 15\ \mu m$ or $5 \times 10^{19}\ cm^{-2}$ water molecules which corresponds to a water vapor mixing ratio of 2×10^{-4}. However, ground-based spectroscopic measurements show the effective temperature of water vapor to be somewhat in excess of the mean atmospheric temperature at the same locations, thus implying excess water amounts in the lower atmosphere. Analysis of the problem by Hess (1976), Farmer (1976), and Flasar and Goody (1976) supported the idea that almost all water vapor on Mars is contained in a comparatively thin (~ 3 km) atmospheric layer near the surface. However, UV measurements on board the Mars 5 orbiter (Krasnopolsky 1979a, Krasnopolsky et al. 1979b) revealed that the aerosol layer formed by ice particles could reach 20 – 35 km and even higher in the evening.

Krasnopolsky and Parshev (1977) and Krasnopolsky et al. (1979) suggested calculating water vapor diffusion in the atmosphere with condensation on the surface and inside the boundary layer taken into account.

It is well known that atmospheric water vapor abundance is a strong function of latitude, season, and local time, and is highly variable from place to place. Therefore, data of observations should be taken carefully. We shall use low latitude data for $L_s = \pm 20°$. Such choice permits the use of the Mariner 9 measurements (surface and atmospheric temperatures, water vapor abundance) after a dust storm, as well as the measurements of Mars 5.

Boundary layer. The average content of water vapor at low latitudes is about 15 μm of precipitable water, which corresponds to its near-surface concentration of several $\mu m\ km^{-1}$. Figure 57 shows the diurnal variation of the surface temperature from the Mariner 9 IR radiometric data for the tropics on February 17, 1972 (Kieffer et al. 1973), as well as the corresponding number densities of saturated vapor. Condensation at the surface occurs during 12 h and 1 h later relative to the sunrise or sunset, due to the surface thermal inertia. During the cold period of a day water is more intensely absorbed by

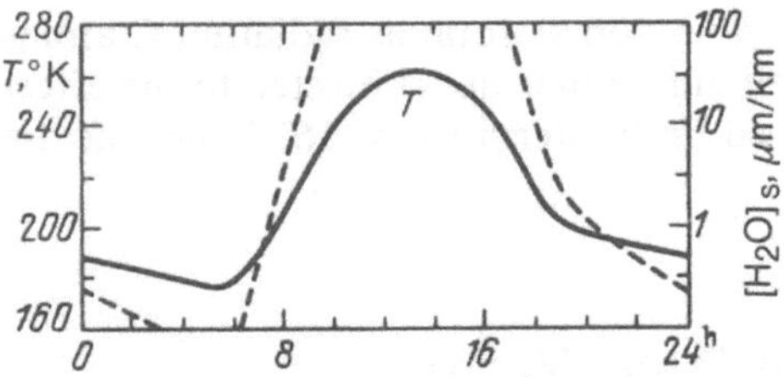

Fig. 57. Diurnal variations of surface temperature and saturated water vapor density (*dashed line*)

the rocks on the surface, during the warm period water returns into the atmosphere. The relationship between adsorption and condensation is not essential for our analysis, and only the latter will be discussed here. Temperature in most of the troposphere is subject to small diurnal variations. Hence, in the nighttime a deep temperature minimum forms near the surface, it gradually rises after sunrise, becoming more and more shallow. This occurs in the lower atmospheric layer about 1 km thick. Obviously, temperature inversion completely vanishes in 4 – 5 h after sunrise (for a more detailed description of a boundary layer temperature profile see Gierasch and Goody 1968, Hess 1976). Due to this, after sunrise the surface frost rapidly evaporates and condenses as ice fog in the temperature inversion region.

Water vapor diffusion in the atmosphere is described by the equation

$$\frac{\partial n}{\partial t} = K\left[\frac{\partial^2 n}{\partial z^2} + \frac{1}{H}\frac{\partial n}{\partial z}\right]$$

where $n = [H_2O]$, K is the eddy diffusion coefficient which is assumed height independent, and H is the scale height. The term with $\partial T/\partial z$ is negligible and is not used in the equation. This is the usual form of the one-dimensional eddy diffusion equation in the absence of sources and sinks in minor component approximation and constant K and H. It may be applied in the height range 0 to 20 km where there is no condensation (but not for the boundary layer). The location of the upper boundary is near the bottom of the aerosol layer according to the Mars 5 measurements. Since the thickness of the boundary layer is small as compared with the scale height, it is not significant for the solution whether condensation is on the surface or in the boundary layer. However, the duration of the condensation process, τ, (we shall refer to it as "night") will now be longer than 12 h: it will be used as a calculation parameter, varying from 12 to 18 h. During the night saturation densities are so low that $n = 0$ can be assumed for $z = 0$. We suppose that with the "day" coming, all the accumulated condensate immediately evaporates and during the day flux at the lower boundary is zero. At the upper boundary the condition of diffusion equilibrium is assumed, i.e., zero flux. The deviations to the actual case from this idealized scheme can be compensated by choosing τ.

It should be mentioned that if the temperature inversion altitude can be much more than 1 km as predicted by Gierasch and Goody (1968), and

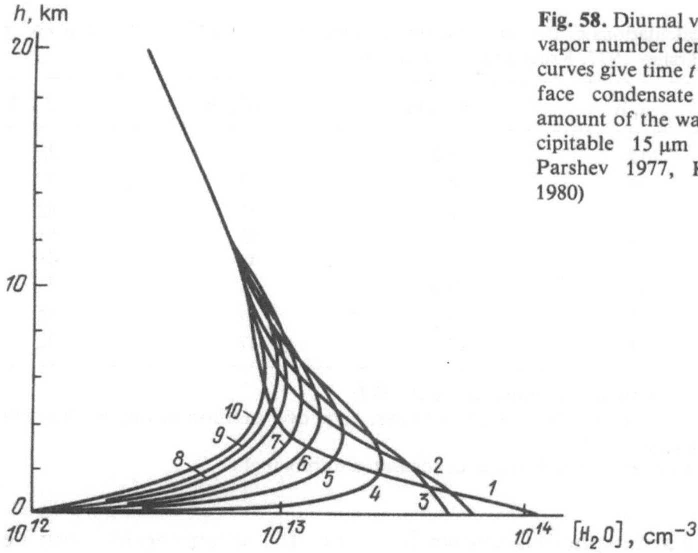

Fig. 58. Diurnal variations of the water vapor number density. The numbers of curves give time $t = n \times 2.4$ h after surface condensate sublimation. Total amount of the water vapor equals precipitable 15 μm (Krasnopolsky and Parshev 1977, Krasnopolsky et al. 1980)

comparable to scale height, the results of calculations given below should be revised.

The results of calculations for $K = 3 \times 10^6\,cm^2\,s^{-1}$, $\tau = 18.00$ h are given in Fig. 58. Ten curves represent altitude profiles of water vapor for different times with a 2.4 h interval. Water vapor distribution in the upper troposphere is subject to small diurnal variations and has a constant mixing ratio. In the lower troposphere diurnal variations are great. During the night (curves 4 – 10) the water density near the lower boundary is very small, and it has a maximum value after ice sublimation (curve 1). In daytime (curves 1 – 3) diffusion water vapor tends to be uniformly distributed. If the boundary between these two areas is defined as level *h*, at which diurnal variations constitute 25%, then $h = 8$ km. The average water mixing ratio in the lower troposphere, with the condensate mentioned above taken into account, is higher than in the upper one. Water distribution can be illustrated by two parts. In the first part it has the constant mixing ratio equal to that in the upper troposphere. The mean temperatures

$$T_{0i} = \int_0^\infty \varrho_i T dz / \int_0^\infty \varrho_i dz \qquad (\varrho_i \text{ is the density of } i\text{-component})$$

of this part and of the atmosphere are equal. The second part represents an excess over this content in the lower troposphere including condensate in the boundary layer, its temperature is determined by temperature distribution in the lower troposphere. Calculation results for different *K* and τ are given in Table 11 together with the maximum quantity of the condensate.

Comparison with the observational results. The excess of the lower layer may be determined from the results of observations, because temperature

Table 11. Results of calculations of the vertical distribution of water vapor for different K and τ (Krasnopolsky and Parshev 1977, Krasnopolsky et al. 1979)

K, $cm^2 s^{-1}$	τ, h	h[a], km	α[b], %	β[c], %
10^6	12	4.2	14	28
	18	5	24	40
3×10^6	12	7	22	42
	18	8	43	53
	21	9.3	48	65
	22.5	9.6	65	77
10^7	12	13.5	27	58
	18	15	47	72

[a] h is the altitude where diurnal variations constitute 25%
[b] α is the fraction of integrated daytime excess of lower layer over uniform mixing profile with respect to total atmospheric water
[c] β is the fraction of total atmospheric water condensated during night

distribution in the atmosphere is known for noon. If the temperature profile from the Mariner 9 data (Fig. 36) is used, then the mean atmospheric temperature is $T_0 = 220$ K; while $T_0 = 204$ K according to Young (1971). The temperature of the near-surface atmosphere is $T_1 = 240$ K and, if $T_0(H_2O) = 225$ K for water vapor, then the excess of the lower layer, α, is equal to 58% for $T_0 = 204$ K and 25% for $T_0 = 220$ K. It may be obtained by the relation $T_0(H_2O) = (1-\alpha)\,T_0 + \alpha T_1$.

When comparing the results of observations and calculations one needs to take into account part of the total atmospheric water vapor amount, β, condensating during the night and determining diurnal variations of water vapor. Ground-based measurements (Fig. 14) show variations by a factor of 2 to 3 which correspond to $\beta = 50\% - 70\%$. With $\alpha = 25\% - 60\%$ the best fit to observations is at $K = (3-10) \times 10^6\ cm^2 s^{-1}$ for $\tau = 12-18$ h (Fig. 58). However, the Viking orbiter measurements (Farmer et al. 1977) showed the tendency of ground-based measurements to overestimate the diurnal variations due to high aerosol extinction at large air masses near both terminators. Then the preferential range of K is $(1-3) \times 10^6\ cm^2 s^{-1}$ with $\alpha \simeq 25\%$ better founded. The diurnal variations by a factor of 1.5 in this case cannot be eliminated by the data of Farmer et al. (1977).

Thus, only about a quarter of the total water vapor constitutes the excess content in the lower troposphere. The other part is distributed in the atmosphere with a constant mixing ratio up to the saturation level in the stratosphere. The water amount in the lowest 4 km is 50% of the total for $K = 3 \times 10^6\ cm^2 s^{-1}$ and $\tau = 12$ h.

The temperature and water abundance are the subject of great latitudinal, seasonal, and local variations, but the sensitivity of results obtained for these variations of input data is not so high. These input data are associated with the parameter τ in our calculation. For example, the temperature decrease is usually connected with the decrease in the water abundance, the former leads to the τ increase, the latter to the τ decrease. So the τ variation is not great, and the dependence of results of calculation on τ is shown in Table 11 (evidently for $\tau = 24$ h no water vapor should be in the atmosphere).

The eddy diffusion coefficient is approximately independent of atmospheric moisture, so such calculations performed for the adequate set of input data should give reliable values of K under conditions corresponding to these input data. The assumptions used in the calculations are (1) the small thickness of the boundary layer as compared to scale height and a clean (without condensation) atmospheric region and (2) one-dimensional approximation. The latter is valid if

$$\left| \frac{\partial [H_2O]}{\partial x} \times \frac{\tau_0 V_x}{[H_2O]} \right| \ll 1 ,$$

i.e., the absolute amount of the scalar product of the horizontal gradient of $[H_2O]$ by wind velocity and $\tau_0 = 24$ h is much less than $[H_2O]$. Of course, it is valid for average values. The results obtained agree with the Viking measurements which tend to give uniform distribution of water vapor in the lower troposphere.

Water vapor dissociation. Photochemistry of the Martian lower atmosphere is first of all the chemistry of CO_2 and H_2O photolytic products and the water photolytic rate is of great importance for it. Figure 59 shows the H_2O photolytic rates used previously in calculations. Let us now analyze the well-known curve 1 obtained by Hunten and McElroy (1970). It is based on the atmospheric temperature profile calculated by Gierasch and Goody (1968), according to which the atmosphere above 15 km has a temperature of 150 K. The total combination of data on the temperature profile of the low latitude atmosphere (Fig. 36) shows that the mean temperature of the atmosphere at heights more than 15 km is about 20° higher, which means that the saturated water vapor density increases by two orders of magnitude. In addition, while

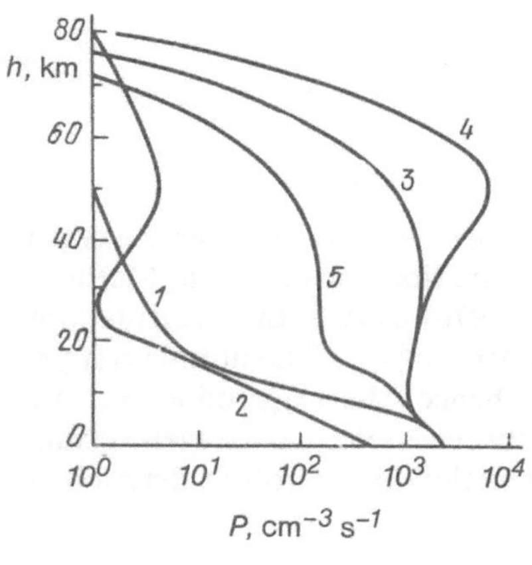

Fig. 59. Water vapor photolytic rates. *Curves 1 and 2* were suggested by Hunten and McElroy (1970) and Izakov and Krasitzky (1977), *curve 3* was calculated with temperature and H_2O profiles from Figs. 36 and 58, *curves 4 and 5* with temperature deviations +10° and −10°, respectively (Krasnopolsky and Parshev 1977, Krasnopolsky et al. 1980)

plotting curve 1 several arbitrary assumptions about spatial (altitudinal and latitudinal) distribution of water vapor were made.

To calculate the water vapor dissociation more realistically, the following characteristics were used as the initial data: absorption cross-sections for CO_2 (Shemansky 1972) and H_2O (Thompson et al. 1963), a solar radiation spectrum (Mount et al. 1980), and a vertical profile of water vapor density (Fig. 58) for three dependences of temperature on height: the standard one shown in Fig. 36 and those with deviations from it by $\pm 10°$. The result of the calculation for the solar zenith angle $Z_0 = 60°$ is given in Fig. 59. The obtained rates of water photolysis are close to curve 1 in the lower troposphere, but exceed it by two orders of magnitude at heights more than 15 km. The variations of the photolytic rate are very strong with changing of the atmospheric temperature.

2.7 Photochemistry of the Lower Atmosphere (Global Average Conditions)

The global averaged photolytic rate and total amount of CO_2 on Mars constitute $10^{12}\,cm^{-2}\,s^{-1}$ (see Fig. 53) and $N = p/mg = 2 \times 10^{23}\,cm^{-2}$, respectively. So the lifetime of CO_2 corresponding to the photolytic rate is 2×10^{11} s or 6000 yr. With other components being neglected, the three body reactions to form O_2 and CO_2 are the only ways of removal of the photolytic products. The formation of CO_2 from O and CO is spin forbidden and its rate coefficient is 10^4 times lower than that of the O_2 formation. Therefore, rather large amounts of O_2 and CO should accumulate in the Martian atmosphere ($\sim 10\%$). However, the measured mixing ratios are two orders of magnitude lower than the expected ones.

To solve the problem, McElroy and Donahue (1972) suggested attributing the increase of CO_2 formation to the following reactions with H_2O photolytic products as catalysts:

$$
\begin{aligned}
& H + O_2 + CO_2 \rightarrow HO_2 + CO_2 \\
& HO_2 + O \rightarrow OH + O_2 \\
& CO + OH \rightarrow CO_2 + H \\
\hline
\text{net}\ & O + CO \rightarrow CO_2 .
\end{aligned}
$$

These reactions should give very low number densities of atomic oxygen below 25 km. To prevent a high rate of O_2 formation above 25 km, McElroy and Donahue (1972) and Liu and Donahue (1976) used in their calculations the eddy diffusion coefficient, K, exceeding $10^8\,cm^2\,s^{-1}$. The atomic oxygen number densities are proportional to K^{-1}, hence, the O_2 production rate should decrease as K^{-2}. The equilibrium between molecular oxygen formation by the three body reactions and its destruction by photolysis determines O_2 number densities.

Parkinson and Hunten (1972) pointed out the possibility of indirect photolysis of O_2 via photolysis of HO_2 and H_2O_2 resulting in its low mixing ratio. In the first case

$$
\begin{array}{ll}
H+O_2+CO_2 & \rightarrow HO_2+CO_2 \\
HO_2+h\nu & \rightarrow OH+O \\
OH+CO & \rightarrow CO_2+H \\
\hline
\text{net } O_2+CO+h\nu & \rightarrow CO_2+O\,.
\end{array}
$$

In the second case

$$
\begin{array}{ll}
2(H+O_2+CO_2 & \rightarrow HO_2+CO_2) \\
HO_2+HO_2 & \rightarrow H_2O_2+O_2 \\
H_2O_2+h\nu & \rightarrow 2OH \\
2(OH+CO & \rightarrow CO_2+H) \\
\hline
\text{net } O_2+2CO+h\nu & \rightarrow 2CO_2\,.
\end{array}
$$

These processes allow the low O_2 mixing ratio at moderate K to be explained. Thus species with odd numbers of hydrogen atoms (H, OH, HO_2) are very active chemically, contrary to those with even numbers (H_2O, H_2). This is valid also for other atmospheric constituents, e.g., odd oxygen O and O_3, odd nitrogen N, NO, NO_2 etc. In spite of some exceptions like HCl and H_2O_2, the latter being much more active than the former, the term "odd" is attributed to atomic species, radicals, and chemically active molecules which are of great importance in the atmospheric photochemistry.

Products of H_2O photolysis and odd hydrogen H, OH, HO_2 accelerate formation of both CO_2 and O_2, the latter by

$$
\begin{array}{ll}
HO_2+O & \rightarrow OH+O_2 \\
OH+O & \rightarrow O_2+H \\
\hline
\text{net } O+O & \rightarrow O_2
\end{array}
$$

and also by

$$
\begin{array}{ll}
O+O_2+CO_2 & \rightarrow O_3+CO_2 \\
O_3+H & \rightarrow OH+O_2 \\
OH+O & \rightarrow O_2+H \\
\hline
\text{net } O+O & \rightarrow O_2\,.
\end{array}
$$

Besides direct photolysis of water vapor, odd hydrogen H is formed by indirect photolysis of H_2O and H_2:

$$O_3 + h\nu \rightarrow O_2(^1\Delta_g) + O(^1D)$$
$$O(^1D) + H_2O \rightarrow OH + OH$$
$$O(^1D) + H_2 \rightarrow OH + H\,.$$

However, the predominant part of $O(^1D)$ is quenched by CO_2. Ozone photolysis also gives excitation of the $O_2(a^1\Delta_g - X^3\Sigma_g^+)$ airglow at 1.27 μm.

The removal of odd hydrogen occurs according to the following reactions

$$H + HO_2 \rightarrow H_2 + O_2$$
$$\qquad\qquad H_2O + O$$
$$HO_2 + OH \rightarrow H_2O + O_2\,.$$

Note that only the latter three reactions out of all catalytic cycles described lead to the sink of odd hydrogen. During its lifetime each H atom participates in a great number of CO_2 and O_2 formation cycles. That is why the presense of odd hydrogen with a mixing ratio of $\sim 10^{-9}$ entails substantial decrease of O_2 and CO amounts.

Current papers on photochemistry of the Martian lower atmosphere differ mainly in adopted values of K and the water vapor photolytic rate. The results presented here are described in the paper by Krasnopolsky and Parshev (1979a). The objective of this paper was time independent photochemical calculations of minor atmospheric constituents up to 60 km at low and middle latitudes under daytime equilibrium conditions best satisfied in the afternoon. The water vapor photolytic rate is taken from Fig. 59 (curve 3) and in our opinion is better founded than other similar estimates.

From calculations of the H_2O vertical distribution, K is equal to 3×10^6 $cm^2\,s^{-1}$ in the lower troposphere (see Chap. 2.6). Interpretation of aerosol layers at 30 km measured by the Mars 5 and Mariner 9 orbiters (Chap. 1.10) gives $K = 3 \times 10^6$ $cm^2\,s^{-1}$ also. Hence, this value may be adopted in an altitude range up to 30 km. K equals 10^7 $cm^2\,s^{-1}$ above 45 km, according to Mariner 9 measurements in the absence of the dust storm. Zurek (1976) obtained an estimate of $K = 5 \times 10^7$ $cm^2\,s^{-1}$ at ~ 40 km from calculations of tides on Mars. Therefore, we assumed the increase of K from 3×10^6 $cm^2\,s^{-1}$ at 30 km to some value at 40–60 km, which was a parameter of the calculations.

Since the time independent problem is under consideration, with diurnal variations of minor compounds being neglected, a correct choice of photolytic rates, i.e., global average, day side global average, or low latitude day side average, should be made. To do this one may calculate a lifetime τ_i of a component i as a ratio of its amount to its total production or loss rate. As will be shown below, $\tau \simeq 1.5$d for odd hydrogen, i.e., significantly smaller than the time of meridional transport. Therefore, the odd hydrogen number densities are subjected to latitudinal and diurnal variations.

If the calculation is carried out for low latitudes curve 3 (Fig. 59) can be used for it and the rate of water photolysis used should amount to half the calculated value under daytime conditions. If the calculation is performed for global average conditions, the additional reduction of this value is necessary which is caused by the decrease of water vapor content at high latitudes. But it is not significant, because first, 80% of the net flux of solar radiation accounts for middle latitudes $\pm 45°$ and the low water content at high latitudes is not very important for global averaging. Second, the main part of water photolysis occurs at altitudes higher than 15 km where smaller variations of temperature and water vapor content can be expected.

Atomic oxygen formed by CO_2 photolysis has $\tau = 7$ h (see below); in calculations the daytime photolytic rate of CO_2 should be used, which exceeds the data of Fig. 53 by a factor of 2 and gives a production rate of 2×10^{12} $cm^{-2} s^{-1}$. Unfortunately, with the intensities of solar UV radiation taken by Krasnopolsky and Parshev (1979a) from measurements of Widing et al. (1970), the CO_2 photolytic rate was overestimated by a factor of $2-3$.

The basic photochemical processes in the Martian atmosphere and their rate constants are given in Table 12. The daytime values of the photolytic rate will be used for the fast processes (R3, R5, R6, R11) and half of those for the slow ones (R2, R10).

Twenty-one reactions of this list reflect the chemistry of odd oxygen O* (O, O_3) and hydrogen H* (H, OH, HO_2, H_2O_2) and eight reactions of odd nitrogen N* (N, NO, NO_2). The influence of nitrogen compounds on O* and H* is relatively insignificant, but their introduction in our calculation makes it more complete, since N* is an aeronomically interesting object in the Martian atmosphere. The nitrogen photochemistry on Mars was considered in more detail by Yung et al. (1977) (see Chap. 1.2), where 50 reactions were used for its description. Our goal in this consideration is somewhat more modest. Therefore, we discuss not all, but only the most important forms of odd nitrogen for which only the most essential chemical processes were chosen. Similarly, the list of the processes with O* and H* can be easily extended, but only the most essential reactions are taken into account and it can be shown that the addition of other reactions does not significantly affect the calculation results.

Let us analyze the choice of some rate coefficients of the reactions. Photolysis of NO (R10) in the lower and middle atmosphere is mainly due to predissociation in δ bands (0,0) 190 nm and (1,0) 183 nm. The photolytic rates calculated by Cieslik and Nicolet (1973) for these bands and with absorption of radiation by CO_2 taken into account are given in Table 12. For the reaction of ozone formation R14 the value was used that had been recommended by Anderson (1976) and the correction was introduced to the higher efficiency of CO_2 as compared with N_2 in this process. The estimates of the R15 rate constant vary from 7×10^{-11} (Hochanadel et al. 1972) to 8×10^{-11} $\exp(-500/T)$ (Anderson 1976). The value close to the average one is in Table 12. In the same manner the values k_{20} (1.5×10^{-12} by Parkinson and Hunten 1972; $4 \times 10^{-11} \exp(-350/T)$ by Anderson 1976), k_{23} (2×10^{-10} by Hochanadel et al. 1972; and 2×10^{-11} by Kaufman 1975) were chosen. We used the only value $k_{16} = 2.6 \times 10^{-11}$ (Phillips and Schiff 1962) obtained with

Table 12. Chemical reactions in the Martian atmosphere and their rate coefficients[a]

Reaction 1	Rate coefficient 2
1. $CO_2+h\nu\rightarrow CO+O$	–
2. $O_2+h\nu\rightarrow O+O$	5.8×10^{-10}
3. $O_3+h\nu\rightarrow O_2(^1\Delta)+O(^1D)$	4.2×10^{-3}
4. $H_2O+h\nu\rightarrow OH+H$	–
5. $HO_2+h\nu\rightarrow OH+O$	2.5×10^{-4}
6. $H_2O_2+h\nu\rightarrow OH+OH$	5×10^{-5}
7. $O(^1D)+CO_2\rightarrow O(^3P)+CO_2$	1.8×10^{-10}
8. $O(^1D)+H_2O\rightarrow OH+OH$	3×10^{-10}
9. $O(^1D)+H_2\rightarrow OH+H$	3×10^{-10}
10. $NO+h\nu\rightarrow N+O$	$3.2\times10^{-6}[\exp(-2\times10^{-16}N_{CO_2})$ $+\exp(-2\times10^{-15}N_{CO_2})]$
11. $NO_2+h\nu\rightarrow NO+O$	4.2×10^{-3}
12. $H+O_2+CO_2\rightarrow HO_2+CO_2$	$2\times10^{-31}(273/T)^{1.3}$
13. $O+O+CO_2\rightarrow O_2+CO_2$	$7\times10^{-31}T^{-1}$
14. $O+O_2+CO_2\rightarrow O_3+CO_2$	$3\times10^{32}\exp(520/T)$
15. $O+HO_2\rightarrow OH+O_2$	2.5×10^{-11}
16. $O_3+H\rightarrow OH+O_2$	$1.3\times10^{-10}\exp(-500/T)$
17. $O+OH\rightarrow O_2+H$	2×10^{-11}
18. $CO+OH\rightarrow CO_2+H$	$2.2\times10^{-13}\exp(-100/T)$
19. $H+HO_2\rightarrow OH+OH$	$4.2\times10^{-10}\exp(-950/T)$
20. $H+HO_2\rightarrow H_2+O_2$	$2\times10^{-11}\exp(-350/T)$
21. $H+HO_2\rightarrow H_2O+O$	$8\times10^{-11}\exp(-500/T)$
22. $HO_2+HO_2\rightarrow H_2O_2+O_2$	$3\times10^{-11}\exp(-500/T)$
23. $HO_2+OH\rightarrow H_2O+O_2$	5×10^{-11}
24. $N+NO\rightarrow N_2+O$	2.2×10^{-11}
25. $NO+O_3\rightarrow NO+O_2$	$2\times10^{-12}T^{1/2}\exp(-1200/T)$
26. $NO+O+CO_2\rightarrow NO_2+CO_2$	$6\times10^{-33}\exp(965/T)$
27. $NO+O_3\rightarrow NO_2+O_2$	$10^{-12}\exp(-1200/T)$
28. $NO+HO_2\rightarrow NO_2+OH$	$2\times10^{-11}\exp(-1200/T)$
29. $NO_2+O\rightarrow NO+O_2$	9×10^{-12}

[a] Photolytic rates are given in s^{-1} for the conditions near the Martian surface. Dimensions of rate coefficients of the second-order and third-order reactions are $cm^3\,s^{-1}$ and $cm^6\,s^{-1}$

$T=300$ K under the assumption that the reaction has an activation energy of about 1 kcal. The adopted rate coefficients of the reactions R15, 16, 23 were confirmed later by laboratory measurements.

The problem of the atmospheric composition model calculation consists now in solving the set of equations:

$$\frac{d\phi_i}{dz}=P_i-L_i,\qquad \phi_i=-K\left(\frac{dn_i}{dz}+\frac{n_i}{H}+\frac{n_i}{T}\frac{dT}{dz}\right)$$

where P_i is the chemical production: i are components; L_i is the loss rate; H is the mean scale height. It is a time independent one-dimensional problem of diffusion of minor components.

Boundary conditions. We considered two types of boundary conditions on the planetary surface. The first assumes there is no interaction of atmospheric components with the surface, $\phi_i(0) = 0$. The second type is another extreme case when the impact of an odd molecule on the surface removes it:

$$\phi_i(0) = -n_i(0)\,V_i/4 \simeq -0.5 \times 10^5\,n_i(0)/\sqrt{\mu_i}\,.$$

Here, V_i is the mean thermal velocity of a molecule; μ_i is the molecular mass. In the second case it is difficult to build a completely correlated composition model since we must know the chemical transformations of molecules impacted on the surface. However, if one assumes the equilibrium of stable components (CO_2, CO, O_2, H_2O, H_2), such a calculation can be carried out.

The conditions on the upper boundary can be taken as the fluxes of H and O atoms and the concentrations of N and NO. The interpretation of the hydrogen airglow observations at 1216 nm results in a flux of hydrogen atoms into the lower atmosphere $\phi_H = -4 \times 10^8$ cm^2 s^{-1} and the molecular hydrogen upward flux from the lower atmosphere will be $\phi_{H_2} = 3 \times 10^8$ cm^{-2} s^{-1}. The downward flux of oxygen atoms at the level of 60 km is equal to the net rate of their formation at heights more than 60 km ($\phi_O = -10^{-12}$ cm^{-2} s^{-1} under daytime conditions). It is more difficult to formulate the conditions for atomic nitrogen and nitrogen oxide. Here, we will use the results obtained by Yung et al. (1977) and take [N] = 1.5×10^5 and [NO] = 5×10^5 cm^{-3}.

The results of calculation for $K = 10^8$ cm^2 s^{-1} with $z > 40$ km are given in Fig. 60.

Atomic oxygen is formed by the photolysis of CO_2 and O_2. Very rapid transformation of ozone to atomic oxygen and vice versa makes it possible to consider them as a single species, odd oxygen O_x. Its sink occurs mainly due to the reactions R15, 16. OH radicals thus formed react with CO to produce CO_2 (R18) or O with giving O_2 (R17) the former predominating below and the latter above 45 km. Since most of the atomic oxygen originates below 45 km and some portion of atomic oxygen produced above 45 km is transported into the troposphere, about 80% of all oxygen atoms form CO_2.

The effective removal of atomic oxygen occurs below 40 km. With oxygen atoms forming, but not being removed above, their downward transport leads to the appearance of a maximum of odd oxygen (O and O_3) near this altitude. If the mixing rate is very high and its time is less than or comparable with the photochemical time of odd oxygen throughout the atmosphere, the odd oxygen maximum should be near the surface. For example, this result was obtained in the calculations by Liu and Donahue (1976) with $K = 4 \times 10^8$ cm^2 s^{-1} being adopted throughout the atmosphere.

Ozone. The ozone to atomic oxygen number density ratio is given by $[O_3]/[O] = (k_{14}/J_3) \times [O_2][CO_2]$. Subscripts indicate the number of the reactions in Table 12. From this relation it is evident that ozone is a predominant species of odd oxygen below 25 km, while atomic oxygen prevails above. The maximum of odd oxygen near 40 km entails a maximum of ozone number densities with a steeper upward decrease. The high value of the eddy

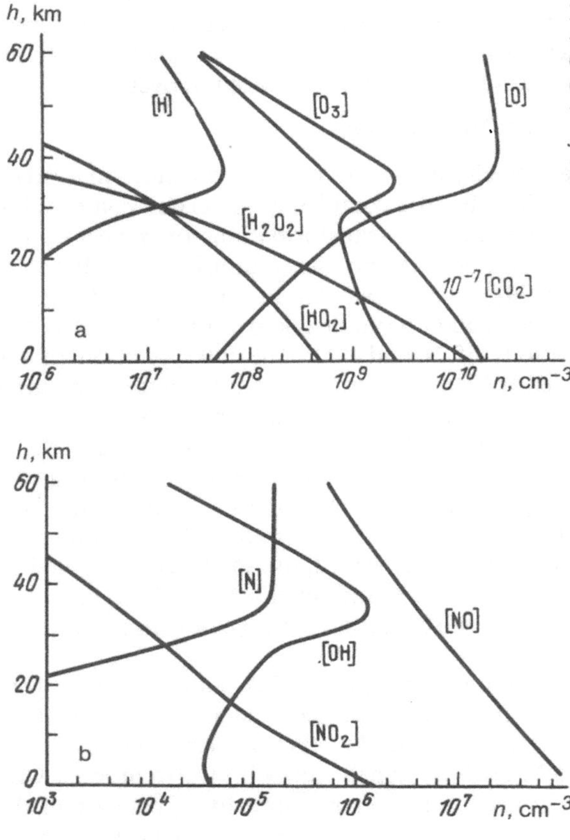

Fig. 60a, b. Chemical composition of the Martian atmosphere up to 60 km. Altitude distributions of O_2, CO, H_2 are similar to CO_2 with $f_{O2} = 1.6 \times 10^{-3}$, $f_{CO} = 2 \times 10^{-3}$, $f_{H2} = 2.5 \times 10^{-5}$ (Krasnopolsky and Parshev 1979a, Krasnopolsky et al. 1980)

diffusion coefficient above 40 km provides the rapid transport of atomic oxygen into the altitude region near 40 km, while a moderate eddy diffusion below 30 km results in the accumulation of atomic oxygen and ozone at 30–40 km, thus making the ozone maximum more distinctive. The higher K is above 40 km, the more narrow and lower the ozone maximum and the higher number density in it. Results of calculations for different K are shown in Table 13. The best fit to the Mars 5 observation near the evening terminator (see Chap. 1.3) is for $K = 10^8\ cm^2\,s^{-1}$. With errors of the measurements taken into account $K = (1-10) \times 10^7\ cm^2\,s^{-1}$ above 40 km. Note that the equilibrium daytime conditions are best applied in the afternoon.

For $K = 10^8\ cm^2\,s^{-1}$ ozone forms a layer 10 km thick having a maximum number density equal to $2.5 \times 10^9\ cm^{-3}$ at 35 km. The ozone minimum is located at 27 km with a number density 3.5 times smaller than the maximum density. This minimum is not so deep for $K = 10^7$ and $3 \times 10^6\ cm^2\,s^{-1}$ and the half-maximum thickness of the ozone layer cannot be determined. The total

Table 13. Dependence of the ozone layer properties on the value of K for $h \geqslant 40$ km

K, $cm^2\,s^{-1}$	10^8	3×10^7	10^7	3×10^6
$O_{3\,max}$, $10^9\,cm^{-3}$	2.5	1.75	0.9	0.63
L[a], km	10	14	–	–
h_{max}, km	25	37	40	54

[a] L is the layer thickness at half maximum

ozone amount calculated coincides with the Mariner spectroscopic upper limit of 3 μm or $7 \times 10^{15}\,cm^{-2}$ for low latitudes. The other upper limit of $[O_3] \times [H] \lesssim 10^{15}\,cm^{-6}$ at 60 km determined by the absence of hydroxyl night airglow (see Chap. 1.5) is consistent with the results of the calculations giving $0.5 \times 10^{15}\,cm^{-6}$ for this product.

The calculated ozone number densities give the O_2 1.27 μm airglow of 7 *MR* for $K = 10^8 cm^2\,s^{-1}$ if the rate coefficient of quenching by CO_2 is $5 \times 10^{-20}\,cm^3\,s^{-1}$ (see Chap. 4.10). This intensity is twice as high as the measured one (3 ± 1) *MR*. The discrepancy may be explained by the fact that the solar UV radiation data (Widing et al. 1970) used in the calculations are overestimated by two- to three fold.

Molecular oxygen forms from O atoms that take part in R17. In our calculation the total rate of this process is equal to $5.3 \times 10^{11}\,cm^{-2}\,s^{-1}$. The direct reaction of O_2 formation R13 provides for a much smaller contribution, $1.65 \times 10^{10}\,cm^{-2}\,s^{-1}$, and the cycle with nitrogen oxydes ending in R29 gives a contribution less by another order of magnitude, $1.65 \times 10^9\,cm^{-2}\,s^{-1}$. So, the fifth part of all the production of O transforms into O_2.

Direct (R2) and indirect (R5, R6) photolysis are the processes of O_2 destruction. Their total rates are equal to 1.15×10^{11}, 1.45×10^{11}, and $2.9 \times 10^{11}\,cm^{-2}\,s^{-1}$, respectively, with $[O_2]/[CO_2] = 1.6 \times 10^{-3}$. Therefore, indirect photolysis of O_2 is prominent here. This is, first of all, because of the great production of odd hydrogen as a result of water photolysis in our calculation. Hence, though the increase of H* should suppress ozone formation; nevertheless, high concentrations of H* in developing the composition model entail high loss rates of O_2 because of photolysis of HO_2 (proportional to H*) and H_2O_2 (proportional to H^{*2}), which allows us to assume higher densities of atomic oxygen in the atmosphere as well as of ozone as a consequence.

The characteristic time of molecular oxygen is 30 yr, during which its complete mixing must occur on the global scale. The obtained mixing ratio $f_{O_2} = 1.6 \times 10^{-3}$ agrees with the results of spectroscopic and mass spectrometric measurements.

Molecular hydrogen forms in the reaction R20; its net rate amounts to $10^9\,cm^{-2}\,s^{-1}$. Some part is transported into the ionosphere where it dissociates:

$$H_2 + CO_2^+ \rightarrow CO_2H^+ + H$$

$$CO_2H^+ + e \rightarrow CO_2 + H\,.$$

The flux of H_2 from the lower atmosphere into the ionosphere is 3×10^8 $cm^{-2} s^{-1}$ (Fig. 50). The remainder dissociates in the lower atmosphere as a result of the reaction with $O(^1D)$. The balance of these processes results in the mixing ratio of $f_{H_2} = 2.5 \times 10^{-5}$ that is in good agreement with the estimate by Anderson (1974). The characteristic time of molecular hydrogen relative to the chemical processes is equal to about 200 yr and relative to escape, 2000 yr.

The water vapor equilibrium in the Martian atmosphere is maintained by the balance of its direct (R4) and indirect (R8) photolysis and the reaction of H_2O formation (R21 and R23). There is a small deficit in this balance equal to about 10^8 $cm^{-2} s^{-1}$ that provides the escape of hydrogen and oxygen.

Nitric oxide in the lower Martian troposphere has in our model concentrations an order of magnitude less than those calculated by Yung et al. (1977). This is because the chemical production of nitric oxide in the region considered is negligible and the transport processes are prominent in determining its number densities and in this case, the higher the transport velocities, the higher the NO concentrations near the surface. Therefore, the difference in the calculation results is explained by the assumed values of the eddy mixing coefficient *K*.

Carbon monoxide has the characteristic time of 6 yr and the mixing ratio of 2×10^{-3}, that are in a good agreement with the experimental data.

The dependence of the total amount of ozone on H_2O *concentrations.* Small variations of atmospheric temperature cause high variations of satured vapor pressure (by a factor of 10 for temperature change by 10°). With falling temperature the rate of H* formation and the rate of atomic oxygen destruction (R15, R16, R17) decrease, thus increasing O_3 number densities. At low atmospheric temperatures the odd hydrogen production is maintained mainly by R9, since H_2 values are still at the same level. The ozone increase makes the rate of R9 higher and the H* rise limits the O_3 increase. Therefore, the dependence of the total amount of ozone on H_2O is not very strong (Fig. 61), but even with a small amount of H_2O, the amount of ozone can exceed by several times the spectroscopic limit for observations on the planetary disk.

H_2O freezing alone cannot provide large amounts of ozone, such as $\sim 10^{17}$ cm^{-2} measured by the Mariner 9 in some regions of the polar caps (Figs. 6, 7).

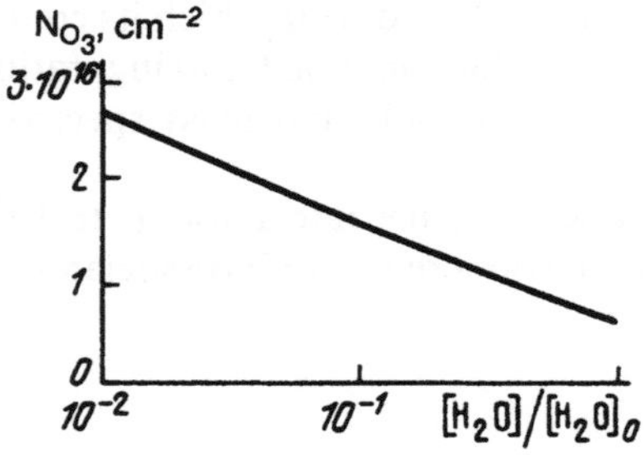

Fig. 61. Total ozone amount vs the water vapor relative abundance in the atmosphere. $[H_2O]_0$ corresponds to Fig. 58, $n = 2$ (Krasnopolsky and Parshev 1979a, Krasnopolsky et al. 1980)

To explain these, Hunten (1974) supposed H_2O_2 freezing which provides an additional sink of odd hydrogen and ozone increase.

The solution for the second type of conditions on the surface, when odd components are removed on it, has been also the subject of our consideration. Here, the concentrations of odd components noticeably decrease in the lowest 3 km. Due to the relatively small value of K this effect does not manifest itself at higher altitudes and only slightly influences the calculated results.

Ozone number densities at 09.30 h LT measured by Mars 5 on the daytime limb of the planet are equal to 10^{10} cm^{-3} for the layer geometry close to that obtained in the calculation. With the possibility of diurnal variations of ozone and the sensitivity of the calculations to the initial data being estimated qualitatively, one may conclude that the assumed scheme of the aeronomical processes in the Martian atmosphere (Table 11) is inconsistent with such high values of $[O_3]$ at 35 km unless the values of k_{16} and k_{17} are very overrated.

What are the ways of solving this contradiction? The fact is that the position of the ozone layer coincides in a surprising manner with the location of the maximum and upper boundary of the aerosol layer formed by particles of H_2O ice. One can assume that during the impact of the oxygen atom on an ice particle, especially on the lattice points where hydrogen atoms are located, the adhesion of the oxygen atom occurs. If the energy of such affinity is about one-tenth of electronvolts, it is stable for thermal molecular impact, but in the case of O_2 ozone forms and the released energy is quite enough for breaking the bond with an ice particle. The advantage of this process over the usual three body reaction of O and O_2 in the presence of CO_2 (R 14) consists here in the fact that the simultaneity of the O and O_2 impact on CO_2 is not required. A simple calculation shows that if the oxygen atom adhesion to an ice particle occurs with an efficiency close to unity, the O_3 number density at 35 – 40 km may be equal to about 10^{10} cm^{-3}.

Generally speaking, it is not necessary for an aerosol particle to take part in such a process. Here, it is important to substitute the three body reaction for the two step process of double collisions. However, it is reasonable to use water aerosol because this procedure helps easily to explain the variability of $[O_3]_{max}$ during a day. It should be noted that the values of $[O_3] \simeq 10^{10}$ cm^{-3} at 35 – 40 km for the considered assumptions must not be close to the average values for the daytime hemisphere, otherwise the equilibrium of most of the components analyzed above can be essentially disbalanced.

Water aerosol particles do not increase ozone densities for $h \lesssim 15$ km, where the three body reaction R 14 is effective enough and $[O_3] \gg [O]$. So the proposed mechanism is not contradictory and may even explain the observed ozone dependence on cloudiness (Barth and Dick 1974) with thick clouds assumed to be in the lower troposphere and thin and very thin ones in the stratosphere and upper troposphere.

The increase of atmospheric reflection at 305 nm which correlates with ozone absorption at middle latitudes (see Chap. 1.3) may be considered as supporting this hypothesis. To check it a correlation of number densities of the mesospheric ozone with those of noctilucent cloud particles may be studied in the Earth's atmosphere.

2.8 Diurnal Variations of Minor Components in the Low Latitude Atmosphere

A model of diurnal variations was calculated by Krasitsky (1978), based on the time-independent model of Izakov and Krasitsky (1977). In this model, diurnal variations of solar UV radiation is described by the time-dependent solar zenith angle. Data on surface temperatures are taken from the Viking measurements. Diurnal H_2O profile variations are not covered in this work. Thermospheric temperature variations are adopted from the two-dimensional model of Izakov and Morozov (1976), which gives an average temperature of 320 K for $F_{10.7\,cm} = 144$. The model under consideration is one-dimensional, i.e., it neglects horizontal transport. Vertical transport is given by the eddy diffusion coefficient $K = 10^5\,cm^2\,s^{-1}$ near the surface and $K = 3 \times 10^7\,cm^2\,s^{-1}$ at the homopause with a linear increase of log K. For the calculations, a day is divided into 50 intervals of different duration which are shorter at sunrise and sunset. Unfortunately, the confused numerical data and illustrations detract from the value of this work. The calculated mixing ratios of O and CO are approximately the same at the ionospheric maximum at 140 km (see Fig. 2 in Krasitsky 1978) and constitute $f_O = 2.3 \times 10^{-2}$ and $f_{CO} = 1.6 \times 10^{-2}$, respectively. Diurnal variations of atomic oxygen are low in the thermosphere up to 200 km. The variations of CO are even higher (by a factor of 2 at 200 km) and are caused mainly by temperature variations. A scale of [O] and [CO] should be multiplied by 10 in Fig. 1 of the paper.

Due to exponential dependence of hydrogen escape on atmospheric temperature, the hydrogen number densities H are subject to greater variations constituting at the base of the exosphere 0.9×10^4 and $7 \times 10^4\,cm^{-3}$ at 10.00 h LT and at midnight, respectively. The average diurnal [H] is $3 \times 10^4\,cm^{-3}$ which is in good agreement with observations.

Odd hydrogen vertical profiles in the lower atmosphere differ greatly at day- and nighttime. Under daytime conditions photolysis of H_2O provides odd hydrogen number densities decreasing upwards with a value of $3 \times 10^8\,cm^{-3}$ near the surface. At night odd hydrogen in the lower atmosphere forms by a downward flux from the upper atmosphere with number densities almost independent of altitude and equal to $2 \times 10^7\,cm^{-3}$ near the surface. Diurnal variations of H_2O_2 are not high, due to its low photolytic rate. With the adopted eddy diffusion coefficient both H and O variations are negligible at 50 – 100 km. However, at $K \simeq 4 \times 10^8\,cm^2\,s^{-1}$ atomic oxygen variations at 40 – 70 km should constitute two orders of magnitude and measurements of these variations could make it possible to estimate K. Figure 62 shows the calculated diurnal variations of [O] and [O_3]. At night ozone forms a maxi-

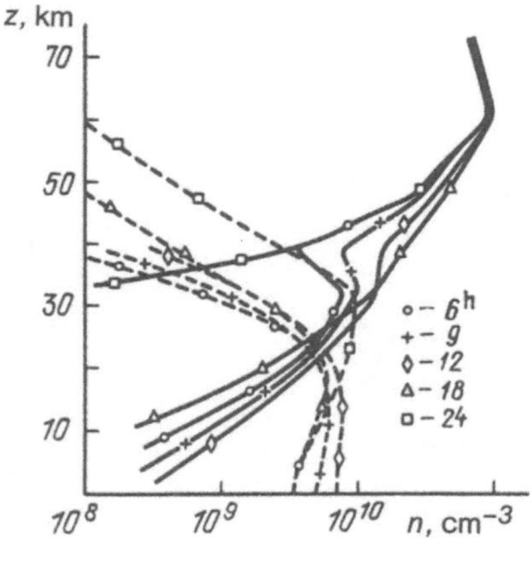

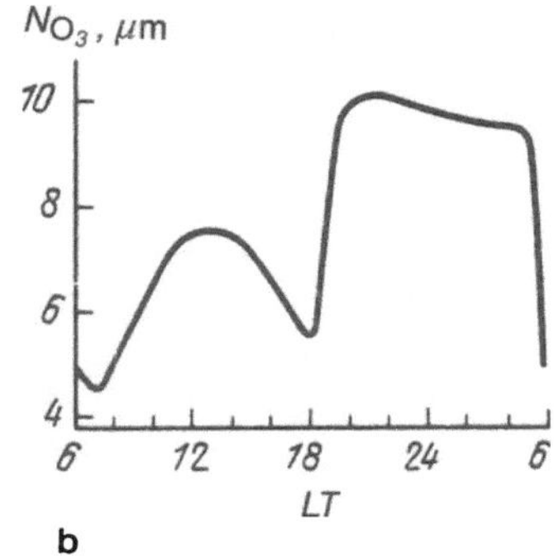

Fig. 62. a Atomic oxygen (*solid lines*) and ozone (*dashed lines*) number densities at various local times. **b** Diurnal variations of the total ozone amount (Krasitzky 1978)

mum at 30 – 35 km with $[O_3] \simeq 10^{10}$ cm^{-3}. In the daytime $[O_3] \simeq 5 \times 10^9$ cm^{-3} up to 25 km with a steep decrease above this altitude. Daytime ozone amounts are higher by a factor of 2 than the spectroscopic limit for low latitudes (3 μm). This difference is considered to be within the limits of measurements and uncertainty in the initial data of the calculations.

2.9 Latitudinal Distribution of Ozone in Different Seasons

The low and middle latitude atmospheric composition model discussed in Chap. 2.7 gives a total ozone amount of 3 μm. With water vapor freezing out it may reach 10 μm. Similar calculations of Liu and Donahue (1976) indicate an ozone increase from 0.92 to 5.2 μm under the same conditions, while the ozone amounts observed by Mariner 9 may be as high as 60 μm (see Chap. 1.3).

Calculations of the atmospheric photochemistry were made by Kong and McElroy (1977a) for different latitudes in two seasons which cover the extreme conditions in winter at 60 – 70° N where maximum ozone amounts were observed. They argue that the large concentrations of ozone may be attributed to the combination of the following factors: low temperatures which allow condensations first of H_2O and then of H_2O_2, thereby reducing the odd hydrogen amount; the increased optical thickness of the ozone layer and a longer photolytic lifetime for O_3 and H_2O_2 especially with a large air mass factor; a more stable lower atmosphere in winter, which inhibits the upward

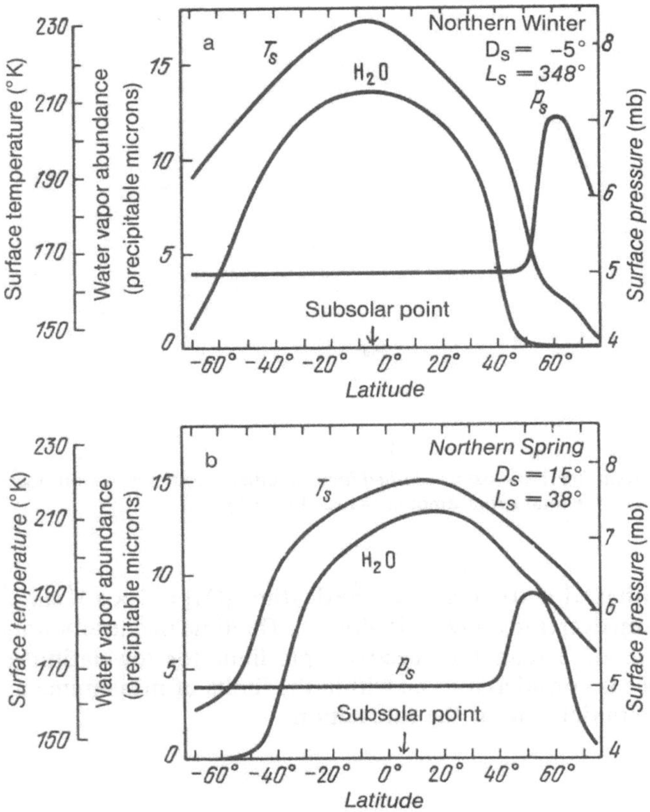

Fig. 63. a Temperature and pressure near the surface and water vapor abundance at various latitudes for winter in the northern hemisphere (L_s = 348°) and **b** spring (L_s = 38°) (Kong and McElroy 1977a)

transport of O_3; large scale meridional winds which could supply additional quantities of odd oxygen from lower latitudes, and an increase in atmospheric pressure in winter with a maximum at 60° N (Woichshyn 1974) which makes the $[O_3]/[O]$ ratio higher and the removal of odd hydrogen more effective due to the following reactions:

$$O + O_2 + CO_2 \rightarrow O_3 + CO_2$$
$$H + O_2 + CO_2 \rightarrow HO_2 + CO_2$$
$$CO + OH \rightarrow CO_2 + H$$
$$H + HO_2 \rightarrow H_2 + O_2$$
$$HO_2 + HO_2 \rightarrow H_2O_2 + O_2$$

with subsequent condensation of H_2O_2.

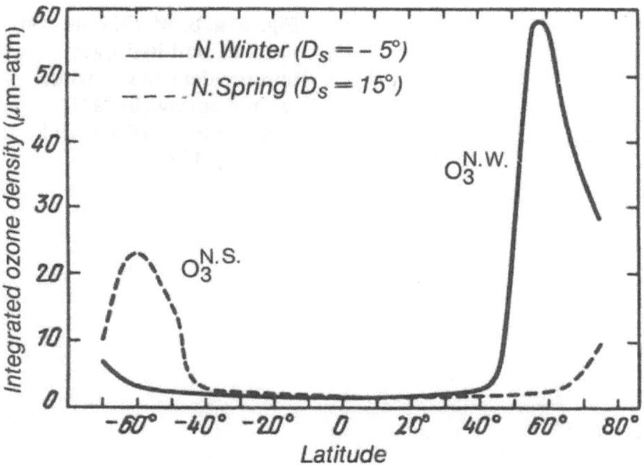

Fig. 64. Latitudinal distribution of ozone amounts for two seasons (*solid line* is for northern winter, *dashed line* is for northern spring) (Kong and McElroy 1977a)

The model under consideration is calculated on the basis of the global average one-dimensional model of Kong and McElroy (1977b) with a chemically passive surface. The atmosphere is divided into latitudinal bands of about 10° in latitude from 70° S to 50° N with three additional bands centered on 58° N, 65° N, and 75° N. Computations have been carried out for two seasons: late winter in the northern hemisphere with a solar declination angle of $-5°$ ($-24°$ at the winter solstice) and a planetocentric solar longitude, L_s, of 348° ($L_s = 0°$ at the vernal equinox), and midspring in the northern hemisphere with a declination of 15° and $L_s = 38°$. The first case describes the Mariner 9 conditions when the largest ozone amounts were detected, and the second corresponds to the significant enhancement of O_3 in the vicinity of the southern polar cap (Fig. 6).

The adopted surface temperatures, pressures, and water abundances at different latitudes are shown in Fig. 63. The curves are based on experimental data obtained before the Viking flights. The relative humidity is assumed to be constant with altitude at high latitudes. At low and middle latitudes it is constant up to 4 km and then sharply drops (this disagrees with the results of Chap. 2.6).

The eddy diffusion coefficient K is assumed to be $2 \times 10^8\,cm^2\,s^{-1}$ at all altitudes and latitudes except 50–75° N in winter and 50–70° S in spring where $K = 10^8\,cm^2\,s^{-1}$ down to 30 km and 5×10^7, 5×10^6, $7.5 \times 10^5\,cm^2\,s^{-1}$ at 20, 10, and 0 km, respectively. The reduced eddy diffusion provides for higher ozone amounts under the extreme conditions.

From the analysis of thermodynamical data on H_2O_2, a dependence of H_2O_2 saturated vapor density on temperature was derived. In a range of 190–140 K this dependence may be given as $n_s = 2.7 \times 10^9 \exp\frac{T-180}{3.4}\,cm^{-3}$.

The calculated amounts of ozone as a function of latitude for both seasons are shown in Fig. 64. They constitute about 2 μm at low and middle latitudes, i.e., lower than the spectroscopic limit, and as high as 60 μm at 60° N in

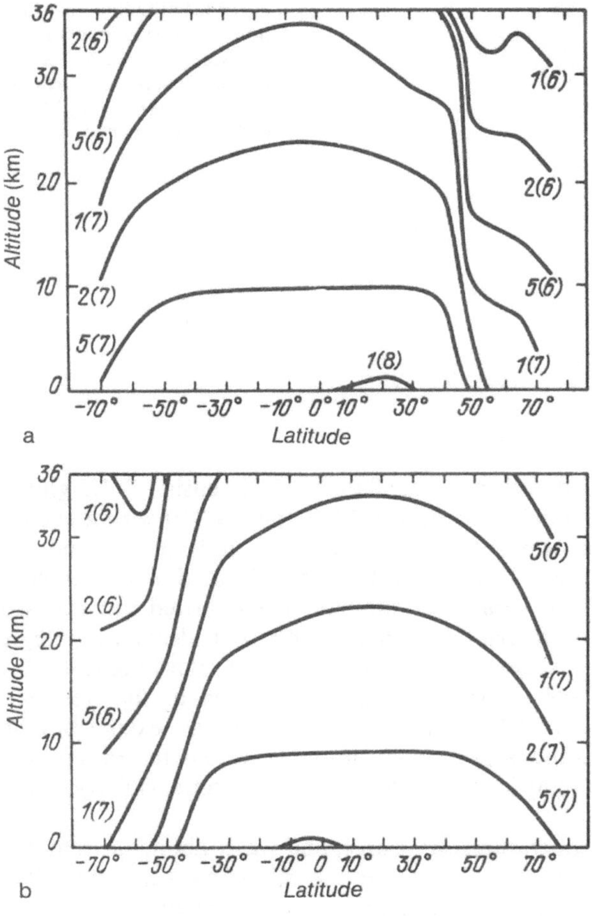

Fig. 65 a, b. Meridional contours of odd hydrogen number densities in winter (**a**) and spring (**b**). *2(7)* means 2×10^7 (Kong and McElroy 1977a)

winter and 23 μm at 60° S in spring. Meridional contours for odd hydrogen number densities are shown in Fig. 65. As is evident from the figure, the number densities at low and middle latitudes exceed the polar ones by one order of magnitude. The fact that this difference is attributed to the condensation of H_2O and H_2O_2 is easy to see when comparing saturated vapor number densities of these species at the temperatures typical of low and polar latitudes: they differ by a factor of ~1000.

The trend of similar contours for odd oxygen $O + O_3$ and ozone (Fig. 66) is opposite to that for H*. Near the surface O_3 has a maximum of 2×10^{11} cm^{-3} at 60° N, while being smaller by a factor of 40 at low latitudes. At 30 km ozone concentrations are almost independent of latitude. The results obtained do not take into account horizontal transport. The decrease of ozone and atomic oxygen concentrations at latitudes more than 60° N in winter is due to the decrease of the CO_2 photolytic rate at large solar zenith angles.

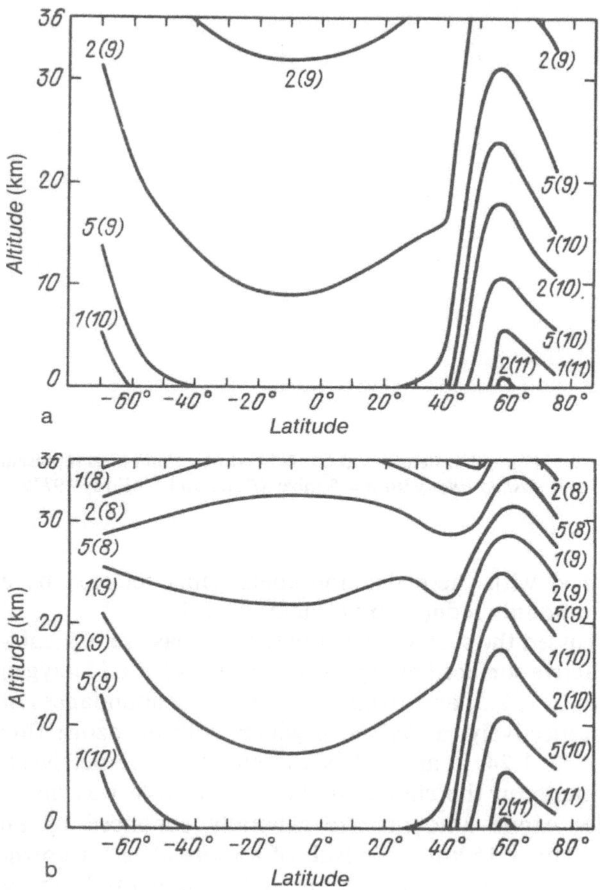

Fig. 66a, b. Meridional contours of odd oxygen (**a**) and ozone (**b**) number densities in winter. *5(9)* means 5×10^9 (Kong and McElroy 1977a)

Figure 67 gives altitude profiles of various forms of odd oxygen and hydrogen at 65° N in winter calculated for surface pressures of 5 and 8.3 mbar. The ozone abundance in the latter case is about twice as high as in the former. The increase of the ozone abundance at high latitudes in winter is caused not only by the decrease of odd hydrogen and the higher atmospheric pressure, but also by the increase of ozone and H_2O_2 lifetimes due to the absorption of solar radiations by large amounts of ozone at large zenith angles. The photolytic rate decreases by a factor of 5 at high latitudes.

Rate coefficients of the main reactions of odd hydrogen removal

$$HO_2 + OH \rightarrow H_2O + O_2$$
$$H + HO_2 \rightarrow H_2 + O_2$$

are assumed to be $2 \times 10^{-10}\,cm^3\,s^{-1}$ and $3 \times 10^{-11}\,e^{-(330/T)}\,cm^3\,s^{-1}$, respectively. To check the sensitivity of the calculations to the chosen values,

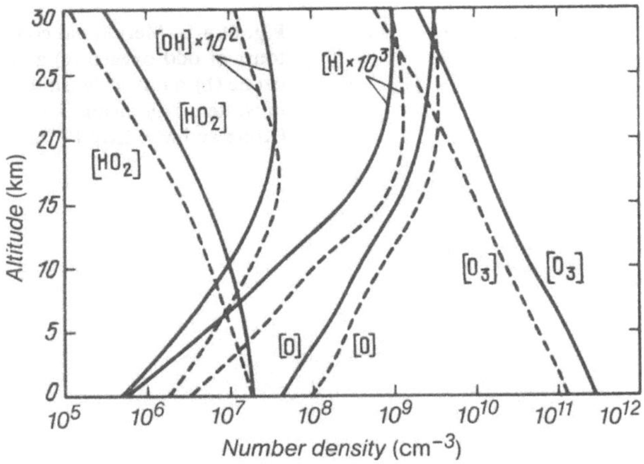

Fig. 67. Altitude profiles of O* and H* number densities at 65° N in winter. *Solid lines* represent calculations with $p = 8.3$ mbar, and *dashed lines* with $p = 5$ mbar (Kong and McElroy 1977a)

calculations were performed with these reaction coefficients reduced by a factor of 4 to give ozone amounts reduced by a factor of 1.5.

In the main model discussed the surface is assumed to be passive. Calculations with the chemically active surface having loss efficiencies for odd oxygen and odd hydrogen equal to 0, 1, 2.5, and 10 cm s^{-1} give ozone abundances of 58, 45, 38, and 31 μm, respectively, at 58° N in winter with an ozone flux absorbed by the surface of (0, 1.24, 2, and 2.7) $\times 10^{11}$ $cm^{-2}\,s^{-1}$, respectively.

The dependence of results on the choice of the eddy profile was investigated. At high latitudes ozone amounts are generally enhanced by an increase of eddy diffusion above 15 km as a result of a more efficient downward supply of odd oxygen. A decrease of K by an order of magnitude above 15 km leads to a reduction in ozone amount by about 25%. A decrease in K below 15 km adopted at high latitudes enhanced ozone amounts by a factor of 1.5, while the increase of K could not affect ozone abundance significantly. It should be borne in mind that this analysis is applicable at high latitudes only.

Calculations with meridional transport taken into account result in the increase of ozone at high latitudes by 6, 32, and 67% with nonlocal sources of odd oxygen being 0.27, 0.65, and 1.5 of the local source, respectively. Thus, the model of the global distribution of ozone on Mars fits the Mariner 9 measurements at low latitudes as well as at 50–75° N in winter and 50–75° S in spring. However, the calculated ozone amounts at 50–75° N in spring turn out to be much smaller than the measured ones (~10 μm). Traub et al. (1979) compared this model with the results of their 1.27 μm airglow measurements. The best agreement is achieved with a rate coefficient of $O_2(^1\Delta_g)$ quenching by CO_2 equal to 3×10^{-20} $cm^2\,s^{-1}$. A twofold discrepancy between the model and the measurements in the winter hemisphere may be explained by the effect of optical thickness and by the fact that the measured amounts in Fig. 6 are maximal and not mean values, as is adopted in the model.

2.10 Seasonal Variations of Atmospheric Composition at a Latitude of 65° N

In the previous section a model of latitudinal variations of minor components in two seasons was described. This model was time independent with appropriate temperatures, solar zenith angles, water vapor amounts, and atmospheric pressure. These values were chosen on the basis of measurements carried out prior to the Viking flights. The main aim of the model was to explain seasonal and latitudinal variations of ozone amounts measured by Mariner 9.

Another approach to this problem was used by Shimazaki and Shimizu (1979). They calculated a time-dependent model of seasonal variations of atmospheric composition at 65° N based on the Viking data.

Further modification of this calculation was made by Shimazaki (1981), who took into account diurnal variations of atmospheric composition. A great improvement of computational technique resulted in one time step of the model being calculated during 0.1 s of computer time. The total number of these steps was 60000, with every step being equal to 12^m to cover the Martian year. Here, we will describe the results of both papers.

Figures 68 and 69 show seasonal variations of surface temperature and pressure measured by Viking 2 and those adopted in calculations. One should note that the latitude of the Viking 2 lander is 48° N and not 65° N as adopted in the calculations. Seasonal dependence of temperature at 65° N deduced

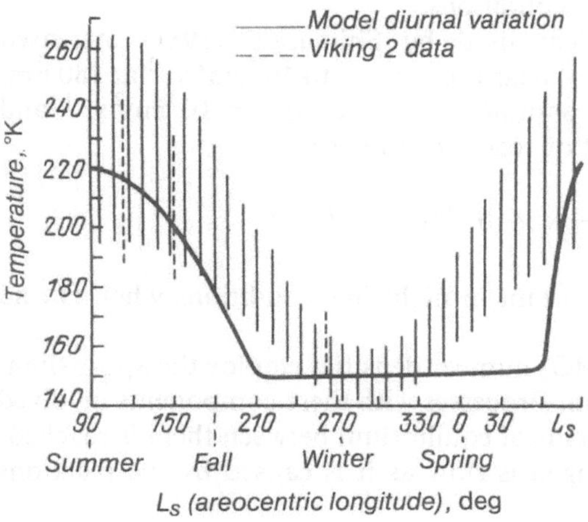

Fig. 68. Diurnal and seasonal variations of the surface temperature measured by Viking 2 at 48° N and adopted in calculations for 65° N (*vertical dashed* and *solid lines*, respectively). *Solid curve* shows seasonal variations of mean surface temperature at 65° measured by infrared radiometers on Mariner 9 and Viking orbiters (Shimazaki 1981)

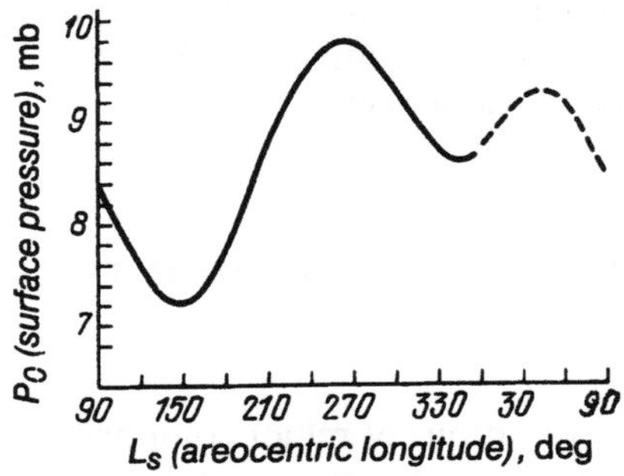

Fig. 69. Seasonal variations of atmospheric pressure adopted by Shimazaki and Shimizu (1979)

from infrared radiometric data aboard the Viking and Mariner 9 orbiters (Kieffer et al. 1977) is also shown in Fig. 66 and differs considerably from that assumed in calculations. The latter has a minimum at $L_s = 290°$, i.e., it is displaced from the winter solstice by 20°. The landing site of Viking 2 is lower by ~2 km than the mean areoid surface, but is close to the mean surface position at the latitude ~65° N. Therefore, the mean pressure exceeds the normal one; its variations caused by the sublimation and condensation of CO_2 in the polar caps are shown in Fig. 67. The temperature was assumed to decrease linearly to 150 K at 60 km.

Calculations by Shimazaki (1981) assume that water vapor amounts are determined only by surface temperature and the dependence used is similar to the curve of saturation, especially at $T < 200$ K. In this case diurnal variations of water vapor amounts should constitute two orders of magnitude, i.e., greatly exceed the real ones (see Chaps. 1.4 and 2.6).

Water vapor abundances assumed by Shimazaki and Shimizu (1979) are in accordance with those measured by the Viking orbiters at ~75° N. The vertical distribution of H_2O adopted has a scale height of 10 km, which adequately approximates daytime conditions.

The eddy diffusion coefficient used by Shimazaki (1981) increased exponentially from $5 \times 10^6\,\mathrm{cm^2\,s^{-1}}$ near the surface to $10^9\,\mathrm{cm^2\,s^{-1}}$ at 180 km. Shimazaki and Shimizu (1979) assumed it to be 10^7 or $10^8\,\mathrm{cm^2\,s^{-1}}$ and independent of height. In the list of reactions the process

$$HO_2 + O_3 \rightarrow OH + 2O_2; \qquad k = 8 \times 10^{-16}\,\mathrm{cm^3\,s^{-1}}$$

was entered. This process is important in nighttime conditions when H and OH are very low.

Calculations of H, OH, and HO_2 number densities employ the approximation of photochemical equilibrium. Processes with these components involved are really very fast, and photochemical equilibrium between them is reached. However, the sink of odd hydrogen is slow as it is caused by the reactions (Chap. 2.7)

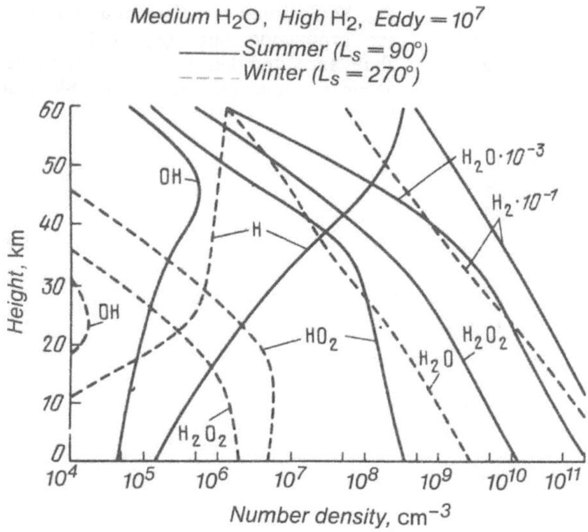

Fig. 70. Atmospheric composition at summer and winter solstices at 65° N (Shimazaki and Shimizu 1979)

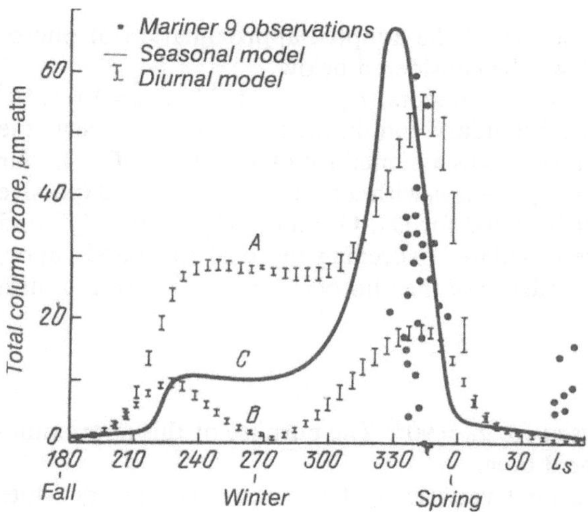

Fig. 71. Seasonal variations of ozone abundance at 65° N (Shimazaki 1981). *Model A* shows calculations with a passive surface, *model B* for the surface sink rate coefficient equal to $\gamma = 1\ \mathrm{cm\,s^{-1}}$, *model C* (Shimazaki and Shimizu 1979) does not involve diurnal variations. *Points* are Mariner 9 observations

$$HO_2 + OH \rightarrow H_2O + O_2$$
$$HO_2 + H \rightarrow H_2 + O_2$$
$$HO_2 + H \rightarrow H_2O + O\,.$$

The photochemical lifetime of odd hydrogen (taken as the ratio of number density to the sum of these reaction rates) constitutes about 2 weeks at 60 km in summer (Fig. 70), while the mixing time is about 24 h with $K = 10^7\ \mathrm{cm^2\,s^{-1}}$ or 3 h with $K = 10^8\ \mathrm{cm^2\,s^{-1}}$. The relation of these times is even much less

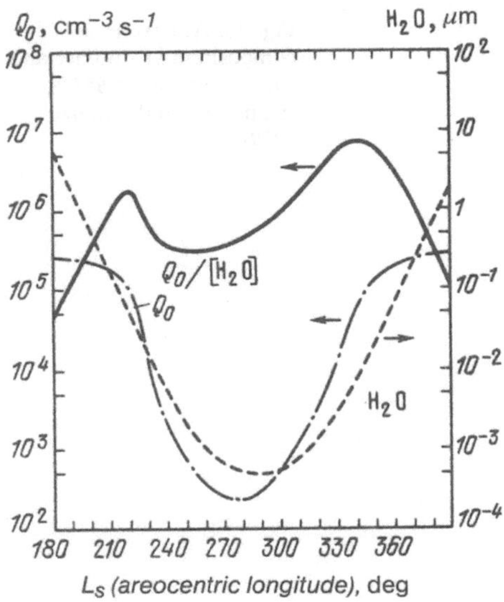

Fig. 72. Seasonal variations of atomic oxygen production rate, Q_O, water vapor abundance, and their ratio proportional to ozone amounts (Shimazaki and Shimizu 1979)

favorable in winter. Consequences of the adopted approximation of photochemical equilibrium for H* will be considered below.

The mixing ratios of stable components, $f_{CO} = 8 \times 10^{-4}$, $f_{O_2} = 1.6 \times 10^3$ and $f_{H_2} = 10^{-5}$, serve as their boundary conditions at the surface. Near the surface photochemical equilibrium was assumed for O, O_3, and H_2O_2. In our opinion the condition of flux, $\phi_i = -\alpha_i n_i$ with a partial case of $\phi_i = 0$ (passive surface) appears to be better founded for O_3, O, and H_2O_2. When H_2O and H_2O_2 number densities were calculated exceeding those of saturated vapor, the latter were used. The dependence of the photolytic rate on the solar zenith angle was given by

$$Q = ae^{-b\tau s z_0}$$

with more accurate values used for $z_0 \simeq 90°$. The purpose of this approximation was to save computational time.

Calculations give the seasonal variations of the ozone column densities shown in Fig. 71. A large peak appears in late winter at $L_s \simeq 340°$ due to the combination of H_2O variations and those of CO_2 and O_2 photolytic rates near the surface (Fig. 72). The former has a minimum at 290°, the latter at 270°, while their ratio, being proportional to ozone abundance, reveals two maxima at 336° and 220°. However, H_2O_2 condensation enhances only the maximum at 336°, while the maximum at 220° is smoothed out and manifests itself only in model *B*.

The main source of odd hydrogen in winter is molecular hydrogen, thus the results of calculations by Shimazaki and Shimizu (1979) for this season do not depend on the humidity assumed. Models *A* and *C* with the passive sur-

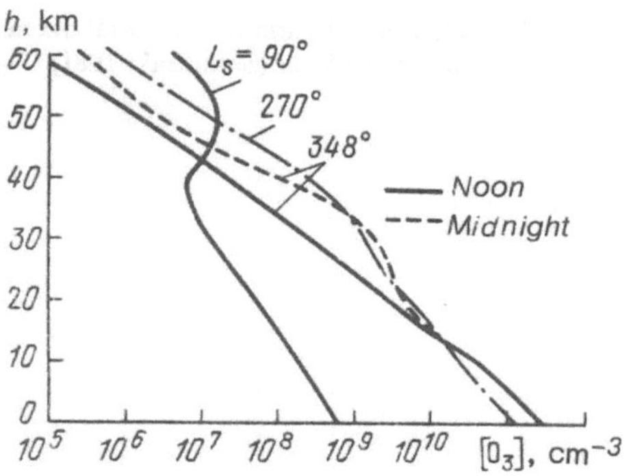

Fig. 73. Ozone number density profiles at 65° N in different seasons. Diurnal variations for $L_s = 90$ and 270° are low and not given in the figure (Shimazaki 1981)

face reproduce well maximal ozone amounts measured in late winter, while model *B* may be close to mean values. However, the measurements (Fig. 6) give ozone content exceeding 3 μm for most of the year from late summer till late spring at high latitudes, while all the models in Fig. 71 feature such a value of ozone amount only from the middle of fall till early spring. In our opinion, this can be attributed to a considerable difference between the assumed and real temperatures (Fig. 68). Indeed, ozone content is 3 μm at $L_s = 200°$ according to all models in Fig. 71. At this time the mean temperature adopted is 190 K (Fig. 68), while the measurements (Kieffer et al. 1977) indicate this temperature at $L_s = 170°$, which shifts the time when the ozone content reaches 3 μm to late summer in accordance with the Mariner 9 measurements. Similarly, a decrease of ozone content to 3 μm in spring should be at $L_s = 70°$ rather than at the 15° calculated.

Figure 73 shows the altitude profiles of ozone number densities predicted for different seasons. Other components are given in Fig. 68 for summer and winter solstices. Due to the approximation of photochemical equilibrium of odd hydrogen its number densities hardly vary with the height, being $3 \times 10^8\ cm^{-3}$ both near the surface and at 60 km. The number densities coincide with the results of our calculations (Fig. 60) near the surface and differ by as much as one order of magnitude at 60 km. Other papers give dependences of odd hydrogen at a height similar to ours. The elevated concentrations of H* at 40 – 60 km (Fig. 70) lead to ozone number densities at these altitudes (Fig. 73) which are underestimated when compared to our calculations and observations aboard Mars 5.

Let us consider now the effect of the surface loss of odd oxygen on the seasonal dependence of ozone content (Fig. 71). With an effective removal velocity of 1 cm s^{-1} and a scale height of 10 km, the lifetime of ozone is 10^6 s relative to this process. Its lifetime relative to atmospheric chemical reactions is 10^3 s in summer and 10^7 s in winter under daytime conditions, while being 10^7 and 10^9 s under nighttime conditions, with the reaction of O_3 with HO_2 as a main removal process. Therefore, the surface activity with $\gamma \simeq 1$ cm s^{-1} is unimportant in daytime in summer. On the other hand, it plays a significant part in other seasons especially in midwinter with very long nights.

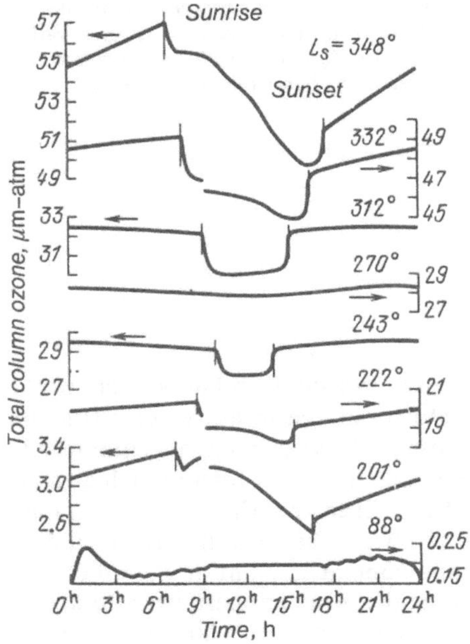

Fig. 74. Diurnal variations of total ozone abundance at 65° N (Shimazaki 1981)

The diurnal variations of ozone content for different seasons are shown in Fig. 74. They are insignificant at solstices when the sun does not rise and set at 65° N (the Martian axis declination is equal to ~25°). Diurnal variations are much smaller than those in the model of Krasitsky (Fig. 62). The curve with $L_s = 201°$ in Fig. 74 may be used for comparison as the closest to the global average conditions of the model of Krasitsky. The difference in the diurnal variations is due to the fact that in the model of Krasitsky considerable ozone amounts are maintained up to ~35 km with large diurnal variations above 15 km, while in the model of Shimazaki, ozone is concentrated near the surface with low diurnal variations.

The profiles of ozone and the atomic oxygen number densities at different time of day and night for fall equinox are shown in Fig. 75. Indeed the diurnal variations are negligible near the surface and are very prominent (of two orders of magnitude) at 35 km.

To explain the results of the Mars 5 measurements, Shimazaki calculated the atmospheric composition at 40 km in the approximation of photochemical equilibrium (Fig. 76) with the atmospheric temperature which determines water vapor densities as a variable. Of course, the accuracy of this approximation provides only order-of-magnitude estimates. From Fig. 74 it is evident that the Mars 5 observations near the evening terminator can be explained at an atmospheric temperature of 145 K, while those at the morning side limb exceed the calculated values at any temperature.

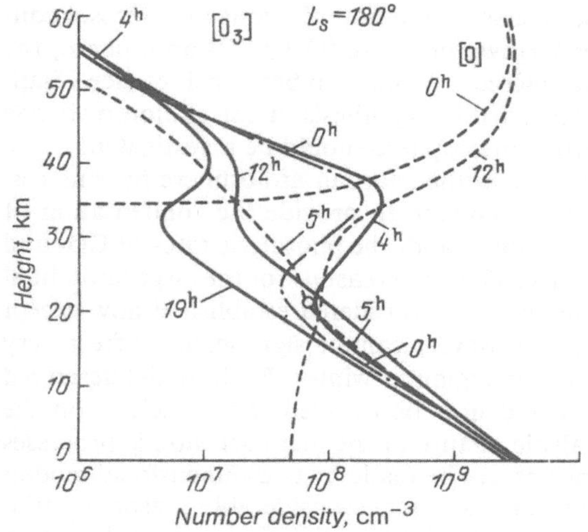

Fig. 75. Ozone and atomic oxygen number densities at different local times in fall equinox (Shimazaki 1981)

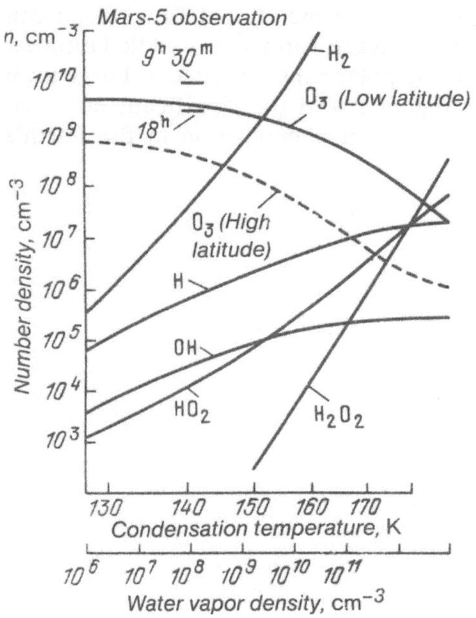

Fig. 76. Atmospheric composition at 40 km calculated by photochemical equilibrium approximation vs atmospheric temperature (Shimazaki 1981)

Shimazaki discussed a possible effect of aerosol particles on ozone number densities (see Chap. 2.7). Quantities and size of aerosol particles measured by Mars 5 produce an active surface of 10^{-7} cm^2 per 1 cm^3 of the atmosphere which is close to that in mesospheric clouds at high latitudes of the Earth. Photolytic rates may be enhanced by about 10% in the presence of aerosol; this effect, along with adsorption of ozone by aerosol particles, reduces the ozone content. There is a weak inverse correlation of ozone and aerosol contents in the Earth's stratosphere. However, there $[O_3] \gg [O]$ and, hence, the hypothesis that aerosol particles increase ozone number densities (see Chap. 2.7) cannot be applied there. To test this hypothesis, a correlation of ozone and aerosol contents in the Earth's mesosphere should be investigated.

So, large ozone amounts appear in the Martian atmosphere in winter at $\sim 60°$ where the temperature is low enough to provide the condensation of H_2O and H_2O_2 and where, on the other hand, the photolytic rates of CO_2 and O_2 are not very low near the surface. The main reasons for the large latitudinal and seasonal variations of ozone may be considered established now except for the aerosol influence which, however, cannot significantly affect very large ozone amounts in the subpolar region in winter. Each model described has some shortcomings which could now be avoided. Nevertheless, on the whole, these models form a realistic picture of the main aeronomic processes in the atmosphere. Further progress can be made by two-dimensional models with horizontal transport taken into account. Considerable seasonal variations of atmospheric pressure as a result of the sublimation and condensation of CO_2 on the polar caps should cause strong meridional flow which can greatly enrich the polar regions with odd oxygen formed at middle latitudes. The local production of odd oxygen in the polar caps is very low. Large local variations of ozone in the subpolar region reveal the significant effect of weather on the ozone amounts, a well-known phenomenon in the Earth's atmosphere.

3 Chemical Composition and Structure of the Venusian Atmosphere and Cloud Layer

3.1 Properties of Aerosol in the Upper Part of the Cloud Layer Deduced from Polarization Measurements

The surface of Venus is concealed by a thick and dense cloud layer which is observed in the visible range as a highly uniform formation. To interpret ground based spectroscopic measurements which for a long time were the main source of information on the chemical composition of the atmosphere, data are needed on the optical properties of cloud and aerosol particles of the outer part of the cloud layer. Besides, these properties may serve as a clue toward the study of the chemical composition of these particles. Important information in this field was obtained by polarization measurements.

According to the theory of light scattering by spherical particles developed by Gustav Mie, the intensity of light scattered by a particle depends on the scattering angle and properties of the particle defined by the three following parameters: $x = (2\pi r)/\lambda$ (r is the particle radius), the real and imaginary part of the refractive index, m_r and m_i, the latter being connected with the absorption coefficient, α, by $m_i = \lambda\alpha/4\pi$. Size distribution of particles of the same origin is given also by three parameters: the mean effective radius a, the effective relative variance b, and the total number of particles per unit volume n. Particle size distribution in terms of these parameters is given by the gamma distribution

$$\frac{dn}{dr} = Anr^{(1-3b)/b}e^{-(r/ab)}$$

where $A = (ab)^{(2b-1)/b}\Gamma^{-1}[(1-2b)/b]$; $\Gamma(x)$ is the gamma function,

$$a = \frac{\int_0^\infty r^3 n(r)\,dr}{\int_0^\infty r^2 n(r)\,dr}, \quad b = \frac{\int_0^\infty \left(1-\frac{r}{a}\right)^2 r^2 n(r)\,dr}{\int_0^\infty r^2 n(r)\,dr}.$$

The maximum of dn/dr occurs at $r_m = a(1-3b)$ and the standard deviation is $\sigma = a[b(1-2b)]^{1/2}$. For this purpose, the Gaussian and log normal distributions can also be used:

$$\frac{dn}{dr} = \frac{n}{\sigma(2\pi)^{1/2}} e^{-\frac{1}{2}\left(\frac{r-a}{\sigma}\right)^2} \qquad \text{(Gaussian distribution)}$$

$$\frac{dn}{dr} = \frac{n}{r(2\pi)^{1/2}\ln\sigma} e^{-\frac{1}{2}\left(\frac{\ln r/a}{\ln\sigma}\right)^2} \qquad \text{(log normal distribution)}\,.$$

By measuring light scattered by an aerosol medium at five scattering angles one can set up five equations and calculate five parameters to describe the physical characteristics of the particles: n, a, b, m_r, m_i. With a larger number of scattering angles the system of equations becomes overdetermined, and the accuracy of the solution may be improved.

This problem is readily solved if the approximation of single scattering is valid, i.e., $\tau \lesssim 0.5$ and $\omega > 0.5$ or $\omega < 0.5$ with an arbitrary τ. Here, τ is the optical thickness and ω is the single scattering albedo. Multiple scattering smooths the phase dependence of the brightness coefficient $R = \pi I/I_0$ (I_0, I are the intensities of incident and scattered radiation) and substantially lowers the accuracy of the solution. Polarization measurements are indispensable in this case. Light components polarized in a source-particle-detector plane and perpendicular to this plane have different conditions of scattering. As a result, polarization of scattered light appears which depends on the scattering angle and the five parameters of the particles described above. However, the polarized component of scattered radiation is due mainly to single scattering. Multiple scattering involves a wide range of arbitrary angles for each single act of scattering and as a result, on the whole, it has a low degree of polarization. Therefore, polarization measurements are very important to study aerosol media with multiple scattering of radiation to deduce the properties of the particles.

Polarization of Venus was measured by Lyot (1929), Coffeen and Gehrels (1969), Dollfus and Coffeen (1970), and Veverka (1971). The most detailed interpretation of these measurements was given by Hansen and Hovenier (1974). We will dwell upon the main results of the latter paper.

The large albedo of Venus in the visible and near infrared regions indicates very low true absorption and large optical thickness of the atmosphere and cloud layer. Thus, $m_i = 0$ and a semi-infinite atmosphere may be assumed. The calculated phase dependence of polarization for various $x = (2\pi a)/\lambda$ and $m_r = 1.4$, $b = 0.05$ are shown in Fig. 77. Similar calculations were made for a set of m_r and b. The best agreement with observations at 550 nm is achieved for $m_r = 1.44 \pm 0.015$, $a = 1.05 \pm 0.1$ μm, $b = 0.07 \pm 0.02$ (Fig. 78). Rayleigh scattering of atmospheric gases was taken into account along with aerosol scattering. A comparison of calculations and measurements at $\lambda = 990$ nm and 365 nm gives $m_r = 1.43 \pm 0.015$ and 1.46 ± 0.015, respectively.

Gas scattering contributes significantly at 365 nm, thus making it possible to derive an atmospheric pressure at a level of $\tau = 1$ where single scattered radiation forms. It equals 50 ± 20 mbar (Fig. 79). The level of $\tau = 1$ defines the upper boundary of the cloud layer and is at 68 km. Good agreement of the

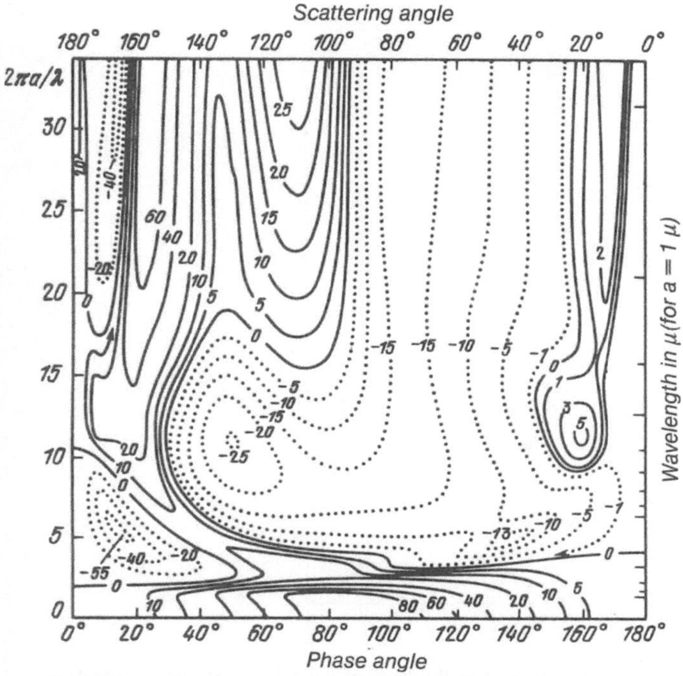

Fig. 77. Percent polarization vs phase angle for various $x = (2\pi a)/\lambda$ with $m_r = 1.4$ and $b = 0.05$. Phase angle is the angle between Sun, object, and observer (Hansen and Hovenier 1974)

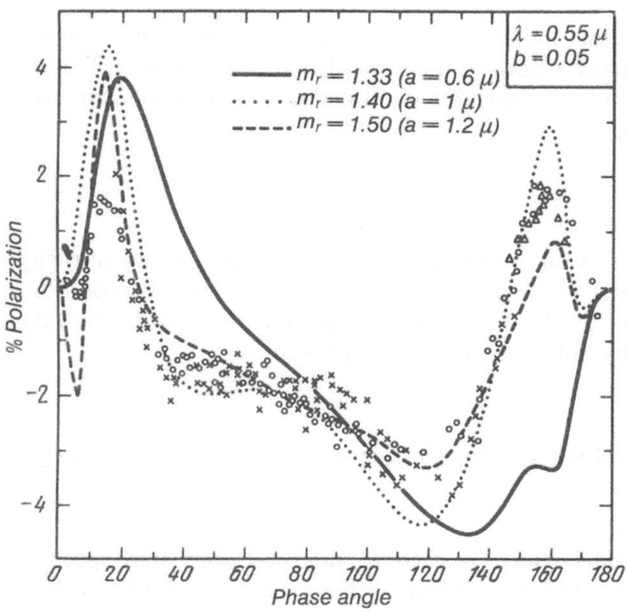

Fig. 78. Calculated (*lines*) and observed (*dots*) polarization of Venus at various phase angles. Calculations were performed for different m_r with a chosen to yield the best agreement with measurements (Hansen and Hovenier 1974). To fit the observations m_r should be between 1.4 and 1.5

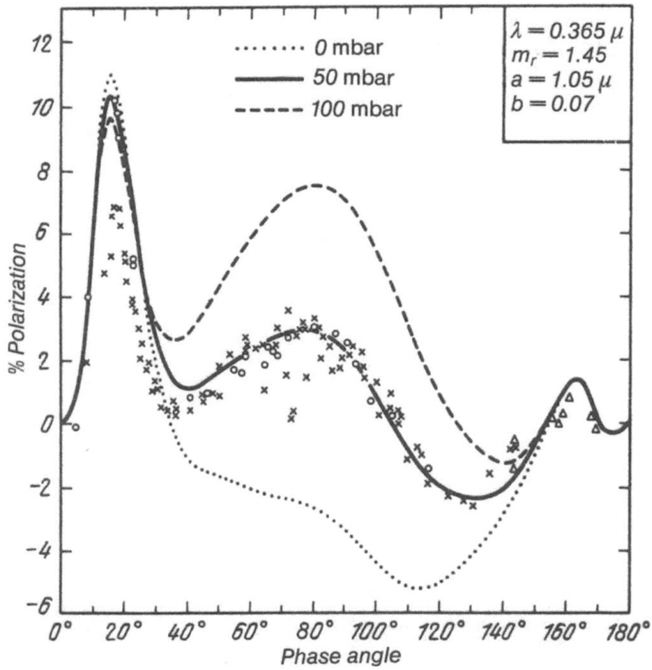

Fig. 79. Determination of atmospheric pressure at the level of $\tau = 1$ by means of polarization measurements at 365 nm (Hansen and Hovenier 1974)

calculations and observations indicates that the adopted approximation of spherical particles is valid and clouds at the upper boundary consist of liquid particles. For liquids with a low absorption coefficient, m_i, the dependence of m_r on wavelength is given by

$$\frac{m_r^2+2}{m_r^2-1} = \frac{A}{\lambda^2} - B\,.$$

The derived m_r are described well by this dependence. The best agreement with the m_r is reached for a water solution of sulfuric acid with a concentration of ~80%. The identification of the cloud particles as sulfuric acid drops was suggested independently by Sill (1972) and Young (1973).

3.2 Interpretation of Spectroscopic Measurements of Escaping Radiation

The simple reflection model which is used to interpret Venusian spectra assumes that clouds form a uniform surface with a given reflectivity. In this case, spectral features observed are due to absorption of the overcloud atmosphere. According to this model, equivalent widths of lines and bands grow with phase angle (see Chap. 1.1). A phase angle is the angle between object-observer and object-Sun directions.

More complicated is the model of an uniform aerosol-gas mixture in the atmosphere. In this model, equivalent widths should decrease slightly with the phase angle which is consistent with observation of CO_2 (see Moroz 1967, 1971). This model is fairly good for absorbers with a height independent mixing ratio. In the general case, this model needs the radiative transfer equation to be solved with the following initial data: the single scattering albedo, ω, the phase function, $\gamma(\varphi)$, the optical thickness of the cloud layer, τ, the reflectivity of an underlying surface, A_s, and the incident and observational angles. Simple approximate, but rather accurate relations for reflectance, R, and transmittance, T, of a plane-parallel optical thick uniform atmosphere was derived by Germogenova (1962) and described by Danielson et al. (1969) and Sobolev (1972). They need the phase function in the form of the Henyey-Greenstein equation

$$\gamma(\varphi)=\frac{1-g^2}{(1+g^2-2g\cos\varphi)^{3/2}}\,,\qquad g=\frac{1}{2}\int_0^{\pi}\gamma(\varphi)\cos\varphi\sin\varphi\,d\varphi\,.$$

Particle properties derived from polarization measurements give $g=0.7$ almost independent of wavelength in the visible range. These relations are

$$R(\mu_0)=r(\mu_0)+\frac{A_S t_L t(\mu_0)}{1-A_S r_L}\,,\qquad T(\mu_0)=\frac{t(\mu_0)}{1-A_S r_L}$$

where μ_0 is the cosine of the incident radiation angle. The other auxiliary values are given by

$$1-\omega=\frac{k^2}{3(1-g)}\,;\quad m=\frac{8k}{3(1-g)}\,;\quad l=e^{-\frac{1.42k}{1-g}}\,;\quad f=\exp\left[-k\left(\tau+\frac{1.42}{1-g}\right)\right]$$

$$r(\mu_0)=1-K(\mu_0)\,m[\tfrac{1}{2}l^{1/2}+f^2/(1-f^2)]\,;\qquad t(\mu_0)=\frac{K(\mu_0)\,mf}{1-f^2}$$

$$r_L=1-m[\tfrac{1}{2}l^{1/2}+f^2/(1-f^2)]\,;\qquad t_L=\frac{mf}{1-f^2}\,.$$

Values of $K(\mu)$ are listed in Table 14. These formulas may be applied for $\tau(1-g)\gtrsim 1$. So one needs ω and g to employ this technique. An exact solution for these quantities is given by the Mie theory for spherical particles and may be used as an approximation for particles with arbitrary shape. This solution needs a computer program to calculate extinction, scattering, and absorption effectivities, K_{ext}, K_{sc}, K_{ab}, defined as ratios of the appropriate cross-sections to πr^2. Then, $1-\omega=K_{ab}/K_{ext}$, and relations for K are given by the approximate formulas derived by van de Hulst (1957) for the case of $1-\omega\ll 1$:

Table 14. $K(\mu)$ (Danielson et al. 1969)

μ	$g=0$	0.25	0.5	0.75	0.875
0	0.433	0.424	0.395	0.333	0.272
0.1	0.540	0.534	0.516	0.484	0.463
0.3	0.711	0.708	0.701	0.692	0.688
0.5	0.872	0.870	0.869	0.869	0.869
0.7	1.028	1.028	1.030	1.033	1.035
0.9	1.182	1.184	1.188	1.192	1.194
1	1.259	1.261	1.265	1.270	1.271

$$K_{\text{ext}} = 2 - \frac{4}{y}\sin y + \frac{4}{y^2}(1-\cos y)$$

$$K_{\text{ab}} = 1 + \frac{1}{2xm_i}\, e^{-4xm_i} + \frac{1}{8x^2m_i^2}\,(e^{-4xm_i}-1)$$

where $y = 4\pi r(m_r - 1)/\lambda$, $x = 2\pi r/\lambda$, $m_i = k\lambda/4\pi$, r is a particle radius, m_r and k are refractive index and absorption coefficient of a particulate matter. $K_{\text{ext}} \simeq 2$ for $y \gtrsim 5$ (large particle approximation) and $K_{\text{ab}} \simeq \frac{4}{3}kr$ for $kr \ll 1$, therefore, $1-\omega = \frac{4}{3}kr$ for large particles. The calculated dependence of g on x and m_r is shown in Fig. 80.

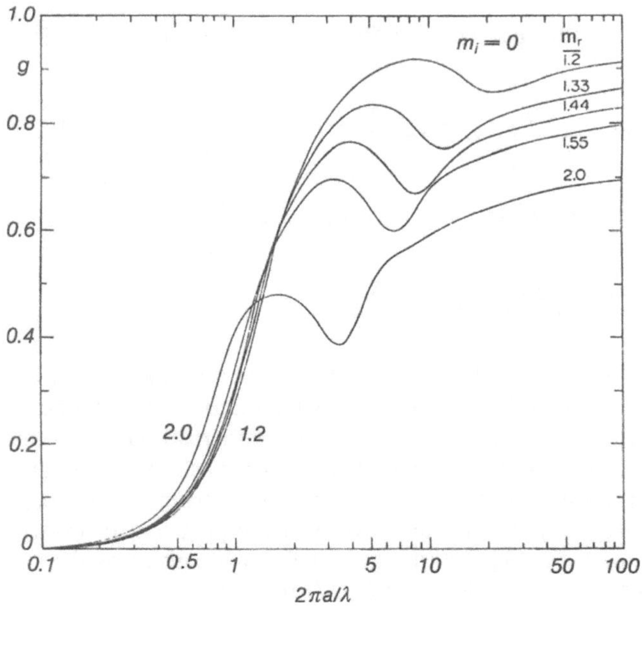

Fig. 80. Asymmetry parameter, g, as a function of effective size parameter, $(2\pi a)/\lambda$. The standard size distribution was used with $b = 0.07$ (Hansen and Travis 1974)

3.3 Spectroscopy in Visible and Infrared Ranges

Visible Range. The curve of the spherical albedo of Venus, A, in a wavelength region of 0.2 to 4 μm as given by Moroz (1981b) and based on measurements of Irvine (1968) and Barker et al. (1975) is shown in Fig. 81. The value of the visual albedo is very large and equal to 0.83 at 600 – 1000 nm. One can suppose that scattering in the cloud layer is conservative, $\omega = 1$, and the difference of A from unity is caused by the finite optical thickness of the cloud layer. For conservative scattering the relations between R and T require the evaluation of indeterminate forms of 0/0 type and with $A_s = 0$ give

$$R(\mu_0) = 1 - T(\mu_0) = 1 - \frac{4K(\mu_0)}{3(1-g)\left(\tau + \dfrac{1.42}{1-g}\right)}.$$

Hence, $\tau \simeq 20$ is in accord with the measurements aboard lander probes (see below).

One may note that $R = \frac{1}{2}$ at $\tau = 4$. This level is interesting in that the cloud layer above and below it contributes equally to the reflected radiation. Though in the model described in Chap. 3.2, the mixing ratio of a possible absorber is assumed to be height independent, deviations from this assumption may be large, and the absorber mixing ratio derived from spectroscopic measurements is attributed usually to the level of $\tau = 4$ (so-called level of line formation) as are polarization measurements to $\tau = 1$.

Spectrum in the 1 – 4 μm Region. Spectroscopy of Venus in the near infrared region is of great importance for investigation of its atmospheric composition. These measurements utilized ground-based telescopes and aircraft to reduce telluric absorption of CO_2 and H_2O. A spectrum with low resolution ($\lambda/\Delta\lambda \simeq 70$) measured by Pollack et al. (1978) from an aircraft at an altitude of 13 km is shown in Fig. 82.

Carbon Dioxide. A great number of CO_2 bands in the region of 1 – 4 μm exist. They were detected for the first time by Adams and Dunham in 1932. A review of measurements up to the end of the 1960's was given by Moroz

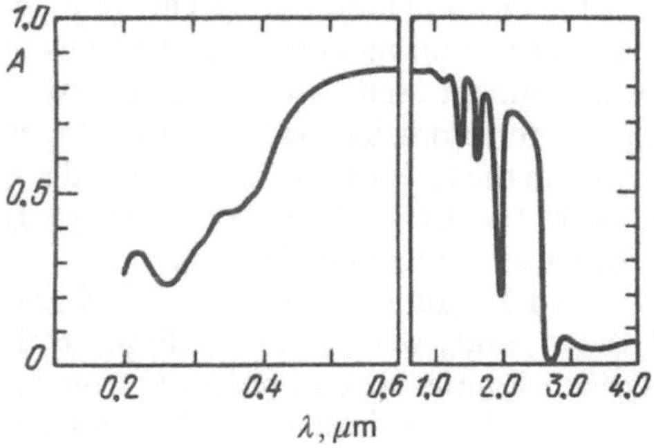

Fig. 81. Albeo of Venus in the range of 0.2 – 4.0 μm (Moroz 1981b)

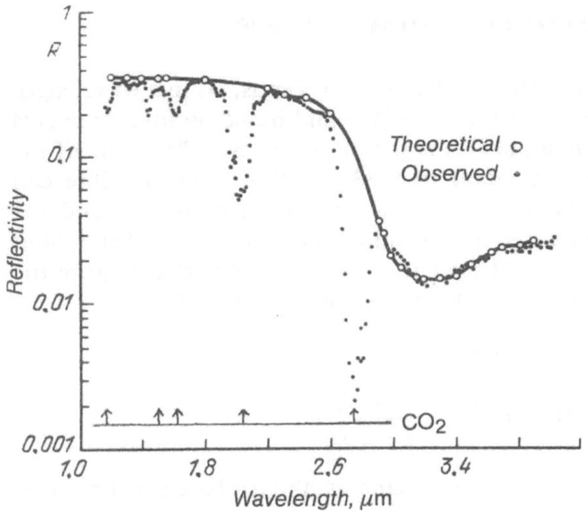

Fig. 82. Reflectivity of Venus at a phase angle of 86°. Here reflectivity refers to the ratio of the observed flux to that which would have been received from the white flat Lambertian surface normally illuminated by the Sun and having an area equal to the illuminated portion of Venus. The positions of CO_2 bands are shown by *arrows*. The theoretical curve fits the observation for drops of sulfuric acid with a concentration of 85% and a mean particle radius of 1 μm (Pollack et al. 1978)

(1971). Observations of weak and saturated bands make it possible to derive the atmospheric pressure and mixing ratio of an absorber (see Chap. 1.1). One should state, however, that this problem for the Venusian atmosphere was not unambiguously solved until successful experiments on spacecraft gave the CO_2 mixing ratio close to unity. A proper correction of a spectroscopic interpretation was made which yielded a CO_2 abundance at the level of line formation of 1 km · atm and atmospheric pressure of 100 – 150 mbar. With the temperature of the upper part of the cloud layer known from infrared observations $T \simeq 230$ K, one can obtain the position of the level of line formation to be ~5 km below the level of $\tau = 1$ ($P = 50$ mbar, $h = 68$ km), i.e., at 63 km. It results in atmospheric pressure and cloud optical thickness increasing from 68 to 63 km, the former by a factor of 2 – 3 and the latter by a factor of 4, which indicates the deviation from the uniform model.

Carbonmonoxide. The (2, 0) band of CO at 2.35 μm was detected first by Moroz (1964), who estimated CO abundance to be 5 cm · atm. A comparison with a CO_2 abundance of 1 km · atm gives the CO mixing ratio, $f_{CO} = 5 \times 10^{-5}$. An earlier detection of CO abundance of 79 cm · atm by Sinton (1962) might be misleading. The CO amounts measured by Moroz were near the threshold of sensitivity of the instrument used. A more perfect spectrometer with high resolution (Michelson interferometer) was used by Connes et al. (1968). In the spectra obtained by them the lines of the 2.35 μm band are readily detected and a comparison with the CO_2 absorption lines gives $f_{CO} = 5 \times 10^{-5}$ which coincides with the value obtained by Moroz.

Water vapor has bands at 1.38, 1.9, and 2.7 μm in the range of 1 – 4 μm. The most reliable measurements of these bands were made by Fink et al. (1972) in Kiuper's laboratory by analysis of Fourier spectra observed from an aircraft. The Venusian spectra were compared with those of the Moon mea-

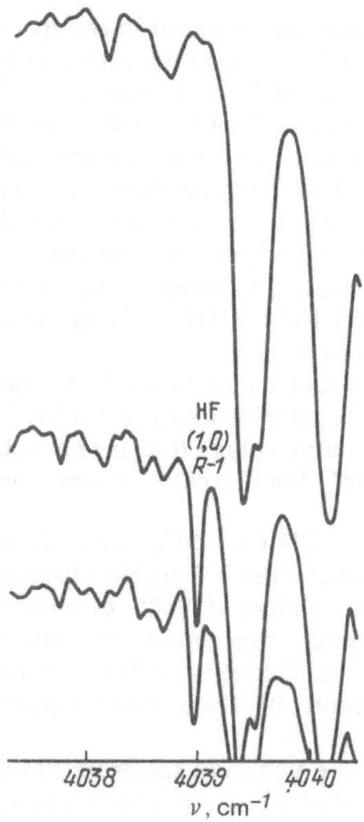

Fig. 83. A line of HF (1,0) band in the Venusian spectrum. The upper curve is the solar spectrum and the lower ones are the Venusian spectra for different times of observations (Connes et al. 1967)

sured in the same flights. Telluric water vapor abundance did not exceed 10 μm precipitable water. Spectra calculated by the simple reflection model fit the measured ones with a water vapor content of 1.8 ± 0.5 μm for the 1.38 μm band, 1.4 ± 0.5 μm for the 1.9 μm band, and less than 2 μm for the 2.7 μm band.

These values are valid for two-way transmission on Venus, i.e., without a correction for the air mass factor. The water vapor mixing ratio was obtained as a ratio of spectroscopic abundances of H_2O for the band at 1.38 μm and those of CO_2 for the band at 1.34 μm (3.6 km · atm without a correction for the air mass factor) to be equal to 0.6×10^{-6}.

Ground-based observations of H_2O search for very weak Doppler shifted satellites with strong telluric absorption lines of H_2O. The most extensive measurements of the 0.82 μm band where made by Barker (1975). The results show a great scatter from <0.5 ppm to 30 ppm. The mean mixing ratio is $\sim 10^{-5}$ or 10 ppm (parts per million) at phase angles smaller than 120°. At large phase angles of 140 – 160° the mean two-way amount of H_2O is smaller by a factor of ~ 5, but it is difficult to interpret in terms of mixing ratio.

HCl and HF. The discovery of these components became possible due to the advance of high resolution Fourier spectroscopy (Connes et al. 1967). The lines of the (2, 0) band of HCl at 1.8 μm were quite pronounced in the spectrum. Their shift from the laboratory-measured lines coincides with a calculated Doppler shift to within 10^{-3} nm. The lines of both isotopes, HCl^{35} and HCl^{37}, were detected with their intensity ratio being 2.4 ± 0.2. The ratio of these isotopes on the Earth is 3.06, thus with the same on Venus the ratio of equivalent widths of the lines should range from $3.06^{1/2}$ to 3.06 depending on atmospheric pressure at the level of line formation. Presuming a terrestrial isotopic abundance ratio the measured intensity ratio gives for the latter a value of 80^{+100}_{-40} mbar; the equivalent widths correspond to an HCl amount of 1.9 ± 0.4 cm · atm in the simple reflection model for the HCl mixing ratio $f_{HCl} = 6 \times 10^{-7}$ with the air mass factor taken into account.

HF was detected by the lines of (1, 0) and (2, 0) bands at 2.45 μm and 1.3 μm, respectively (Fig. 83). Their equivalent widths indicate $f_{HF} = 5 \times 10^{-9}$. Among all components observed in the Venusian atmosphere this one has the lowest mixing ratio. One should note that HF lines greatly exceed the threshold sensitivity of the spectrometer used.

Spectroscopic Limits. Upper limits of CH_4, CH_3Cl, C_2H_2, and HCN mixing ratios of 10^{-6} were obtained by Connes et al. (1967) from the absence of appropriate absorption lines and bands. Kuiper (1968) estimated $f_{NH_3} < 3 \times 10^{-8}$ and $f_{COS} < 10^{-8}$. Many attempts were made to detect molecular oxygen by the band at 762 nm. Traub and Carleton (1974) gave the upper limit, $f_{O_2} < 10^{-6}$, which was improved recently by Trauger and Lunine (1983): $f_{O_2} < 3 \times 10^{-7}$.

Sulfuric Acid. Apart from molecular bands the Venusian spectrum has a broad depression centered at 3.2 μm. The absence of a fine structure suggests that aerosol particles of the cloud layer and not gaseous components are responsible for this depression. In a detailed study of the nature of the depression, Pollack et al. (1978) have shown that out of all possible candidates (drops of Hg, H_2O, water solutions of HCl, H_2SO_4, HNO_3, condensates of H_2O, hydrocarbons, C_3O_2, solid particles of NH_4Cl, $FeCl_2 \cdot 2H_2O$, Hg_2Cl_2, $(NH_4)_2SO_4$, silicates, and carbonates) the best agreement with the measurements is observed for drops of sulfuric acid water solution with a concentration of 85% and a mean particle radius of 1 μm (Fig. 82).

The Spectral Region of 8–22 μm. The most detailed observations in this range were carried out by Samuelson et al. (1975) with a Fourier spectrometer having a resolution power, $\lambda/\Delta\lambda$, of 1000. Thermal radiation of the cloud layer dominates here. To compensate telluric absorption, spectra of Venus and the Moon were compared with $T = 360$ K and emissivity $\varepsilon = 1$ assumed for the latter.

To deduce data on the cloud layer from the measurements, the spectrum was modeled with profiles of temperature and CO_2 concentration obtained from the Mariner 5 radio-occultation experiments (Fjeldbo et al. 1971) and with the uniform model of the cloud layer adopted. The single parameter of this model is the volume extinction coefficient σ at a given level (60 km in this case). The measured spectrum as a dependence of $\sigma(\lambda)$ at 60 km is shown in

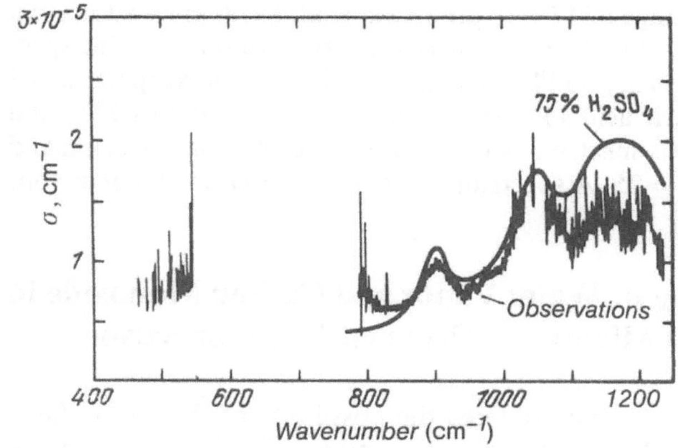

Fig. 84. Venusian spectrum in the range of 8 – 22 μm referred to wave number scale and given as a dependence of the volume extinction coefficient at 60 km. The *smooth curve* represents calculations for drops of sulfuric acid with a concentration of 75%, a radius 1 μm, and a number density of 600 cm at 60 km (Samuelson et al. 1975)

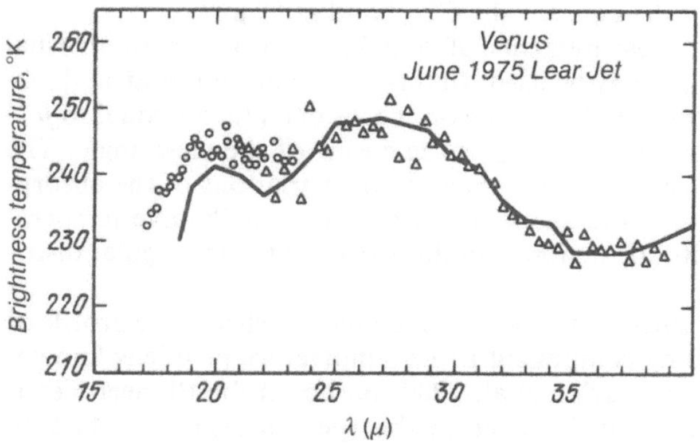

Fig. 85. Spectrum of Venus in the range of 17 – 38 μm: observations and calculations for drops of sulfuric acid with a concentration of 75% and radius of 1 μm (Reed et al. 1978)

Fig. 84. Due to strong telluric absorption, ground-based measurements are impossible in the 5 – 8 μm and 13 – 18 μm ranges. This is the reason why the data in the range of 550 – 780 cm^{-1} are absent in Fig. 84.

The extinction coefficient is a product of the number of particles and their extinction cross-section. The latter may be calculated by means of the Mie theory with the complex refraction index and size distribution known. Such calculations for sulfuric acid drops having a concentration of 75% and size distribution deduced by Hansen and Hovenier (see Chap. 3.1) fit the observations with $n = 600\ cm^{-3}$ at 60 km. A fair agreement of the calculated and measured spectra is the important confirmation of sulfuric acid being present in the cloud layer. Some deviations may be explained by differences in particle size and concentration of sulfuric acid solution in particles, as well as by the presence of aerosol of other chemical compositions.

Spectroscopy in the range of 17 – 38 μm (Reed et al. 1978) gives additional evidence for the presence of sulfuric acid aerosol in the cloud layer. The spectrum of Venus measured is in good agreement with the calculated spectrum of aerosol particles of sulfuric acid solution having a concentration of 75% and radii of 1 μm (Fig. 85). This spectrum was measured from aircraft at 14 km and has a resolution of $\lambda/\Delta\lambda = 25$. A spectrum of Mars was used for comparison.

3.4 Remote Sounding of Water Vapor and Carbon Monoxide in Far Infrared and Microwave Regions. Radio Spectrum

Water Vapor. In Chap. 1.8 we considered the possibility of the atmospheric temperature profile being deduced from measurements of a shape of a band in the thermal infrared region if a species to which the band belongs is mixed uniformly. This is exactly the case for the CO_2 band at 15 μm.

For similar purposes measurements of angular distribution of thermal radiation at one wavelength may be used. In this case, unit slant optical thickness refers to different altitudes depending on the air mass factor equal to μ^{-1} in a plane-parallel atmosphere, where μ is the cosine of the view angle. The range of μ is typically smaller than the range of the variations of the absorption coefficient, therefore, band shape measurements usually give temperature profiles of a wider altitude region than those deduced from angular distribution.

Both kinds of measurements may be also explored to retrieve the profile of a mixing ratio of a radiating component if the temperature profile is known.

An infrared radiometer (Taylor et al. 1980) on board the Pioneer Venus orbiter consisted of five channels covering the spectral region of 11.5 to 15 μm, i.e., the short part of the 15 μm band of CO_2, a pressure modulated photometer, a near infrared channel, and a far infrared channel centered at the 45 μm rotation band of water vapor. With the temperature profile retrieved from the 11.5 – 15 μm channels, the 45 μm channel was used for water vapor measurements (Schofield et al. 1982).

Brightness temperature of the atmosphere should increase at large view angles for a positive temperature gradient and decrease for a negative one, the so-called limb brightening and limb darkening. Mean limb darkening curves at 45 and 11.5 μm measured at two locations with solar fixed coordinates of 25° N, 225° E (night, solar zenith angle, z_0, is about 145°) and 25° N, 345° E (noon, $z_0 \simeq 30°$) are shown in Fig. 86. The 11.5 μm channel for comparison was intended to minimize the role of CO_2 absorption. Three sources of atmospheric capacity are considered: CO_2 and cloud particles for both channels and water vapor for the 45 μm channel. Cloud particle size distribution with an effective radius of 1.05 μm and an effective variance of 0.07 containing 75% H_2SO_4 is assumed (Chap. 3.1). Calculations by the Mie theory yield in 45/11 μm channels an optical depth ratio of 0.36. Qualitatively this ratio

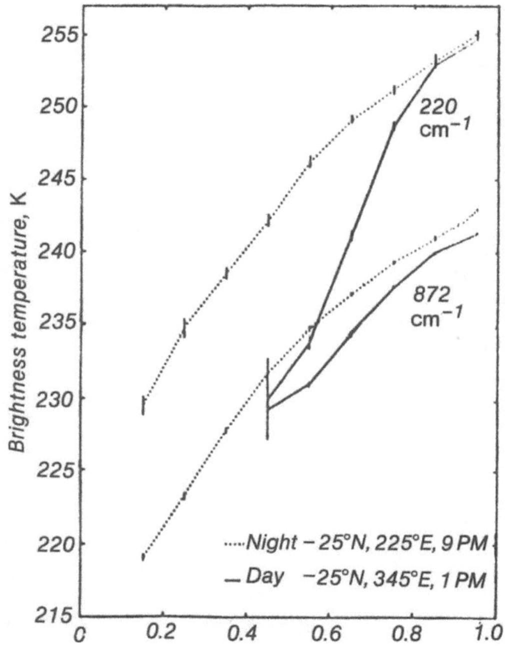

Fig. 86. Mean limb darkening curves at 45 and 11.5 μm for the day side (*solid lines*) and night side (*dotted lines*) (Schofield et al. 1982)

may be predicted by comparison of parameters $x = (2\pi r)/\lambda$ with both being under unity.

Limb darkening is much steeper at 45 μm than at 11.5 μm in daytime. With the atmospheric temperature profile taken into account (see below, Chap. 3.13) one can see that radiation in the 11.5 μm and 45 μm channels forms at 65 and 62 km, respectively, for $\mu = 1$. A deeper formation of 45 μm radiation means that the effect of water vapor absorption is smaller than the difference in cloud optical thicknesses. At $\mu = 0.45$ both channels see the level of 70 km. The 11.5 μm channel is sensitive mainly to cloud particles, thus, giving $\tau_{cloud} \simeq \mu \simeq 0.45$ at 70 km. Then $\tau_{cloud} \simeq 0.36 \times 0.45 = 0.16$ at 45 μm, and with $\tau_{cloud} + \tau_{H_2O} \simeq \mu = 0.45$ at 70 km for the 45 μm channel, one obtains $\tau_{H_2O} \simeq 0.3$, i.e., the water vapor amount is equal to ~80 μm of precipitable water above 70 km. The solving of appropriate equations results in 38.9 ± 2.4 μm above unit cloud optical depth at 11.5 μm which corresponds to ~100 mbar or 65 km and in the water vapor mixing ratio profile shown in Fig. 87b. The simple consideration above should help to understand the results obtained, which seem very unusual.

Limb darkening curves at night (Fig. 86) are similar for both channels though displaced by 10 – 12°. This means that aerosol is responsible for atmospheric opacity in both channels and the water vapor mixing ratio is much lower than in daytime. Figure 87a shows the retrieved profile of f_{H_2O} for nighttime resulting in a water vapor amount of 3.7 ± 0.9 μm, as well as sen-

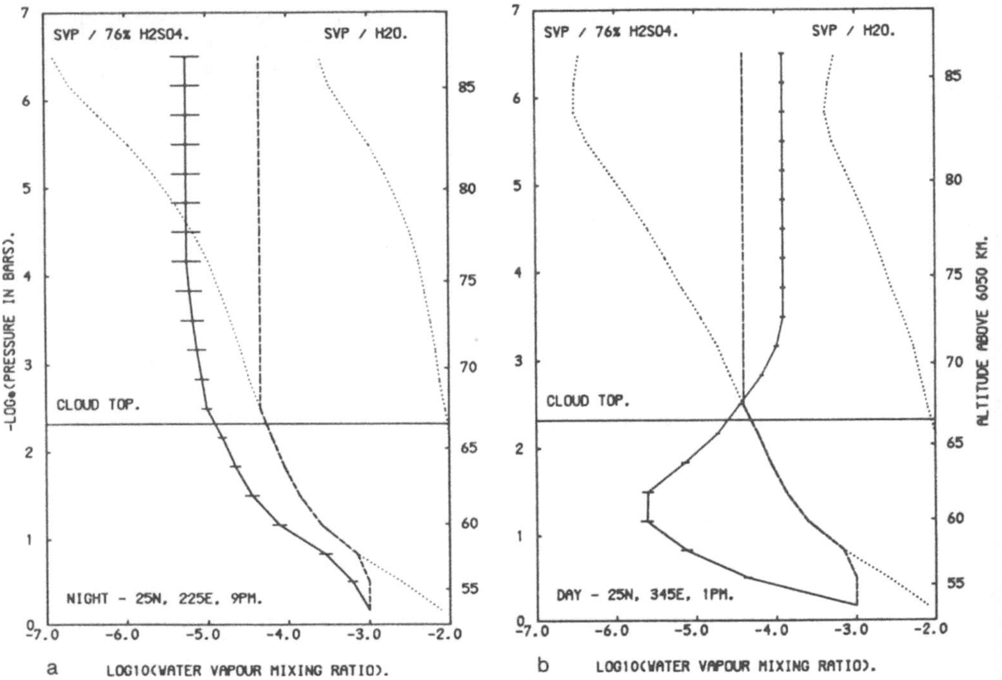

Fig. 87. a Water vapor mixing ratio profile at 25° N and 225° E (9 P.M. local time). The retrieved profile (*solid line*) is compared with the first guess profile (*dashed line*) and profiles corresponding to the saturated vapor pressure of water over both 75% sulfuric acid and water (*dotted lines*). **b** The same for 25° N and 345° E (1 P.M. local time) (Schofield et al. 1982)

sitivity of the profiles obtained to the assumed initial profiles and mixing ratio at a lower boundary of 53 km.

The derived distribution of water vapor amounts at low latitudes is given in Fig. 88. Daytime amounts exceed 10 μm from 09.00 h to 17.00 h LT, being typically smaller than 3 μm in nighttime. To obtain this map, as well as the profiles in Figs. 86 and 87, a great number of limb darkening curves measured at fixed solar coordinates were averaged, and these mean curves were used to retrieve water vapor amounts.

Cloud unit optical depth for 11.5 μm is at 100 mbar in Fig. 87, while it is at 50 mbar in visible and near infrared regions. It is caused by $(2\pi r)/\lambda \gg 1$ in the latter case with the approximation of large particles resulting in an extinction cross-section of $2\pi r^2$ instead of $\sim \pi r^2$ for $(2\pi r)/\lambda \sim 1$. One precipitable micron above 100 mbar is equivalent to a mixing ratio of about 2×10^{-6}. So daytime maximal mixing ratios could reach 100 ppm.

The results obtained are most sensitive to the adopted cloud model through the ratio of cloud optical depths at 45/11.5 μm (Fig. 89). For example, with the first mode of aerosol particles taken into account (see below,

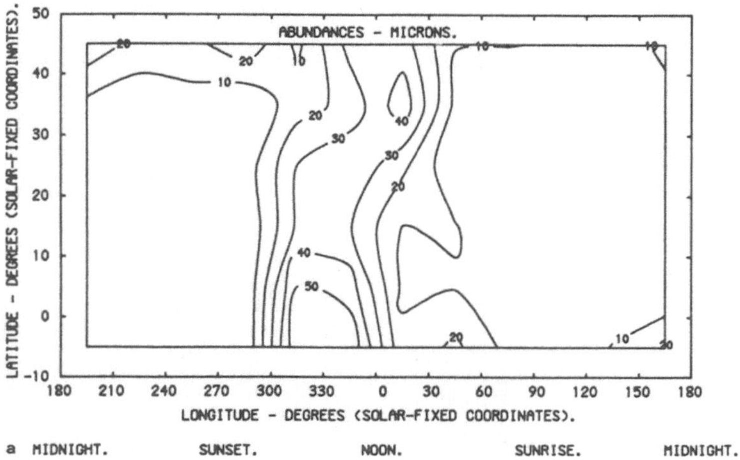

Fig. 88. Variations of the water vapor column abundance (in μm) above 65 km. One precipitable micron above the cloud unit optical depth is equivalent to a mean mixing ratio of approx. 2 ppm (Schofield et al. 1982)

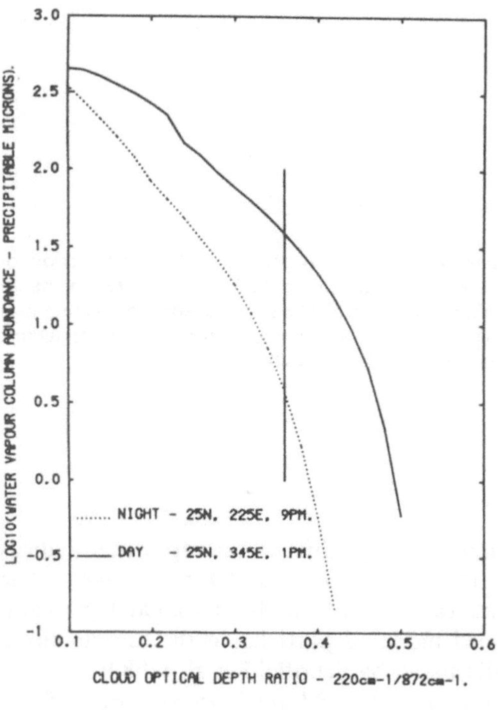

Fig. 89. The dependence of the retrieved water vapor column abundances on the cloud optical depth ratio at 45/11.5 μm for typical day side (*solid line*) and night side (*dotted line*) conditions. The standard cloud model is indicated by the *vertical solid line* (Schofield et al. 1982)

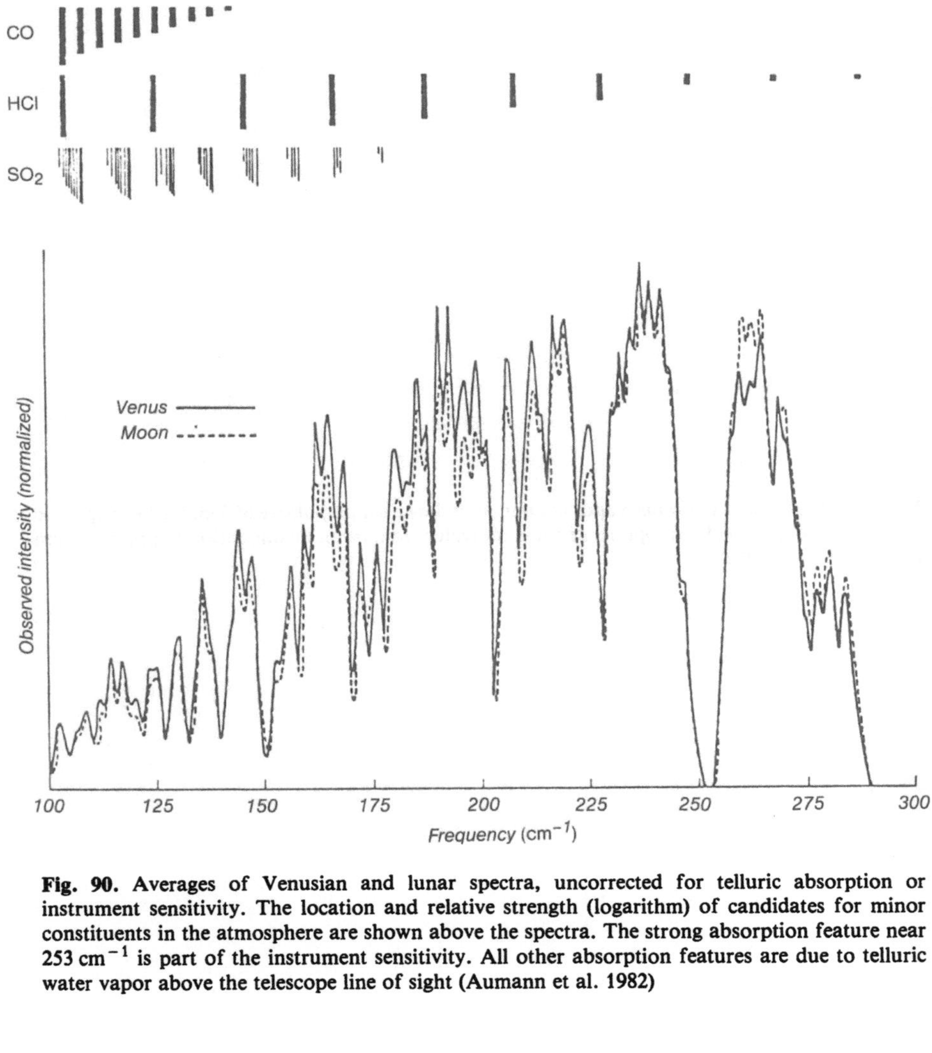

Fig. 90. Averages of Venusian and lunar spectra, uncorrected for telluric absorption or instrument sensitivity. The location and relative strength (logarithm) of candidates for minor constituents in the atmosphere are shown above the spectra. The strong absorption feature near 253 cm^{-1} is part of the instrument sensitivity. All other absorption features are due to telluric water vapor above the telescope line of sight (Aumann et al. 1982)

Chap. 3.7) this ratio is 0.39; daytime and nighttime water vapor abundances are smaller than in Figs. 87 and 88 by a factor of 1.7 and 4, respectively. The sensitivity of retrievals to the altitude variation of the cloud optical depth ratio and a possible influence of the third mode of particles in the middle cloud layer below 56 km have not been discussed by Schofield et al. (1982).

Spectroscopic measurements in the range of 37 to 91 μm were made from G. Kiuper Airborne Observatory by Aumann et al. (1982). The instrument had

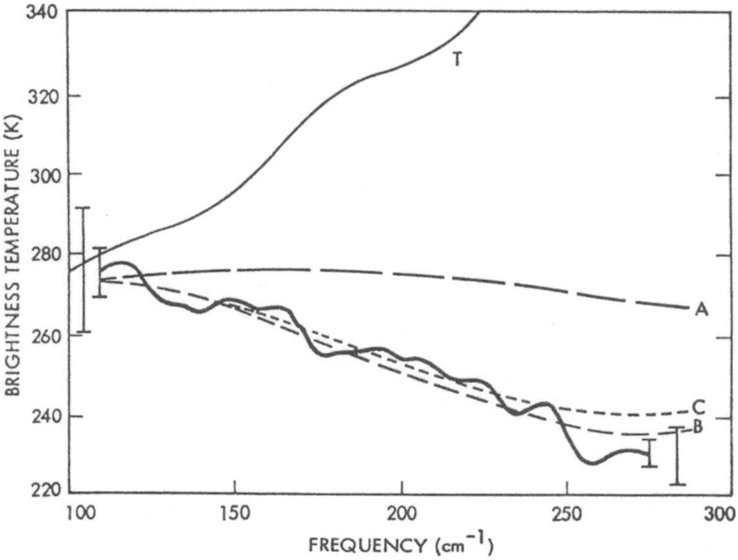

Fig. 91. Fit of various models to the spectral data. The spectrum labeled *T* represents a model with no clouds; that of *A* includes the clouds detected by various PV experiments, and those of *B* and *C* include large particles of $r \gg 1$ μm with number densities of $\sim 1\ cm^{-3}$ (Aumann et al. 1982)

a resolution power of $\lambda/\Delta\lambda \simeq 1000$, and observations were conducted at a flight altitude of ~13 km when the Venusian phase angle was 60°. Calibration was made against a lunar spectrum measured in the same flight with a lunar surface temperature and emissivity assumed to be 340 ± 10 K and 0.9 ± 0.04, respectively. Comparison of the normalized Venusian and Moon spectra uncorrected for telluric absorption and instrument sensitivity is shown in Fig. 90. Measurements were made during the Pioneer Venus orbiter operation in March, 1979. The amount of telluric water vapor along the line of sight measured by a special instrument during the same flight was 17 ± 2 μm. The calibrated spectrum (Fig. 91) agrees well with the data of Reed et al. (1978). Some features of the spectrum are of the order of error bars and have not been discussed. Figure 91 also represents some calculated spectra. The spectrum labeled *T* indicates the absorption due to the CO_2 collision-induced dipole transitions only, with a temperature profile measured by Pioneer Venus and without any contribution of cloud opacity. The spectrum *A* includes also the effect of cloud particles measured by various Pioneer Venus instruments. It disagrees with the observed spectrum and imposes the constraint that any additional infrared opacity must be below 40 mbar in the upper cloud, probably due to the presence of large particles of $r \gg 1$ μm with number densities of about 1 particle cm^{-3}. However, these particles should be and were not measured by the particle size spectrometer (see Chap. 3.7) and it turns out to be the main difficulty in this explanation.

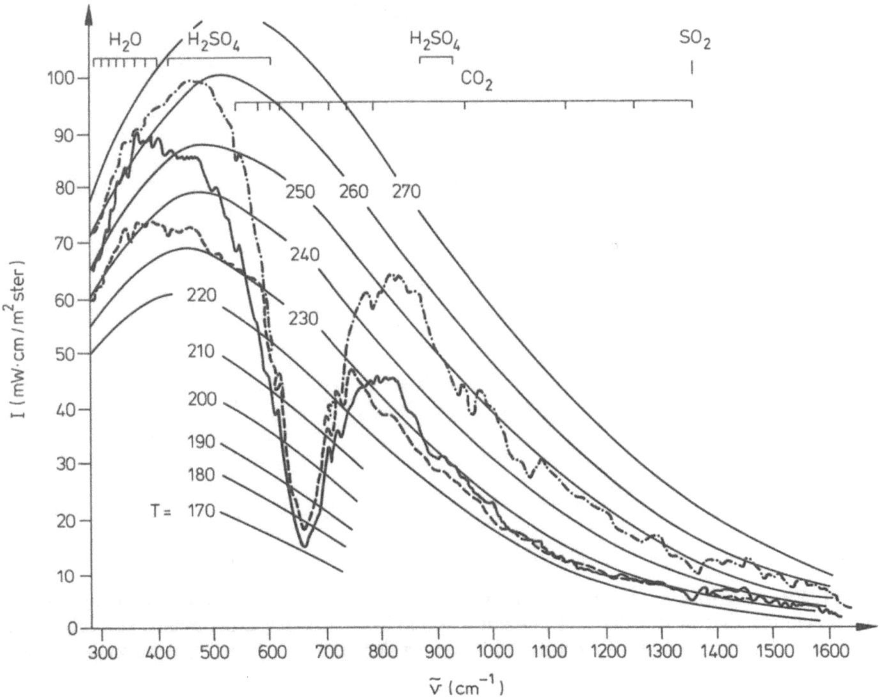

Fig. 92. Spectra measured by the V 15 orbiter (Ertel et al. 1984). Positions of absorption bands are shown. Smooth curves are black body spectra given for comparison

So, both series of far infrared (radiometric and spectrometric) observations on board the Pioneer Venus orbiter and the aircraft, respectively, indicate a source of opacity in the upper cloud layer. The spectrometric measurements appear to eliminate water vapor as a possible source, because H_2O absorption is equal in magnitude in the Venusian and Moon spectra.

V15, 16 observations were carried out by a Fourier spectrometer in the range of 6 – 36 μm with a resolution power of 200. The measured spectra (Fig. 92) demonstrate the strong CO_2 15 μm band and the bands of H_2O, H_2SO_4, and SO_2. This experiment was aimed at studying the thermal regime and the dynamics of the mesosphere and global distribution of the absorbers (Ertel et al. 1984). The preliminary interpretation of the data obtained shows that the afternoon-night asymmetry of water vapor distribution is less pronounced as compared to that measured by Schofield et al. (1982). Mean values of $f_{H_2O} = 30$ ppm at 58 km and $f_{SO_2} = 2$ ppm at 66 km were deduced from the bands at 30 μm and 7.5 μm, respectively. Reference levels of thermal radiation are $\tau = 1$, contrary to those of scattered radiation at $\tau(1-g) = 1$. Both calculations based on the properties of the clouds and the brightness temperatures at 30 μm and 7.5 μm indicate the above altitudes.

Carbon Monoxide vertical distribution was deduced by Wilson et al. (1981) from measurements of the shape of the CO rotational line $J = 0 \rightarrow 1$ at

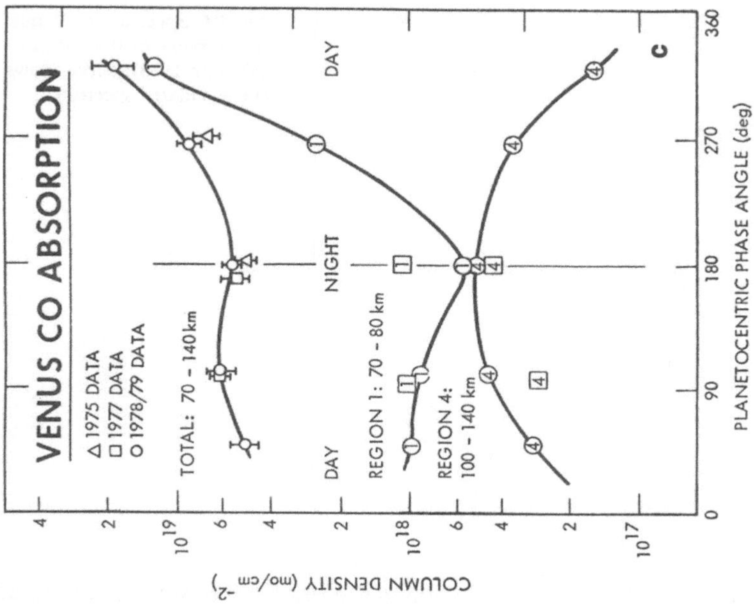

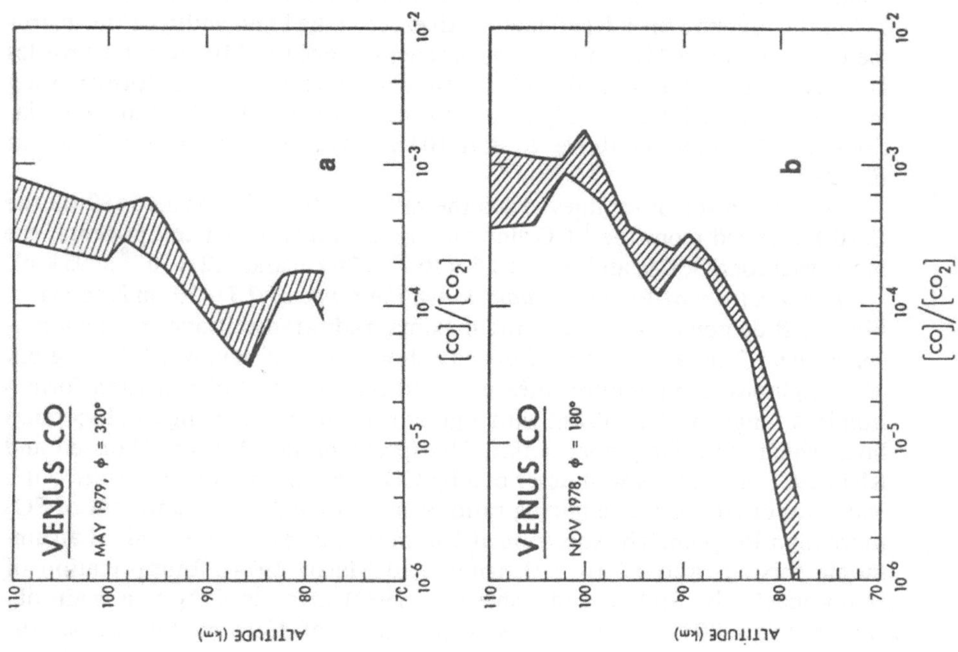

Fig. 93. a CO mixing ratio profiles at phase angles of 320° (morning) and **b** 180° (midnight in the center of the disk). **c** Phase angle variations of CO amounts above 70 and 100 km and near 75 km (Wilson and Klein 1981)

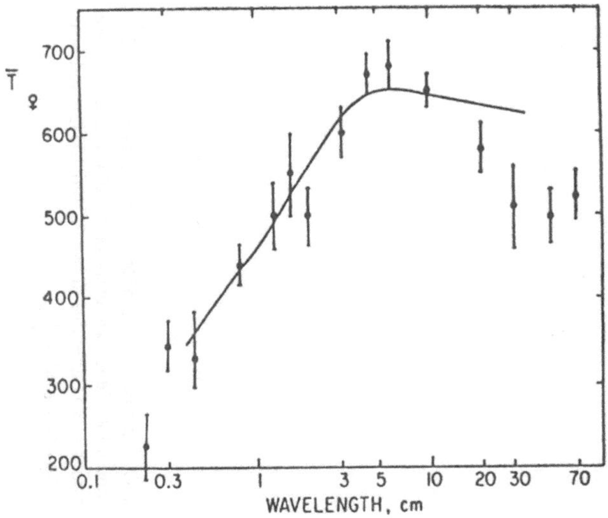

Fig. 94. Spectrum of Venus in the range of 0.3 – 70 cm (Kuzmin 1983); curve shows the calculated spectrum

2.6 mm (see Chap. 1.2). The instrument had the resolving power, $\lambda/\Delta\lambda$, equal to 3×10^6. Continuous radiation forms in this range by CO_2 opacity at ~50 km and has a brightness temperature of 340 K according to the measurements. The absorption of the continuum by CO forms near 100 km where the temperature is ~170 K. CO profiles obtained refer to 78 – 110 km. Below 75 km the pressure broadened line width exceeds the band width of the instrument, while above 110 km Doppler broadening occurs. Mixing ratio profiles at phase angles of 320° and 180° are shown in Fig. 93 a, b, the former being morning and the latter midnight in the center of the disk. Phase angle variations of CO amounts above 70 and 100 km and near 75 km are shown in Fig. 93 c.

The search for other lines yields the ratio of $^{12}C/^{13}C = 85 \pm 15$ (89 on the Earth) deduced from the ^{13}CO line and upper limits of ten components, the most strict for SO, CS, and HCN ($< 5 \times 10^{-8}$ at 70 km and $< 2 \times 10^{-8}$ at 95 km).

The spectrum of Venusian radiation in the range of 0.3 – 30 cm is shown in Fig. 94. It demonstrates clearly the maximal radiotransmittance of the atmosphere near 7 cm and enhanced opacity shortward and longward this region. The brightness temperatures measured indicate the level of radiation formation in a range of 15 to 50 km with a great pressure broadening of absorption lines which look like a continuum. The spectrum near 1.35 cm (Janssen and Klein 1981, Fig. 95) is well described by CO_2 absorption only and rules out a water vapor tropospheric mixing ratio of more than 3×10^{-3} and that of SO_2 more than 180 ppm. The spectrum at 3.8 cm results in an upper limit of an ammonia mixing ratio of 10^{-5} (Kuzmin and Marov 1974). Interpretation of radio-opacity by Steffes and Eshleman (1982) involves CO_2 as a main absorber below 30 km, SO_2 with a mixing ratio of 150 ppm throughout the

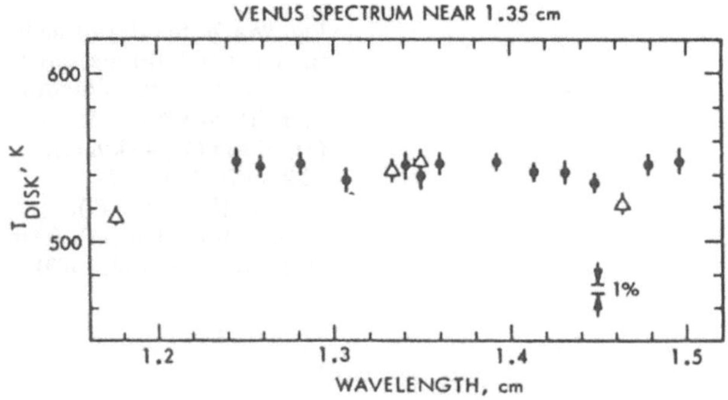

Fig. 95. Spectrum of Venus near 1.35 cm (Janssen and Klein 1981)

troposphere, and gaseous sulfuric acid with a mixing ratio of ~40 ppm above 35 km. The contributions of these components to optical thickness at a wavelength of 1 cm are 10.6, 5.2, and 3.6, respectively. The radio spectrum in Fig. 95 cannot give an unambiguous identification of these species. Interferometric measurements at 3.4 mm (Good and Schloerb 1983) show a steep decrease of f_{SO_2} above 40 km.

3.5 Optical Measurements in the 0.45 – 1.2 μm Range from Venera Landers

Measurements of atmospheric chemical composition by chemical detectors were made first by Veneras 4, 5, 6 (Vinogradov et al. 1970) and later by Veneras 13, 14 (Surkov et al. 1983a). These measurements resulted in a CO_2 mixing ratio of 97 ± 3%, while they failed to detect nitrogen, oxygen, and ammonia. Electrolytic, manometric, and other detectors were used to measure water vapor abundance. Analysis of Veneras 4, 5, 6 data with subsequent laboratory experiments (Andreichikov 1978) yields a water vapor mixing ratio decreasing from 2% at 55 km to ~1000 ppm at 40 km and 100 ppm at 20 km. The last Veneras give 2000 ± 1000 ppm at 46 – 50 km.

Data from Veneras 9 – 14 (V9 – 14). Photometric and low resolution spectrometric techniques were used on board these probes to measure water vapor abundance and to search for other absorbers. V9, 10 instruments (Moroz et al. 1976) provided measurements with three filters centered at the (211) band of H_2O at 820 nm, the (005) band of CO_2 at 870 nm and continuous radiation at 800 nm. Both bands feature equal absorption with $f_{H_2O} \simeq 10^{-3}$. These measurements yield $f_{H_2O} \simeq 3 \times 10^{-4}$ at 25 – 45 km (Ustinov and Moroz 1978).

Further modifications of these instruments result in spectrometers for a range of 0.43 – 1.17 μm with a resolving power of $\lambda/\Delta\lambda \simeq 30$ (Ekonomov

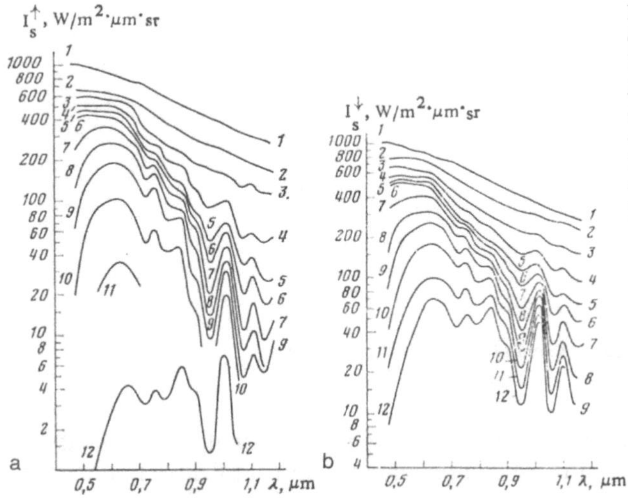

Fig. 96a, b. Spectra of nadir (**a**) and zenith (**b**) measured by V14 at various altitudes: I_0/π (1), 62 km (2), 55 km (3), 52 km (4), 49 km (5), 40 km (6), 25 km (7), 16.5 km (8), 10 km (9), 4.5 km (10), 1 km (11), 0 km (12) (Moshkin et al. 1983)

et al. 1979, Moroz et al. 1979) utilizing circular variable filters. A scanning photometer with four broad band ($\sim$100 nm) filters was used to study angular distribution of light, while the spectrometer was directed upward vertically. These instruments were installed on board the V11, 12 landers. Instead of the scanning photometer the V13, 14 spectrometers had a scanning adapter to measure spectra at different viewing angles. A near UV photometer with a band pass from 320 to 390 nm measured UV absorption in the cloud layer (Moshkin et al. 1983).

Spectra of zenith and nadir at different altitudes measured by V14 are shown in Fig. 96. The results obtained by the last four Veneras are very similar. They show attenuation of the continuum radiation by a factor of 3 at the altitudes from 62 to 49 km, relatively high transparency of the atmosphere at 49–24 km, and a more optically dense atmosphere below. The spectra show several CO_2 and H_2O bands with equivalent widths increasing at low altitudes. Another feature of the spectra is absorption shortward of 550 nm.

Below we will consider the results of the interpretation of H_2O bands measured by V11–14 and continuum radiation and blue absorption for V11, 12 observations. The interpretation of continuum radiation and blue absorption for V13, 14 is now under way. The UV photometric data will be discussed in Chap. 3.8.

Properties of the aerosol component were deduced mainly from observations with the scanning photometer (Golovin et al. 1981, 1982, Moroz et al. 1981, Ekonomov et al. 1983). The measurements began at 63 km with zenith light intensity being smaller than that from the lower hemisphere. According to calculations they are equal for $3(1-g)\tau \gtrsim 2$. Hence, $\tau \lesssim 2$ at 63 km with the Henyey-Greenstein parameter $g = 0.7$ (Chap. 3.2) for the upper part of the cloud layer. Photometric profiles measured inside an optically thick atmosphere are interpreted as follows. The difference between downward and upward light fluxes $F_{\downarrow} - F_{\uparrow}$ at the altitude h is a flux, absorbed by the atmosphere layers below h, the so-called net flux. A derivative of this flux is the energy flux absorbed per unit length, i.e.,

$$\frac{d(F_{\downarrow} - F_{\uparrow})}{dh} \simeq 2(F_{\downarrow} + F_{\uparrow})\,\sigma(1-\omega)$$

where $\sigma(1-\omega)$ is the true absorption coefficient, σ is the extinction coefficient, and ω is the single scattering albedo. The factor of 2 is due to the mean cosine of the incident angle being $\sim\frac{1}{2}$.

The net flux decreases total luminosity per unit volume of the low lying atmosphere. This decrease is proportional to the scattering coefficient $\sigma\omega$ and unisotropy of scattering, i.e.,

$$\frac{d(F_{\downarrow} + F_{\uparrow})}{dh} = A(F_{\downarrow} - F_{\uparrow})\,\sigma\omega\,.$$

With $\omega \simeq 1$ and the Henyey-Greenstein phase function (see Chap. 3.2), Ustinov (1977) found $A = \frac{3}{2}(1-g)$. Angular distribution of radiation in an optically thick atmosphere (Sobolev 1972) is approximated by

$$I(h, \cos z) = \tfrac{1}{2} f_0(h) + \tfrac{3}{2} f_1(h) \cos z\,.$$

Integration of this relation over the upper and lower hemispheres yields

$$F_{\downarrow} + F_{\uparrow} = \pi f_0\,, \qquad F_{\downarrow} - F_{\uparrow} = 2\pi f_1\,.$$

The substitution of the latter expressions in the above relations gives

$$\sigma(1-\omega) = \frac{1}{f_0}\,\frac{df_1}{dh}\,, \qquad \sigma(1-g) = \frac{1}{3f_1}\,\frac{df_0}{dh}\,.$$

This solution of the inverse problem of optical sounding in a thick atmosphere was first used by Ustinov (1977) as a partial case of a more complicated matrix differential equation.

Having determined $f_0(h)$ and $f_1(h)$ from pairs of equations for two viewing directions at each h, one can deduce σ, ω, g.

The problem of optical sounding is ambiguous in principle, i.e., two relations determine three values and one needs an assumption on one of them. The more the two cosines of z differ, the higher is the accuracy of the solution. Measurements with $z = 0$ and 130° were used for interpretation. The results are shown in Fig. 97 with $g = 0.7$. Size spectra of cloud particles, measured by Knollenberg and Hunten (1980) on board the Pioneer Venus large probe yield g in the range of 0.6 – 0.8 at different altitudes. The V 11, 12 curves in Fig. 97 confirm the presence of three cloud layers with the upper cloud layer above 57 km having an optical thickness of $\tau = 7$ and 6 for V 11 and V 12, respectively; the middle cloud layer at 57 – 51 km with $\tau = 9$ and 12.5; the lower cloud layer at 51 – 48 km with $\tau = 12$ and 20. Only the upper limit for ω,

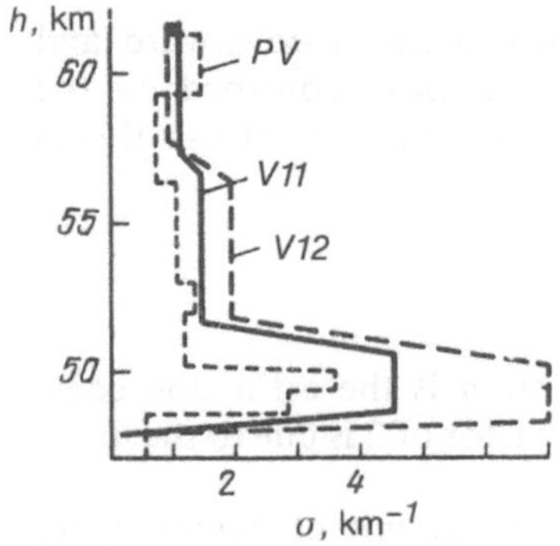

Fig. 97. Aerosol extinction coefficient in the Venusian cloud layer (Golovin et al. 1981)

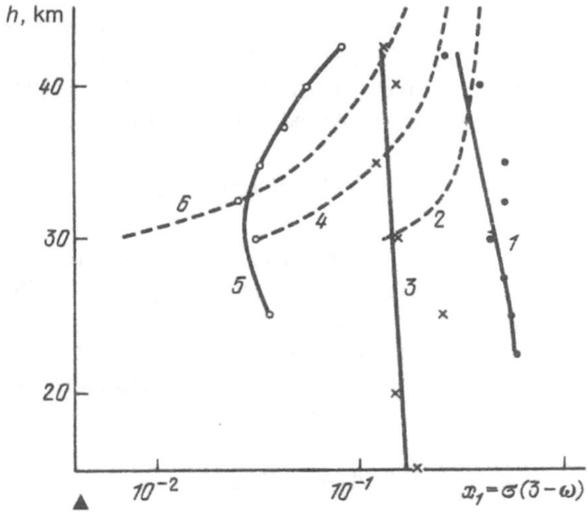

▲ Fig. 98

Fig. 98. Subcloud aerosol extinction coefficient $\sigma(3-\omega)$ measured by Venera 11 and 12 (*solid* and *dashed lines*, respectively) at various wavelengths: 1, 2: 0.49 μm, 3, 4: 0.7 μm, 5, 6: 1 μm (Golovin et al. 1981)

$1-\omega \leqslant 5\times 10^{-4}$, may be derived in the middle and lower cloud layers in the range of 0.49 – 0.73 μm.

Figure 97 shows the extinction coefficients varying from $1.3\ \text{km}^{-1}$ in the upper layer to $5-7\ \text{km}^{-1}$ in the lower one. Aerosol was detected and measured below 48 km as well (Fig. 98). Its contribution exceeds that of gaseous Rayleigh scattering down to ~20 km and ~30 km at the locations of V 11, and V 12 landing, respectively, being of the order of $\sigma \simeq 0.1\ \text{km}^{-1}$ and with wavelength dependences of σ varying with altitude. These dependences result in a particle size comparable with wavelength, i.e., $(2\pi r)/\lambda \lesssim 1$ or $r \lesssim 0.1$ μm. The extinction coefficient of small particles with $(2\pi r)/\lambda \ll 1$, i.e., $r \lesssim 0.03$ μm, is known to be $\sigma(\lambda) \sim \lambda^{-4}$ while the measure spectra can be approximated by λ^{-3}, therefore 0.03 μm $\lesssim r \lesssim$ 0.1 μm. More precise values of r were derived from the measured $\sigma(\lambda)$ by the Mie theory. It was assumed that $g = 0$ under the clouds, which holds for small particles.

The aerosol measurements below the cloud layer are summed up as follows. According to V 11 data, $r = 0.08 - 0.1$ μm at 40 km, 0.05 – 0.07 μm at 35 km, and smaller than 0.04 μm at 25 km. The extinction coefficient at a wavelength range of 0.6 – 0.7 μm with minimal true absorption is almost height independent from 45 to 20 km (Fig. 99). From V 12 measurements, $r = 0.2 - 0.3$ μm at 45 – 40 km and 0.2 μm at 35 km. A steep decrease of aerosol density occurs at 40 – 30 km. A similar interpretation of V 9, 10 yields $r \lesssim 0.1$ μm at 30 – 45 km and the lower boundary of aerosol at 30 – 25 km (Fig. 99).

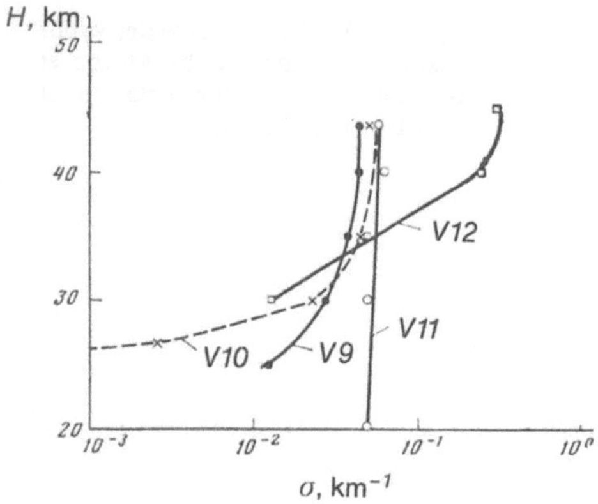

Fig. 99. Subcloud aerosol extinction coefficient at 0.65 μm measured by V9 – 12 (Golovin and Ustinov 1982)

The measurements cannot yield the refractive index of small aerosol particles. Its value of $m_r \lesssim 1.8$ was derived only for V12 data at 45 km where $r \simeq 0.25$ μm. With $m_r = 1.5$ adopted and the extinction coefficients known, one can obtain the aerosol particle number density, n. Extinction coefficients almost coincide for all four probes at 35 km. Calculations give $n \simeq 10^4\ \mathrm{cm}^{-3}$ and an aerosol mass mixing ratio, f_a, of 5×10^{-9} at this altitude with an aerosol matter density of $2\ \mathrm{g\ cm}^{-3}$.

Water Vapor (Moroz et al. 1979, 1983, Golovin et al. 1982). To calculate absorption in H_2O bands with optical properties of the aerosol medium known from continuum radiation measurements, one should have a dependence of the water vapor absorption coefficient on temperature and pressure for different wavelengths inside the band, assuming the altitude profile of H_2O mixing ratio and performing radiative transfer calculations. Varying the assumed H_2O distribution to fit the observations, one can get the real H_2O mixing ratio profile. This calculation presents two difficulties: the absence of the absorption coefficients of H_2O bands in the region of $\lambda < 1$ μm for the essential range of temperatures and pressures, and a complicated structure of the H_2O band with a large number of lines and a strong dependence of the absorption coefficient on wavelength and on H_2O abundance. The latter was discussed briefly in Chap. 1.1.

Moroz et al. (1979) measured the absorption coefficients of H_2O near infrared bands for p and T close to those in the lower atmosphere of Venus. The data obtained were set in a computing program of radiative transfer for the Elsasser model of band (a band with equally spaced lines with locally equal intensities). The calculation assumed a two flux approximation. Figure 100 shows the dependences of $f_{H_2O}(h)$ adopted in calculations. The best fit to observations is with the curve *A 1*, i.e., $f_{H_2O} \simeq 2 \times 10^{-4}$ at 40 – 60 km decreasing to 2×10^{-5} near the surface. Interpretation of all four probe data yields similar dependences of f_{H_2O}. Moroz et al. (1983) give a more detailed profile of f_{H_2O} based on V13, 14 measurements (Table 15). These experiments give a

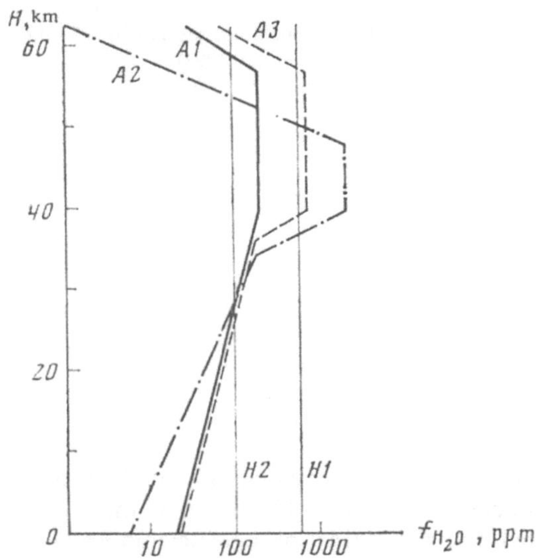

Fig. 100. Models of the water vapor mixing ratio adopted by Moroz et al. (1983) for the interpretation of the water vapor bands

Table 15. Water vapor mixing ratio vs height derived from V 13, 14 optical measurements (Moroz et al. 1983)

Height range, km	54 – 50	50 – 40	40 – 20	10 – 0
H_2O, ppm	500 ± 200	200^{+200}_{-100}	100^{+100}_{-50}	20^{+30}_{-10}

more reliable profile f_{H_2O} due to the fact that the spectral measurements were carried out in both upward and downward directions.

Blue Absorption. The spectra measured are distinguished by a strong decrease of intensity in a range of 600 nm to 450 nm, especially in the lowest 30 km. Two processes may underlie this phenomenon: Rayleigh scattering in the lower atmosphere and true absorption. Simple calculations by means of the asymptotic formulas (Chap. 3.2) yield low values of true absorption up to 8 km and enhanced absorption at 8 – 30 km with $1-\omega \simeq 10^{-3}$ and 10^{-2} at 450 nm in the altitude ranges mentioned. A more thorough interpretation of the scanning photometric data from V 11, 12 results in $1-\omega \simeq 2 \times 10^{-3}$, 5×10^{-3}, and 10^{-3} at altitudes 0, 12, and 30 km respectively, in a wavelength interval of 450 to 550 nm. Sanko (1980) has found that this distribution is close to that of sulfur allotropes S_3 and S_4 in the Venusian atmosphere, with the total gaseous sulfur mixing ratio independent of height. He compared temperature dependence of the gaseous sulfur absorption coefficient with that of number densities of various sulfur allotropes S_n ($n \leqslant 8$) in the same laboratory measurements and concluded that S_3 and S_4 were responsible for absorption in the range of 450 – 600 nm. If absorption at 500 nm is caused by S_3, then the mixing ratio of S_2, which is a predominant allotrope in the conditions of the Venusian atmosphere below 40 km, is $f_{S_2} = 8 \times 10^{-8}$, while the same interpretation at 450 nm yields $f_{S_2} = 2 \times 10^{-8}$ as well as the data of the scanning photometer in the range of 450 – 550 nm. These values have been obtained on the assumption of thermochemical equilibrium between sulfur allotropes. We shall see below (Chap. 4.7) that this assumption is invalid here, so these results should be revised. To sum up: if true absorption at 450 – 550 nm is caused by sulfur vapor, then the S_3 allotrope is responsible for it, and with a S_3 absorption cross-section of 1.6×10^{-17} cm^2 at 490 nm, the measured absorption gives S_3 number densities of 6×10^9, 7.5×10^9, and 4×10^8 cm^{-3} at 0, 12 and 30 km, respectively.

With chlorine Cl_2, bromine Br_2, and nitric dioxide NO_2 considered as possible sources of absorption, one can derive the upper limits on their mixing ratios: $f_{Cl_2} < 10^{-7}$, $f_{Br_2} < 2 \times 10^{-10}$, $f_{NO_2} < 5 \times 10^{-10}$ (Golovin et al. 1982, Krasnopolsky and Parshev 1979). The absence of an oxygen absorption band at 762 nm leads to $f_{O_2} < 3 \times 10^{-5}$ (Krasnopolsky and Parshev 1979).

3.6 Mass Spectrometric and Gas Chromatographic Measurements in the Lower Atmosphere

Mass Spectrometry on Board V11 – V14 (Istomin et al. 1979, 1980, 1983). Mass spectrometers at these lander probes have analyzed a mass range of 11 – 138 (105 at V 11, 12) with a resolution of ~35. They were switched on at 25 km and operated till landing. Twenty-two samples of atmospheric gas were analyzed and 176 spectra measured by V 11, 12 (250 by V 13, 14). A very sophisticated inlet system was used to provide precise measurements in the dense atmosphere.

The instrument had two operational modes, the first for measurements of chemically active species and the second for noble gases which were enriched inside the instrument by removing chemically active components, including CO_2 and N_2, by activated titan. The sensitivity of the instruments on board V 13, 14 was higher than that of V 11, 12 by a factor of 2 in the chemical mode and by a factor of 40 in the noble gas mode. This was achieved by the increasing of the atmospheric sample volume in both modes up to the optimum level which was defined from V 11, V 12 data. The detection limit of the V 13, 14 instruments was about 1 ppb in the noble gas mode. A spectrum in the noble gas mode is shown in Fig. 101.

A comparison of mass peaks at 14 in the atmospheric and laboratory spectra gives for the nitrogen mixing ratio a value of $4 \pm 0.3\%$. The 15 and 14 peak ratio shows the nitrogen isotopic ratio close to that on Earth. Similarly, peaks at 12 and 13 formed by dissociation of CO_2 in the instrument ion source yield $^{13}C/^{12}C = (1.1 \pm 0.02) \times 10^{-2}$, which agrees with the Earth's ratio. Com-

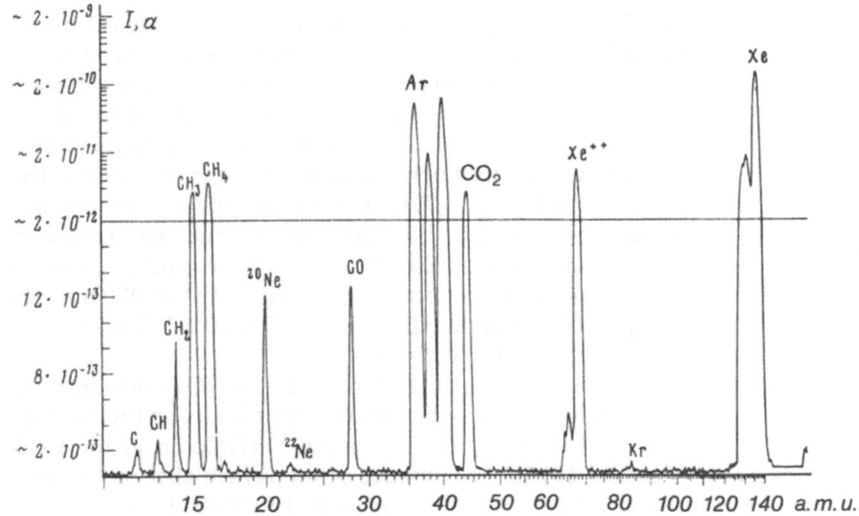

Fig. 101. The mass spectrum measured by V13 at 25 km in the noble gas mode (Istomin et al. 1983). Xenon is calibration gas

parison of mass peaks at 32 and 46 results in an oxygen isotopic ratio, $^{18}O/^{16}O$, with the former produced by dissociation of $^{12}C^{16}O_2$ and the latter formed by $^{12}C^{16}O^{18}O$. This ratio exceeds the Earth's by $4 \pm 2\%$. Though the difference constitutes only two mean errors, Istomin et al. (1980) attribute it to contribution of $^{12}C^{34}S$. The presence of CS with a mixing ratio of 4×10^{-3} explains the 46 and 32 mass peaks, with 70% of the latter formed by CO_2 dissociation and the rest by CS dissociation. Here equal isotopic ratios of $^{18}O/^{16}O = 2.04 \times 10^{-3}$ and $^{34}S/^{32}S = 4.5 \times 10^{-2}$ are assumed on Venus and the Earth. The CS mixing ratio obtained seems too high and is difficult to explain by thermochemical or photochemical processes in the lower atmosphere (see Chap. 4.5).

A peak at 64 was not detected, and Istomin et al. (1980) concluded that $f_{SO_2} < 25$ ppm, with equal sensitivity of the instrument to both CO_2 and SO_2. This assumption may not be valid due to a higher adsorption of SO_2. On the other hand, there is some theoretical explanation of SO_2 being a minor sulfur-bearing compound in the lowest scale height (see Chap. 4.5). The ^{18}O atoms formed by CO_2 dissociation account only for 70% of the peak at 18, the rest being atmospheric water vapor with a mixing ratio, f_{H_2O}, equal to 100 ppm.

The argon mixing ratio was measured to be 100 ppm with an $^{40}Ar/^{36}Ar$ isotopic ratio of 1.11 ± 0.2 instead of 300 as on the Earth. The ^{36}Ar and ^{38}Ar are primary isotopes, while ^{40}Ar is formed by radioactive decay of ^{40}K. The ratio of primary isotopes, $^{36}Ar/^{38}Ar$, coincides with that on the Earth, where it is 5.35. From V11, 12 measurements the neon mixing ratio was found to be 12 ± 4 ppm. The V11, 12 mass spectrometers featured low sensitivity to this gas due to its high ionization potential. The V13, 14 mass spectrometers were much more sensitive and result in $f_{Ne} = 8$ ppm with an $^{20}Ne/^{22}Ne$ isotopic ratio of 12.15 ± 0.1. The measurements by the same instrument on the Earth gave a ratio of 10.07 ± 0.35 (from 9.24 to 10.3 according to other measurements). Thus, this ratio on Venus is intermediate in value between the Earth and solar wind (13.7 ± 0.3). The data on neon take into account some contribution of double ionized CO_2 and ^{40}Ar to the peaks at 22 and 20. The upper limit of $^{22}Ne/^{21}Ne \geqslant 15$ is derived from the absence of the peak at 21. Peaks of krypton near 84 were poorly detected in individual V11, 12 spectra, but were well observed after the averaging of the spectra. Istomin et al. (1980) give a krypton mixing ratio of 600 ± 200 ppb with isotopic ratios close to those on the Earth. The last Veneras, V13, 14, yield a much smaller krypton abundance, about 35 ppb. The authors of the experiments think that the krypton mixing ratio measured by V11, 12 is a result of laboratory calibration by xenon with some krypton as impurity. As we discussed before, krypton was detected by the averaging of the spectra. This procedure was not used for the V11, 12 calibration spectra, so this impurity was disregarded. For V13, precautions had been taken to reduce the background of krypton caused by the laboratory calibration of the instrument. As for V14, the instrument was calibrated by pure ^{136}Xe (99.99%) and had almost no background at the peaks of krypton and normal xenon with masses of 129, 131, 132. For the mixing ratio of normal xenon, only an upper limit of 20 ppb is available now.

Istomin et al. (1983) raised the problem of mass spectrometric sensitivity to minimal abundances of noble gases. i.e., krypton and xenon. Calibration is usually made with their mixing ratio ranging from 10 to 100 ppm, and the measured values in the atmosphere may be underestimated due to the gas adsorption by parts of the instrument.

Mass Spectrometric Measurements in the Lower Atmosphere by Pioneer Venus (Hoffman et al. 1980). The Pioneer Venus (PV) large probe neutral mass spectrometer analyzed mass spectra of atmospheric species in the range of 1 to 208 mass units with a rather high spectral resolution which allowed multiplets (for example, CH_4 and ^{16}O) to be separated. The instrument was switched on at 65 km, and its inlet system was blocked by an aerosol particle at 50 km. The mass spectrum of this particle has strong peaks at 64, 48, and 18 referred to SO_2, SO, and H_2O, respectively. A laboratory imitation of this spectrum results in the composition of the particle being a sulfuric acid solution with a concentration of 85%. At 28 km the particle evaporated, but trace amounts of its products inside the mass spectrometer contributed to the lower atmospheric spectra as well.

Donahue et al. (1982) studied the mass peaks of the droplet at 18 and 19 and deduced a ratio $HDO/H_2O = (3.2 \pm 0.4) \times 10^{-2}$ or $D/H = (1.6 \pm 0.2) \times 10^{-2}$, far higher than a terrestrial ratio of 1.5×10^{-4}. This implies that initial hydrogen abundance on Venus was several orders of magnitude larger than now and deuterium enrichment is due to preferential escape of hydrogen (see Chap. 4.4). The isotopic peaks of $H^{35}Cl^+$, $H^{16}O^+$, the peak of $^{19}F^+$, and the single ionized peaks of $^{38}Ar^+$ and $H^{37}Cl^+$ should be used to account for the contributions of $H^{18}O^+$, $^{38}Ar^{++}$, and $H^{37}Cl^{++}$ to the mass peak at 19 (von Zahn et al. 1983). High spectral resolution of the instrument facilitates this procedure. However, the data reduction leading to the recommended deuterium enrichment has not been published.

The argon mixing ratio was found to be $f_{Ar} = 70^{+50}_{-30}$ ppm with isotopic ratios similar to those measured by Venera mass spectrometers.

Abundances of other noble gases were measured relative to argon, thus including error of argon determination. Neon data are $f_{Ne} = 10^{+20}_{-6}$ ppm and $^{20}Ne/^{22}Ne = 14^{+6}_{-3}$. The peak at 3 may be caused by H_3^+, HD^+, and $^3He^+$. The former two components are contaminated with terrestrial hydrogen from the methane calibration gas (von Zahn et al. 1983), thus reducing the $^3He/^4He$ ratio to less than 3×10^{-4} (1.4×10^{-6} on the Earth). Krypton was not detected at first; however, a more thorough analysis by Donahue et al. (1981) indicated trace amounts of krypton with a mixing ratio of 50 ppb and an upper limit of xenon, f_{Xe}, of less than 120 ppb.

The isotopic ratio of carbon was measured by peaks of CO_2 at 45 and 44 to be $^{13}C/^{12}C = (1.19 \pm 0.06) \times 10^{-2}$ close to that on Earth. Similarly, the ratio $^{15}N/^{14}N = (0.37 \pm 0.08) \times 10^{-2}$ was deduced from the 29/28 peak ratio (Hoffman et al. 1979). Correlation of the peaks at 28 ($^{12}C^{16}O + ^{14}N_2$) and 30 ($^{12}C^{18}O$) results in $f_{N_2} = 4 \pm 2\%$. As carbon oxide formed by CO_2 dissociation inside the instrument is far more abundant than the atmospheric one, the latter cannot be measured. The isotopic ratio of oxygen coincides with that on the Earth ($^{18}O/^{16}O = 2 \times 10^{-3}$). The peak at 34 after the subtracting of ^{34}S produced by dissociation of SO_2 features an altitude dependence similar to the peaks of ^{36}Ar and the peak at 33. It led Hoffman et al. (1980) to the conclusion that the peaks at 34 and 33 are formed by H_2S and HS, respectively. Then, $f_{H_2S} = 3 \pm 2$ ppm in the lowest scale height and ~1 ppm in the cloud layer. The similarity of the peak 30 and ^{36}Ar altitude profiles reveals that the former is caused by atmospheric gas, ethane with a mixing ratio of ~2 ppm being the most probable candidate. Other estimates obtained are upper limits: $f_{SO_2} \leq 10$ ppm at 55 km and ≤ 300 ppm at 24 km, $f_{H_2O} < 10^{-3}$, $f_{O_2} < 30$ ppm, $f_{COS} < 3$ ppm above 24 km, $f_{HCl} + f_{Cl_2} < 10$ ppm, $f_{Hg} < 5$ ppm. As soon as all noble gas mixing ratios were derived by comparison of their peaks with that of ^{36}Ar for both Venera and PV data, the difference in ^{36}Ar mixing ratio (50 ppm from V11–14 and 30 ppm from PV) became a subject of discussion. According to Istomin et al. (1983), the ratio of peaks at 4 and 44, equal to 2% inside the PV instrument corresponds to a helium background mixing ratio of 14% for a helium leak from inner parts of the spacecraft. Laboratory simulation of this situation with a prototype of the Venera spectrometer showed a decrease in the instrument sensitivity by 20–25% to argon due to large amounts of helium inside the instrument. Thus, some part of the discrepancy could perhaps be explained.

Gas Chromatography of the Atmospheric Composition by V12–14. If one puts a sample of atmospheric gas at the end of a tube filled by a sorbent (a kind of powder with great gaseous adsorption on the surface of particles) and then isolates this end from the atmosphere and produces an excessive pressure of a carrier gas, the latter forms a flow transporting the atmospheric mixture to the other end of the tube. The time of this transport depends on adsorption, i.e., is determined by properties of the sorbent and chemical nature of the gas. Hence, a separation of components of the atmospheric mixture occurs at the output of the device. A detector insensitive to the carrier gas, but sensitive to admixtures in it may be developed. For example, rather low energy electrons can ionize atmospheric gases, but not helium which has the highest ionization potential. Therefore, with helium as a carrier gas, the ionic current is a measure of a quantity of admixture, while the time of its appearing measured from the moment of sampling, the so-called retention time, is indicative of its chemical nature. Such a system of chemical analysis of a gas is called a gas chromatograph. This idea was used first in experiments of the Russian scientist M. S. Tzwet in 1904. In spite of the simplicity of the idea the gas chromatograph is a rather complicated device with many valves, separators, chromatographic columns (tubes with sorbents), gas sample systems, detectors, etc. Gas chromatographs make it possible to analyze a gas mixture during a short period of time (several minutes) and to detect components with mixing ratios of the order of one-tenth ppm and larger for atmospheric pressure of the order of bars. No changes of chemical composition of atmospheric gases occur inside gas chromatographs, contrary to mass spectrometers with unwanted electron impact dissociation inside ion sources. However, gas chromatographs fail to analyze at low atmospheric pressures, e.g., in upper atmospheres, due to the fact that their sensitivity is proportional to atmospheric pressure. Some components cannot be separated by a given sorbent, so systems of several columns with different sorbents are used. Also, gas chromatographs cannot separate isotopes. Minor components with mixing ratios of about 10 ppm produce deviations of the detector signal smaller than 1%, therefore gas chromatographs are not as impressive as mass spectra.

The V 12 gas chromatograph made eight analyses of atmospheric composition from 42 km down to the surface (Gelman et al. 1979). Mixing ratios of nitrogen, argon, and sulfur dioxide were found to be height independent and equal to $2.5 \pm 0.3\%$, 100 ± 30 ppm, and 130 ± 35 ppm, respectively. The carbonmonoxide mixing ratio equals 30 ppm above 36 km with a gradual decrease to 17 ppm at 12 km. Some contamination of the instrument by water vapor lowered its sensitivity to this component; however, atmospheric and instrument water vapors differ slightly in their retention times, thus, for the former, estimates give $f_{H_2O} = 50^{+50}_{-30}$ ppm. Oxygen was not detected with an upper limit of 20 ppm. After computer processing the chromatographs show traces of Kr, COS, and H_2S, but no qualitative estimates have been made.

Improved versions of this device were used on board the V 13, 14 (Mukhin et al. 1983), which included two columns with neon ionization detectors and one with an electron capture detector (Fig. 102). Instrument detection limits evaluated by preflight calibration is better than 1 ppm for N_2, H_2, O_2, Kr, CH_4, SO_2, H_2S, COS, and about 3 ppm for Ar, CO, H_2O. Unfortunately,

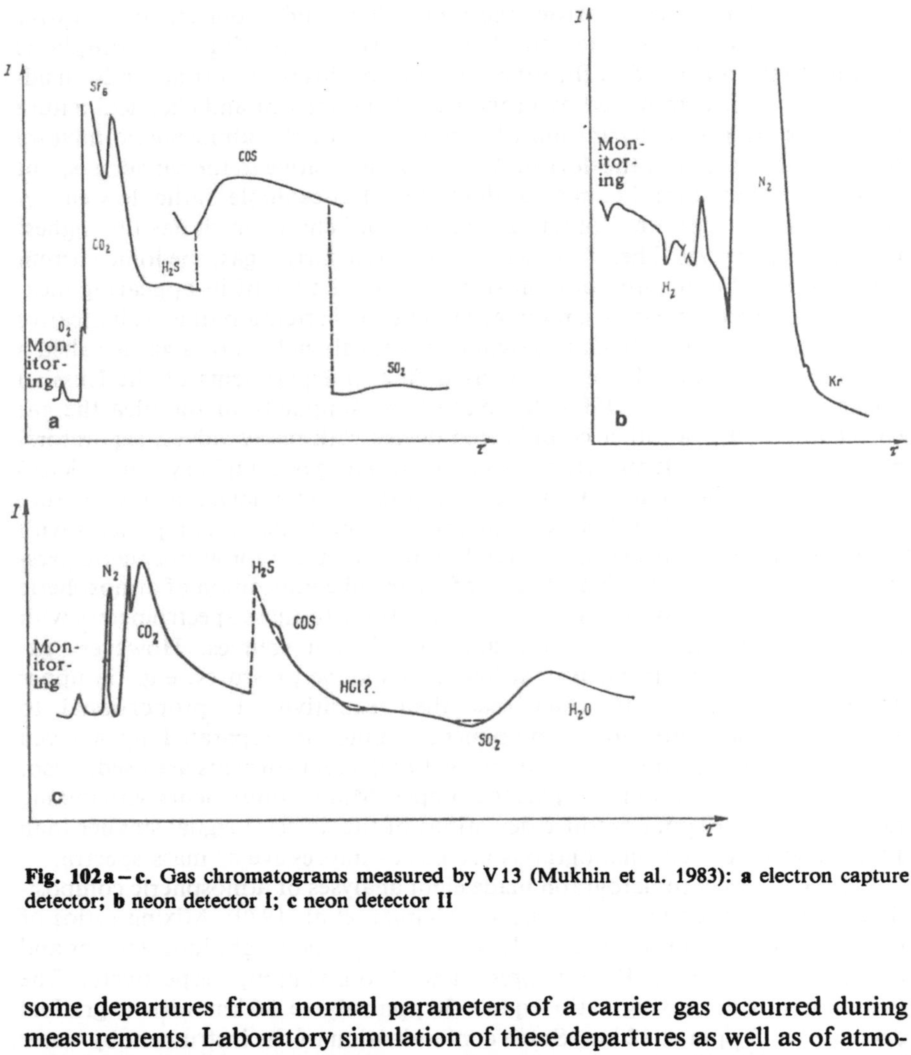

Fig. 102a–c. Gas chromatograms measured by V13 (Mukhin et al. 1983): **a** electron capture detector; **b** neon detector I; **c** neon detector II

some departures from normal parameters of a carrier gas occurred during measurements. Laboratory simulation of these departures as well as of atmospheric composition, temperature, and pressure to reproduce the measured chromatographs yield mixing ratios of some components (Table 16). SF_6 was tentatively identified by an unexpected peak in Fig. 102a with the following properties of a species: chemical stability up to 500°C to materials of the instrument including sorbents, insolubility in water, and high detection limit by the electron capture detector. Mukhin et al. have proposed either N_2O or SF_6 with mixing ratios of 1% and 0.2 ppm, respectively, to satisfy these requirements. Obviously, $f_{N_2O} \simeq 10^{-2}$ should be rejected and $f_{SF_6} \simeq 2 \times 10^{-7}$ was recommended. Further laboratory simulations are needed to reveal other components and altitude profiles of the detected ones.

Table 16. Mixing ratios of components measured by V13, 14 gas chromatographs (Mukhin et al. 1983)

Component	Mixing ratio, ppm	Altitude range, km
H_2	25 ± 10	58 – 49
O_2	18 ± 4	58 – 35
Kr	0.7 ± 0.3	49 – 37
H_2O	700 ± 300	58 – 49
H_2S	80 ± 40	37 – 29
COS	40 ± 20	37 – 29
SF_6	0.2 ± 0.1	58 – 35

Gas Chromatography Aboard the PV Large Probe (Oyama et al. 1980). During its descent in the Venusian lower atmosphere three samples at 51.6, 41.7, and 21.6 km were studied. Nitrogen was measured with the highest accuracy, its mixing ratios equal 4.6 ± 0.14, 3.54 ± 0.04, and $3.41 \pm 0.01\%$, respectively. Its variation exceeds error bars, thus indicating deviations from a uniformly mixed atmosphere. This fact can be explained only by the formation of nitrogen-bearing species with mixing ratio equal to $\sim 1\%$, which seems improbable. The experiment shows large and variable water vapor mixing ratio: $f_{H_2O} < 600$ ppm at 51.6 km and $f_{H_2O} = (5.2 \pm 0.7) \times 10^{-3}$, and $(1.35 \pm 0.15) \times 10^{-3}$ at 41.7 and 21.6 km, respectively. The former two values correspond to saturated water vapor above a sulfuric acid solution with a concentration of 87% and 89% at atmospheric temperatures. However, saturated water vapor density varies greatly with concentrations of sulfuric acid solution and temperatures. Therefore, in our opinion, this coincidence cannot be considered as confirming the measurements.

The first and the second analyses result in an oxygen mixing ratio, f_{O_2}, of 44 ± 25 and 16 ± 7 ppm, respectively. Other gases detected are argon ($f_{Ar} = 67 \pm 2$ ppm), SO_2 (180 ppm), and neon (4 ppm). Upper limits derived from the third sample are 1 ppm for hydrocarbons, 2 ppm for H_2S and COS, and 10 ppm for hydrogen.

3.7 Physical Characteristics of the Cloud Layer

We will discuss now such properties of the cloud layer as number densities and sizes of particles, their phase (liquid or solid), refractive index, and absorption coefficient of their matter, geometry of layers, etc. On the basis of these data one can draw some conclusions on the nature of the cloud layer and specifically on its chemical composition. An apt example is provided by polarimetric observations (Chap. 3.1) which show the upper part of the cloud layer consisting of drops of sulfuric acid solution. Three techniques were utilized by lander probes to investigate the physical properties of the cloud layer: photometry, nephelometry, and particle size spectrometry. Photometric measurements aboard Venera probes were discussed in Chap. 3.5.

Nephelometric Measurements of Veneras (Marov et al. 1978, 1980, 1983). According to Chap. 3.1 scattering properties of a group of particles may be described by five parameters: mean radius and its variance, particle number density, refractive index, and the absorption factor of their matter. Therefore, measurements of light scattered by some cloud volume at five angles can yield these parameters. This idea is employed in the nephelometer, in which a narrow beam illuminates some volume of atmospheric medium to measure the intensities of light scattered at different angles. As ground-based photometric measurements reveal very low true absorption in the Venusian clouds in the visible range, four parameters are sufficient, and V9, 10 nephelometers measured light scattering at four angles of 4°, 15°, 45°, and 180°.

The V9, 10 observations reveal three cloud layers with different properties. The upper layer has its bottom at 58 km and consists of particles having a mean radius, r, increasing downward from 1 to 1.3 μm with a rather narrow size distribution and a refractive index, n, of 1.46 at $\lambda = 0.92$ μm. Evidently these particles are detected by polarimetry (Chap. 3.1) and consist of sulfuric acid. The upper cloud layer features an altitude heterogeneity, and number density of particles can vary in it by a factor of 1.5 at points spaced by about ~1 km. Throughout 58 to 57 km, n changes from $1.46^{+0.01}_{-0.12}$ to 1.42 ± 0.015 and the effective radius increases to 1.6 μm with further increase up to 1.9 ± 0.2 μm at the bottom of the middle cloud layer at 51 km. The middle cloud layer is the most optically thick and located at 57 – 51 km. Here the particle size distribution is broad enough and does not rule out two kinds of particles of different nature. The refractive index may be as low as 1.33 (which is the value of liquid water) and may reach as high as 1.42 if true absorption by particles occurs with an imaginary component of the refractive index, $m_i = \lambda\alpha/4\pi$, of 3×10^{-3}, where α is the absorption factor of particle material. In the lower cloud layer at 51 – 49 km the main radius is ~1.5 μm; however, a single mode distribution seems inappropriate in this case. The values of the refractive index, m_r, range from 1.4 to 1.43. Like the upper cloud layer, the lower one is rather nonuniform in height. The subcloud haze of small particles with $r = 0.13 \pm 0.03$ μm was observed at 49 – 32 km.

These results were supplemented by measurements with back scattering nephelometers aboard V11, 13, 14 which utilized only one angle of scattering, given specified properties of particles to derive their number densities. These properties were assumed to be the same as measured by the V9, 10 nephelometers. Altitude dependences of the cloud extinction coefficient observed by all Venera nephelometers are shown in Fig. 103 and the optical thicknesses of different cloud layers are given in Table 17.

Particle Size Spectrometry Aboard Pioneer Venus (Knollenberg and Hunten 1980). The spectrometer has a laser and a beam splitter to form three narrow beams. Shadows of particles intersecting the beam were projected on

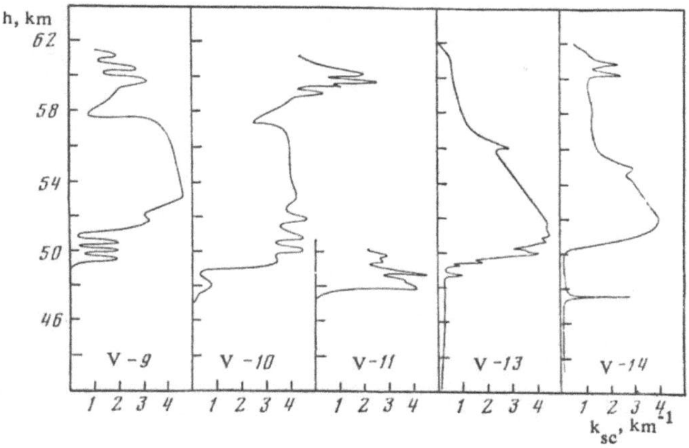

Fig. 103. Altitude profiles of the cloud extinction coefficient measured by V9–14 nephelometers (Marov et al. 1983)

Table 17. Optical thicknesses of different parts of Venusian clouds measured by the Venera nephelometers (Marov et al. 1978, 1980, 1983)

Cloud layer	V9	V10	V11	V13	V14
Upper	7	17	–	3 ± 1	12 ± 4
Middle	26	28	–	22 ± 2	10 ± 1
Lower	2	8	8	6 ± 1	5 ± 1
Subcloud haze	<3	<3	–	<2	<3

three linear detectors, each having 12 photodiodes with diameters of 2.5, 10, and 25 μm, respectively. A particle was recorded with a shadow covering more than two thirds of a photodiode area. The number of photodiodes shadowed simultaneously gave the particle size. Thus, three linear arrays provided measurements in size ranges of 2.5–25 μm, 10–100 μm, and 25–250 μm, respectively. Particle sizes in a range of 0.5–2.5 μm were evaluated from signals of the 2.5 μm photodiodes measured. Typical particle size distribution measured appears to be rather nonuniform, having three maxima (Fig. 104), which suggest three kinds of particles, each described by a standard distribution. Convolution of the measured spectra by these distributions is shown in Fig. 104. However, the data on the first mode are not sufficient to make an unambiguous choice of parameters and cannot reveal its number density. Yet contribution of the smallest unmeasured particles to the extinction coefficient and mass loading should be small even for the first mode, so the uncertainty of these values does not exceed 20%. And there is a considerable deviation of the measured distribution from the adopted one at $r \gtrsim 10$ μm.

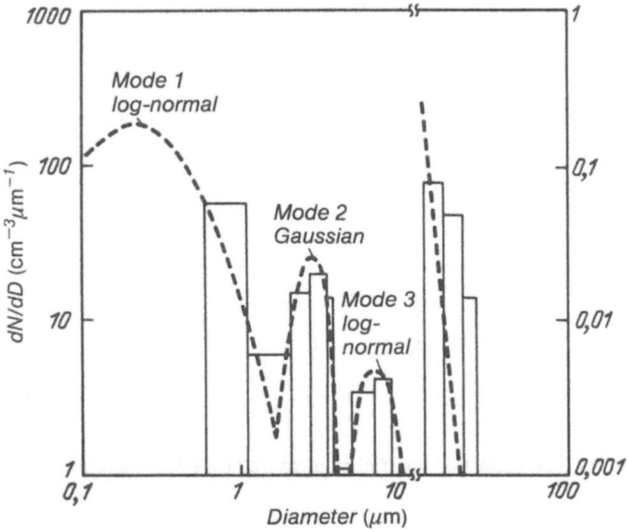

Fig. 104. Size spectrum of cloud particles at 54 km. *Solid line* represents PV measurements and *dashed line* approximates the measured spectrum by the three mode distribution. Size spectrum is the dependence of $\frac{dN(D)}{dD}$, where $N(D)$ is the number density of the particles having a diameter exceeding D (Knollenberg and Hunten 1980)

Trimodal size distribution is typical for the Venusian clouds and though the mean effective radius of every mode and its variance are slightly different at different altitudes, one can consider main cloud characteristics for every mode separately. Altitude dependences of number density, extinction coefficient, and mass loading for every mode are shown in Fig. 105 a, b, c. This figure confirms the presence of three cloud layers with boundaries at 57, 50, and 48 km revealed by the V 9, 10 nephelometers. The upper cloud layer is above 57 km and consists of the first and second modes of particles with radii of 0.2 and 1 – 1.2 μm, respectively. Maximum of mass loading is located at 60 km with small downward decrease of the second mode and a substantial decrease of the first mode. Optical thickness and column mass loading of the upper cloud layer at 65 – 57 km are $\tau = 7$ and $M = 10^{-3}$ g cm^{-2}, respectively, with a contribution of the first mode of 30% and 10%.

The third mode with $r \simeq 3.5$ μm appears in the middle cloud layer at 57 – 50 km. These large particles form the main contribution to the extinction coefficient and mass loading. The number density of the third mode at the tail of distribution, $r \simeq 10 - 20$ μm, measured by the 2.5 – 25 μm and 10 – 100 μm arrays, appears to differ very much. Detailed geometric analysis reveals a substantial deviation of particle shape from the spheric one. Actually, particles of rectangular shape 4 – 8 μm thick with an arbitrary length can be measured by the channel of 2.5 – 25 μm and not by the 10 – 100 μm channel. To fit the observation the third mode particles should be nonspheric crystals with an aspect ratio of 3:1 and a mean length of ~20 μm. Optical thickness of the middle cloud layer is $\tau = 0.34 + 3.5 + 6.1 = 11$ and column mass density is $M = (0.2 + 4.7 + 40) \times 10^{-4} = 4.5 \times 10^{-3}$ g cm^{-3}. Summands show contributions of the modes.

The lower cloud layer at 50–48 km is detached from the middle one by a gap 0.3 km thick with the second and the third modes being absent, while the first mode features a monotonic downward increase throughout the middle and lower cloud layers with a scale height equal approximately to the atmospheric one. In the lower cloud layer mass loadings of the second and the third modes are larger by factors of 2 and 7, respectively, than in the middle cloud layer. All data on the third mode are given for spherical particles.

Apart from the three main cloud layers, the measurements reveal two very thin layers 100 and 20 m thick at 47.5 and 46.3 km, respectively, consisting of particles of the first and second modes. The layer at 47.5 km is very close to the lower cloud boundary and may be detached from it. The size of the second mode particles in the layer at 46.3 km ($r = 1.5$ μm) exceeds that in main clouds and the nature of this layer is not clear.

Particles of the first mode with $r \simeq 0.1 - 0.15$ μm were detected also in the region from the lower cloud boundary at 48 km to 31 km. Optical thickness of this subcloud haze is 0.25; its column mass loading being 10^{-5} g cm^{-2}.

The second mode of cloud particles has the most narrow size distribution and is identified well as particles which are responsible for polarization of light reflected by the planet and which are essentially drops of a sulfuric acid solution (Chap. 3.1). The first mode particles may have the same composition. Their production has a maximum at 60–65 km where absorption of solar near UV radiation occurs (Tomasko et al. 1980). Their number density decreases below 60 km, perhaps, as a result of conversion to the second mode. In the lower cloud layer and the subcloud haze the particles of the first mode may serve as condensation nuclei for the second and the third modes and are not sulfuric acid drops. Maybe some part of them consists of sulfates.

PV Nephelometric Observations (Ragent and Blamont 1980). The particle size spectrometer does not provide data on particle refractive indices, m_r, important for the determination of the particle chemical composition. Therefore, the measurements of size spectra were supplemented by nephelometric measurements at one scattering angle. This is a back scattering nephelometer which allows m_r to be derived by means of the Mie theory, given spherical particles and known size distribution. For several kinds of particles only a mean value of m_r or m_r or one mode with these or other modes adopted may be obtained.

Though particle size measurements were carried out on the PV large probe, it is reasonable to assume the identity of the structure and size distribution at different locations. Then nephelometric measurements by other PV probes provide data on variations of cloud vertical structure. To calculate the back scattering (the real angle was 172°) cross-section of particles, the Mie

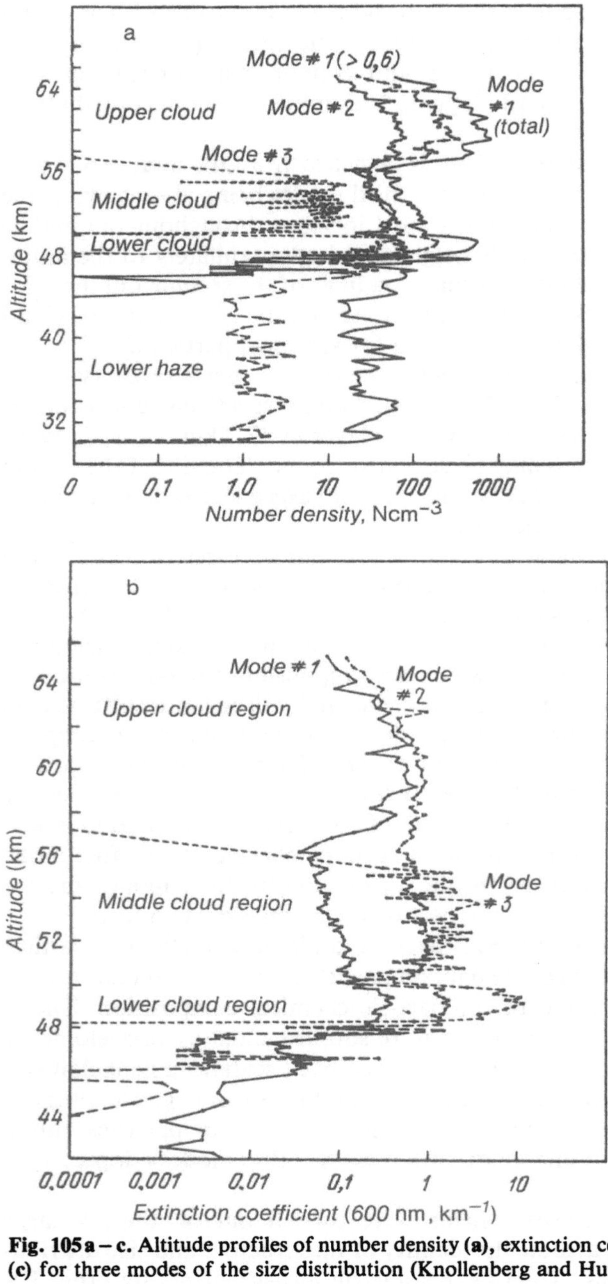

Fig. 105 a – c. Altitude profiles of number density (**a**), extinction coefficient (**b**), and mass loading (**c**) for three modes of the size distribution (Knollenberg and Hunten 1980)

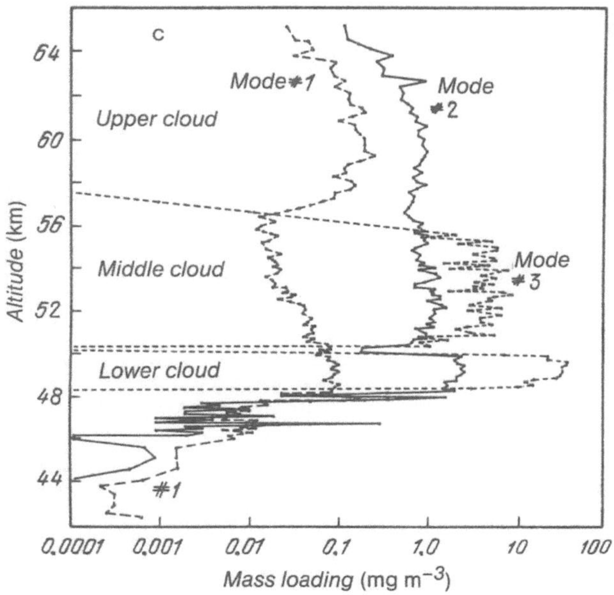

Fig. 105c

theory should be used. This cross-section is approximately proportional to the geometric cross-section with a steep increase with m_r. The dependence on m_r is more complicated for true absorption (imaginary refractive index m_i). PV nephelometric observations were carried out at $\lambda = 900$ nm.

The results of measurements at four PV probes are shown in Fig. 106a. This figure confirms the multilayer structure of the Venusian clouds discovered by Venera probes. Ragent and Blamont (1980) denoted these layers by *A, B, C, D*.

According to the large probe observations, the *D* layer is located down to 57 km and corresponds to the upper cloud layer. Fitting of calculations to the observations results in $m_r = 1.44$ for both modes of particles. This fitting is very good at 60 – 57 km, while the calculated scattering cross-section is smaller than the measured one above 60 km. It may be attributed to very small particles with $r < 0.3$ μm which are not measured by the particle size spectrometer. If one puts in calculation data on the upper haze (see below) for small particles, good agreement with observations may be obtained with $n_r = 1.44$ for a total altitude range of 64 – 57 km.

The *C* layer is the middle cloud layer at 56.5 – 50 km. It contains large particles of the third mode with low number density and gives rise to great statistical error of the measurements. To improve statistics, calculations and observations were compared for the whole layer without altitude dependence of its component taken into account. The best fit is a mean $m_r = 1.39$.

The *B* layer at 49.5 – 47.5 km is the lower cloud layer with the third particle mode expected to prevail. However, their contribution turns out to be

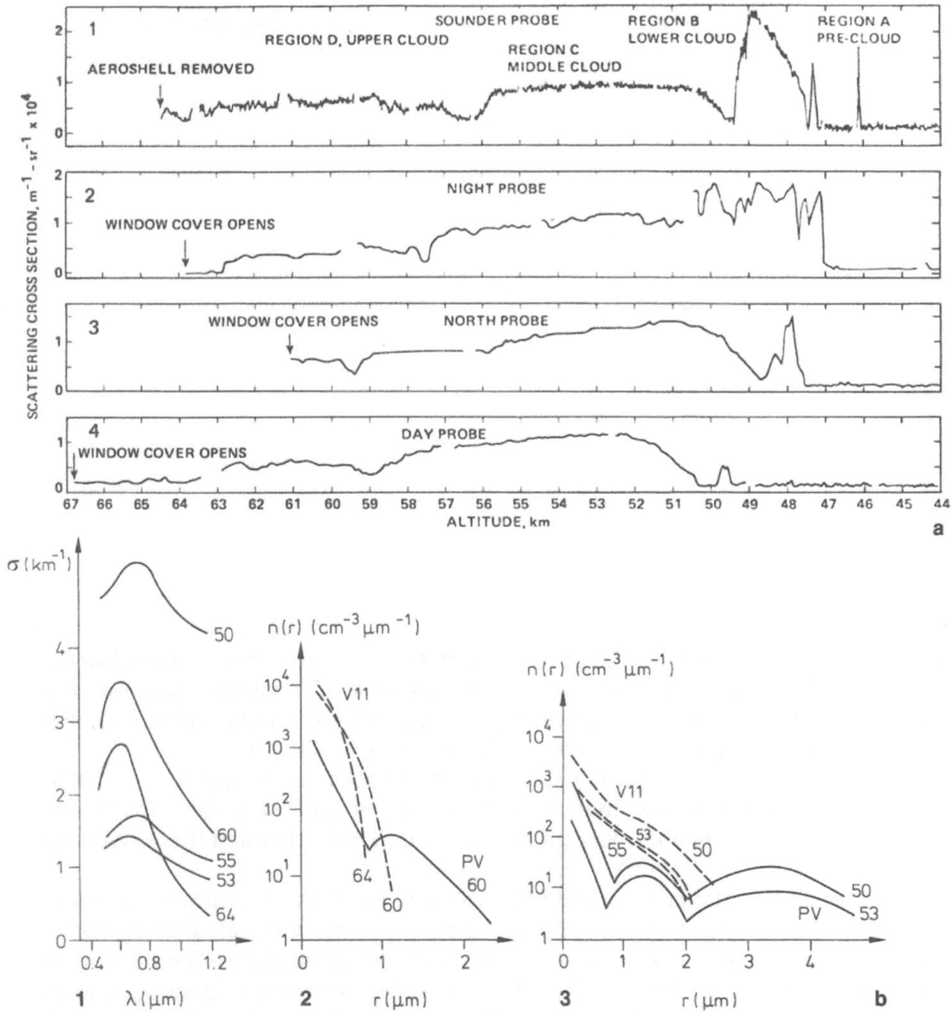

Fig. 106. a Back-scattering cross-section vs height measured by the nephelometers on board four PV probes (*1–4*) (Ragent and Blamont 1980). **b** *1* (Aerosol extinction coefficient vs wavelength; *2, 3*) particle size spectra derived from the curves in (*1*). *Numbers* near the curves show the altitudes to which a curve refers to. Result of the PV particle size spectrometer are also shown (Titov 1984)

small. This may be solved assuming true absorption by these particles. Then the third mode may have the refractive index ranging from $1.33-10^{-3}i$ to $1.44-8\times10^{-3}i$ with $m_r = 1.4$ for the first and the second modes. With low true absorption of this layer measured by photometric experiments the first value seems preferable. Another solution of the problem is nonspherical particles. Calculations can be carried out only for some specified shapes of particles (e.g., cylindrical) and until now have not been performed.

The *A* region consists of thin detached layers detected also by V9, 10 nephelometers. There may be three or less of them and their thickness is tens and hundreds of meters. Calculations fit the observations with $m_r = 1.44$ for the particles in these layers and in the bottom of the *B* layer.

The results discussed have allowed Ragent and Blamont (1980) to conclude that the first and the second modes throughout the clouds are concentrated sulfuric acid solution drops. The observed altitude variations of m_r are caused by variations of temperature and concentration of the solution. The authors avoid definite conclusions on the composition of the third mode and assume sulfuric acid and/or any salt. After sulfuric acid evaporation small particles of the salt appear to form a subcloud haze at 48 – 32 km measured by the Veneras and the PV particle size spectrometer.

Comparison of measurements by the four probes results in variations of the cloud layer properties. The upper boundary of the clouds is located at 63 km according to the night and day probe data and above 65 km according to the large probe. Probably the variations of the upper boundary height are responsible for the observed UV contrasts. The boundary between the *D* and *C* regions was detected at all probes and may be the stable feature in the atmosphere; its altitude varies in a range of 56 – 60 km (see, however, V13, 14 measurements in Fig. 103). Temperature of this feature is close to the freezing point of the sulfuric acid solution with a concentration of 80%. Thus, this feature can be related to a phase transition. The *C* region has an enhanced scattering coefficient and is about 6 – 8 km thick. The *B* layer is very variable and very weak according to the northern probe data and almost absent according to the day probe data. The same refers to thin layers at 45 – 47 km.

Photometry. The results of the Veneras photometric experiments are described in Chaps. 3.5 and 3.8. Here we consider the cloud particle size distribution deduced from these measurements.

Titov (1984) analyzed the measured spectra at wavelengths free from contamination of the CO_2 and H_2O bands. With radiative transfer calculations applied to the spectra he obtained curves of the aerosol extinction coefficient versus wavelength at various heights (Fig. 106b). These curves were used to derive the particle size spectra by means of the integral equation

$$\int_{r_{min}}^{r_{max}} K\left(\frac{2\pi r}{\lambda}, m\right) \pi r^2 \frac{dN}{dr}\, dr = \sigma(\lambda) .$$

Here K is the extinction factor, r is the particle radius, m is the refractive index, $\sigma(\lambda)$ is the measured extinction coefficient in the cloud layer, and dN/dr is the particle size distribution to be found. Tikhonov's method was utilized to solve this equation. The extinction coefficient spectra are sensitive to particles having $2\pi r/\lambda$ in the range from ~1 to ~10. Particles with $2\pi r/\lambda < 1$ have very low extinction factors and contribute only slightly, those with $2\pi r/\lambda > 10$ have extinction factors independent on wavelength and can be derived with very low accuracy. The size distributions obtained (Fig.

106b, 2 and 3) were compared with those measured by the PV particle size spectrometer. These distributions ignore large particles and exceed the measured ones by a factor of ~3 in the size range of 0.2 – 2.0 µm. Combination of the data with V11 nephelometry results in the refractive index of 1.44 – 1.53 in the upper and middle cloud layer and of 1.33 in the lower cloud layer.

Four photometric experiments were carried out by the PV probes.

Photometers for wavelengths of 365 and 530 nm with band widths of 40 and 70 nm were included as parts of the nephelometers at all four probes. Unfortunately these photometers had leaks at 750 and 950 nm for the 365 nm channels and at 925 nm and for the 530 nm channels. The relative sensitivities in the bands were 1:0.28:0.33 and 1:0.085, respectively (Ragent and Blamont 1980). Due to low intensity of near UV radiation in the Venusian clouds, it seems very difficult to separate the UV contribution to the measured radiation.

Another photometer (Tomasko et al. 1980) measured radiation at 0.59 – 0.67 µm, 0.4 – 1.0 µm, and 0.4 – 1.8 µm in five directions aboard the large probe. This experiment confirms the data on the cloud structure from the particle size spectrometer and the nephelometer with some additions and changes. The atmospheric optical thickness at 64 km not measured by the other in situ instruments is $\tau = 4$, and the solar radiation at this height is attenuated by a factor of 2. The low cloud layer optical thickness is smaller by a factor of 2.5 than that measured by the particle size spectrometer on the assumption of spherical particles. This is an additional confirmation of the presence of nonspherical particles in the lower cloud layer. Some absorption of solar radiation occurs in the upper cloud layer and below 35 km. The middle and lower cloud layers absorb remarkably little sunlight.

Net radiation measurements (i.e., the difference of total radiation fluxes from upper and lower hemispheres) were carried out at all four PV probes (Suomi et al. 1980). These measurements indicate a source of infrared opacity in the upper cloud layer and above, probably small haze particles. High infrared opacity of the lower cloud layer indicates that large particles consist of sulfuric acid. The water vapor mixing ratio may be as high as tenths of a percent near the cloud bottom and as low as a few ppm near the surface. It may vary from place to place by two orders of magnitude. The net flux data for the North and Night PV probes were analyzed by Marov et al. (1984) to derive the altitude dependence of water vapor mixing ratio. They obtained that f_{H_2O} reaches the maximum of 30 ppm and 600 ppm near 40 km at the North and Night probes, respectively, with a steep decrease of f_{H_2O} down to 25 km. Below 25 km the contribution of water vapor to the infrared opacity of the atmosphere is negligible.

On the Presence of Large Solid Particles in the Clouds. We discussed above that some inconsistencies of the particle size of spectrometric, nephelometric, and photometric observations, can be resolved by the hypothesis of nonspherical shape of the Mode 3 particles. Toon et al. (1984) argue that a discontinuity in the size spectra at $r \simeq 2.5$ µm which forms the separate Mode 3

(Fig. 104) coincides with the transition from the first to the second photodiode arrays and may be the instrumental artifact caused by an error or shift in the calibration. Therefore the part of the size spectrum in the range of $r \simeq 2.5-5$ μm may be incorrect, while the measured particles with $r > 5$ μm form a tail of the Mode 2 sulfuric acid droplets. This hypothesis removes also the inconsistencies of the data of the different instruments.

The particle size spectrometer properties and data were reexamined by Knollenberg (1984) to check both suggestions. According to his analysis, the shift or error in the calibration to remove the Mode 3 is too large to be accepted, but the evidence of high aspect ratio solid particles is not as strong as for Mode 3. However, the only way to reduce the optical thickness of the lower cloud layer by a factor of more than 2 is to assume the solid particles. We feel that this problem needs further investigation.

Radio-Occultation Data on the Lower Cloud Layer. Some additional atmospheric absorption caused by clouds was measured by the V9, 10 orbiter radio-occultations (Timofeeva et al. 1978). More detailed study of this phenomenon was based on PV radio-occulations (Cimino 1982). A dual frequency radio system operated at wavelengths of 13 and 3.5 cm. Both phase and amplitude analyses of the measurements were carried out, the former resulting in atmospheric refractive index data connected with pressure-temperature profiles, the latter yielding atmospheric absorption. Several error sources were taken into account to rectify absorption data. After that gaseous atmospheric absorption was calculated assuming $f_{H_2O} = 200$ ppm, f_{SO_2} varying from 100 ppm at 2 bar to 1 ppm at 0.3 bar and H_2SO_4 equilibrium vapor pressure with f_{H_2O} adopted and temperature known. The difference between the measured and calculated values of absorption was analyzed as an aerosol absorption. Approximation of the results at two wavelengths by power law yields the dependence $\lambda^{-1.2 \pm 0.2}$, while λ^{-2} and $\lambda^{0.2}$ dependences are expected for liquid and solid particles, respectively. Cimino (1982) considered a model with a solid dielectric particle core and a spherical shell of sulfuric acid around it. In this case, power depends on the ratio of the core to particle diameters, q, fitting the observation for $q = 0.97$. Thus, the large particles of the lower cloud layer are solid with a very thin coating of sulfuric acid.

The value of q obtained was used to derive mass loading of the lower cloud layer and its variations (Fig. 107). Mass loading is the greatest at high latitudes and the smallest at middle latitudes with a slight increase at the equatorial region. This behavior can be explained by a cooler atmosphere at high latitudes. A mean mass loading at middle latitudes is close to that measured by the PV particle size spectrometer (Fig. 105 c).

Upper Haze. TV images of the planetary limb from Mariner 10 (O'Leary 1975) prove aerosol layers to be near 85 km. Near UV and visible spectra of scattered radiation at the day side limb have been measured by the V9, 10 orbiters (Krasnopolsky 1979c). Single scattering approximation for $\tau < 0.3$ with simulation of the main cloud layer as a Lambert surface with reflection of Venus was used for their interpretation. Fits of calculated spectra with

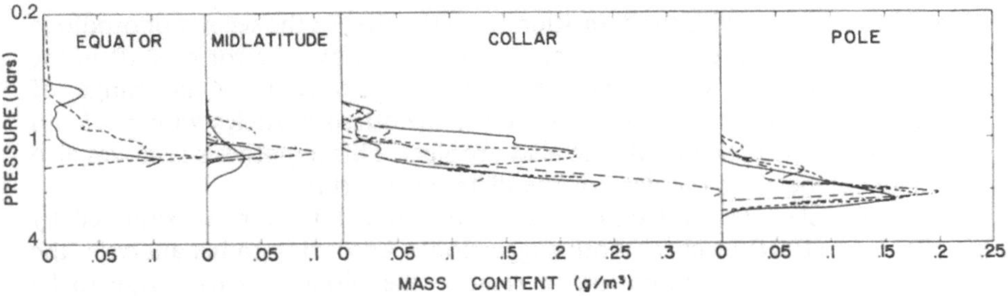

Fig. 107. Mass loading of the lower cloud layer measured by radio-occultation technique on board the PV orbiter (Cimino 1982)

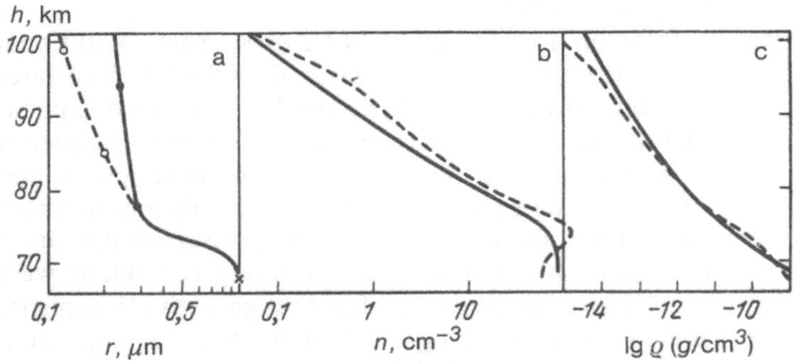

Fig. 108a–c. Radii (**a**), number densities (**b**), and mass loadings (**c**) of the upper haze (Krasnopolsky 1979c)

$m_r = 1.46$ which is appropriate to 80% solution of sulfuric acid at $T \simeq 200$ K to the observed ones yield mean particle radii. Then particle number densities and mass loading are derived from the measured intensities. The results are shown in Fig. 108a, b, c. The curves are extrapolated to $r = 1$ μm at 68 km according to ground-based polarimetry (Chap. 3.1). So the mean particle radius varies from 1 μm at 68 km to 0.1–0.2 μm at 100 km, number density from 100 to 0.1 cm^{-3}, mass loading from 10^{-9} to 10^{-15} g cm^{-3}, and the sulfuric acid volume mixing ratio from 1 ppm to 5 ppb, respectively. Some spectroscopic upper limits are deduced at 77, 85 and 94 km.

The upper haze been detected also by a PV polarimetric experiment (Kawabata et al. 1980). To explain polarization measured at 270, 365, 550, 690, and 935 nm one needs to assume a haze consisting of small particles with $r = 0.23 \pm 0.04$ μm and $m_r = 1.45 \pm 0.04$ at 550 nm above the main cloud deck consisting of particles with $r = 1$ μm and $m_r = 1.44$. Optical thickness of the upper haze at 365 nm is very variable and can be as high as 0.8 in polar and subpolar regions, while the typical value at low latitudes is 0.06. Long-term observations (Kawabata et al. 1984) show a gradual decrease of the upper haze optical thickness in 1978–1983 (see below, Chap. 3.8).

High resolution limb scans at 365 and 690 nm were obtained by the PV polarimeter (Lane and Opstbaum 1983). The mean brightness coefficient at 93 km is close to the detection limit threshold of the instrument, 1.5×10^{-2}. At ~ 84 km a slant optical thickness was ~ 1 so a vertical one is $\sim 10^{-2}$. Therefore, a mean aerosol scale height equals 2 km with variations from 1 to 3 km. Detached haze layers were observed at some scans. Phase angle dependence of the brightness coefficient at large slant optical thickness shows the appropriate range of the mean particle radius to be $0.15-0.4$ μm for 365 nm measurements and larger by a factor of 2 for 690 nm. The recommended value is $r \simeq 0.2$ μm and $m_r = 1.3-1.5$.

All the results discussed are in good agreement despite different techniques used. Some information on the upper haze at $15-35$ mbar (~ 73 km) was obtained by infrared heterodyne spectroscopy of CO_2 lines at 10.86 and 10.33 μm (Deming et al. 1982). This method utilizes a CO_2 laser with a frequency ν_0 as a heterodyne of the detector to convert a frequency ν of the measured radiation to the differential frequency $\nu - \nu_0$ with an appropriate analysis of the signal by ultrahigh frequency radiotechnique. Thus, very high resolution profiles of CO_2 lines can be measured and vertical distribution of temperature may be derived from them. However, the temperatures obtained differ significantly from those measured by the PV orbiter infrared radiometer (Taylor et al. 1980). The difference can be explained by a haze at $15-35$ mbar with $\tau \simeq 0.6$ and $\tau \simeq 1.4$ at day and night sides, respectively, in low latitude regions. We should note that $\tau \simeq 0.6$ at 35 mbar correlates well with $\tau \simeq 1$ at 50 mbar deduced from the ground-based polarimetry (Chap. 3.1).

3.8 Ultraviolet Absorption in the Cloud Layer

Ground-Based Observations. The yellowish colour of Venus which has been observed for a long time is due to a decrease in the planetary albedo at wavelengths shorter than 500 nm. The most detailed and reliable spectroscopic measurements at 300 to 500 nm were carried out by Barker et al. (1975). They show that the albedo at 330 to 380 nm is reduced by nearly one-half in comparison with the visible region. The absorption seems to be continuous and has a weak phase angle dependence (Fig. 109). At 300 to 330 nm, the albedo curve has a more complicated structure. Barker's (1979) measurements of the albedo curve with a high spectral resolution of about 10^{-2} nm indicated the presence of SO_2 absorption bands (Fig. 110). The SO_2 abundance calculated for these bands by the simple reflection model (see Chap. 1.2) is quite different for different bands and decreases sharply with increasing phase angle. At phase angles more than 90° the SO_2 bands disappear and the corresponding calculated mixing ratio does not exceed 2×10^{-8}. The phase angle dependence accounts, primarily, for failure in the search for SO_2 in earlier ground-based and rocket experiments.

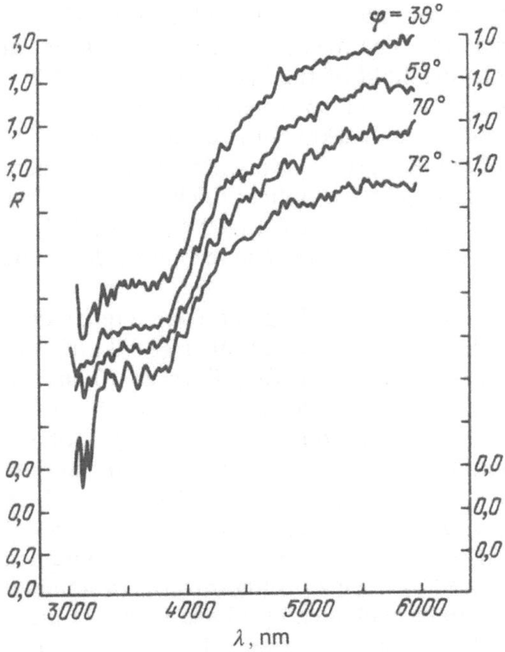

Fig. 109. Relative reflectivity of Venus in the range of 300 – 600 nm at various phase angles (Barker et al. 1975)

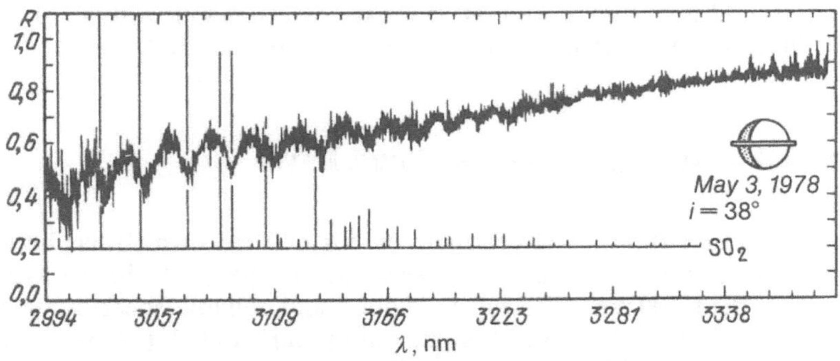

Fig. 110. High resolution spectrum of Venus in the range of 300 – 340 nm. Positions of the SO_2 bands as well as the geometry of observations are shown (Barker 1979)

Spectroscopic Measurements on Board the Pioneer Venus Orbiter. An UV spectrometer for the 110 to 360 nm region with a spectral resolution of 1.3 nm was carried by the Pioneer Venus orbiter (Stewart et al. 1979). The spectrometer operated both in a spectral and photometric mode when measurements were conducted at a selected wavelength. The latter provided the monochromatic images of Venus obtained by scanning the planetary disk due to the spinning of the orbiter.

An analysis of observations of the reflected radiation from Venus was made by Esposito et al. (1979). The atmosphere is rather bright at high

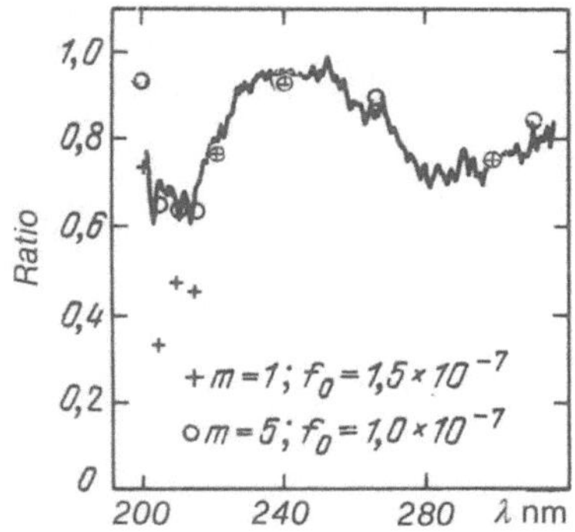

Fig. 111. Ratio of the Venusian spectra measured at the dark and bright regions compared with two models for the vertical distribution of SO_2: uniform mixing ratio model (*crosses*) that for the scale height of CO_2 to SO_2, ratio of five (*circles*) (Esposito et al. 1979)

latitudes; its reflectivity can be adequately described by a homogeneous scattering model (Hansen and Hovenier 1974, see Chap. 3.1). According to this model, the cloud layer consists of sulfuric acid droplets with a radius of about 1 μm mixed uniformly with the gaseous atmosphere. At middle and low latitudes the atmosphere becomes darker even in spectral intervals without SO_2 bands. To explain the measured reflectivity and limb darkening outside SO_2 bands Esposito et al. (1979) introduced two wavelength independent absorbers. The first leads to a decrease in a single scattering albedo of droplets to $\omega = 0.98$, the second represents a pure absorbing layer with $\tau = 0.2$ at 75 mbar (about 66 km), where the atmospheric scattering optical thickness equals 1.5. The absorption at 75 mbar agrees well with the photometric measurements of sunlight extinction by Tomasko et al. (1980) (see Chap. 3.7).

The SO_2 absorption was detected by a comparison of spectra of dark and bright regions measured at similar solar zenith angles and the orientation of the instrument optical axis. Their ratio (Fig. 111) exhibits the main features of the wavelength dependence of the SO_2 absorption coefficient, though it gives different SO_2 amounts for different wavelengths just as Barker's spectra. This indicates that the simple reflection model cannot be used and makes it possible to obtain the vertical SO_2 number density profile because weak and strong bands form at different altitudes in the atmosphere. Besides the absorbing layer and the cloud layer of $\tau = 20$ with a droplet radius of 1 μm, a conservative scattering lower atmosphere of $\tau = 64$ was introduced in the calculation. The SO_2 distribution was given by its number density at 40 mbar (70 km) and scale height. The agreement between measurements and the calculations corresponds to $f_{SO_2} = 10^{-7}$ at 40 mbar and $H_{SO_2} = 1$ km (Fig. 111). Thus, the ratio of the atmosphere scale height to that of SO_2 is equal to 5.

To study SO_2 variations, a large number of PV ultraviolet spectra was analyzed by Esposito and Gates (1981). They found the SO_2 scale height varying in a range of 1 to 3 km and a low latitude mean SO_2 mixing ratio of 20 ppb at 40 mbar.

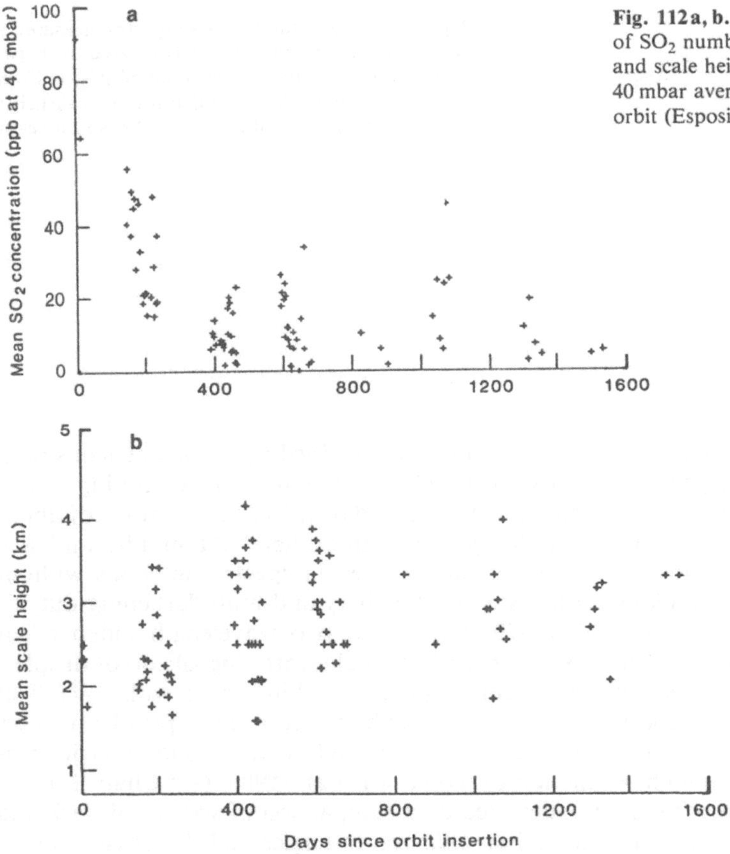

Fig. 112a, b. Time variations of SO_2 number density (**a**) and scale height (**b**) at 40 mbar averaged for every orbit (Esposito 1984)

A steep decrease of the UV albedo occurs below 200 nm due to the strong CO_2 absorption.

Long-term variations of the global mean SO_2 mixing ratio at 40 mbar (Fig. 112) and its scale height were recently observed (Esposito 1984). These data show a gradual decrease of f_{SO_2} from ~100 ppb at the end of 1978 to ~10 ppb at the end of 1982 which is accompanied by a small increase of the scale height. The upper haze optical thickness deduced from the PV polarimetric measurements (Kawabata et al. 1984) demonstrates similar behavior. This may be correlated with almost simultaneous discovery of SO_2 in the upper clouds in 1978 by several observers after failures in previous attempts. Ground-based polarimetric observations made in the late 1960's indicate the absence of the upper haze also (see Chap. 3.1). However, there is evidence of this haze being much thicker in 1959 (Dollfus et al. 1979). Photochemical transformation of SO_2 to H_2SO_4 which is a material of the upper haze particles, occurs in and above the upper cloud layer (Chap. 4.9), resulting in a correlation of f_{SO_2} at 70 km and the upper haze optical thickness.

All these data may imply episodical injection of SO_2 into the region above the cloud tops with a period between injection of decades. Esposito (1984) considers two possible explanations:

1. the SO_2 mixing ratio may differ by an order of magnitude above and below the stable layer at 40 – 50 km (Seiff et al. 1980). A large scale episodical mixing in this layer would result in a massive injection of SO_2 into the middle and upper cloud layers where atmospheric mixing is much more intense;
2. a volcanic eruption forms a convective plume of heated, emitted, and entrained gas which can rise to the middle and upper cloud layers with subsequent mixing in these layers. Calculations show that a thermal output of the eruption should exceed 10^{16} W if the convective plume formed reaches 50 km. This heating rate is about 10 – 100 times larger than that from the strongest terrestrial eruptions. Almost the same difference exists for stratospheric aerosol production after a large volcanic eruption: 2×10^{14} g of the Venusian polar haze and 10^{13} g of stratospheric haze on the Earth.

We should draw attention to the fact that the total amount of sulfur dioxide above 50 km exceeds 10^{16} g and its substantial change can be hardly produced by a volcanic eruption. However, it will be shown in Chap. 4.12 that even very small variations of SO_2/H_2O at 50 km with a time scale comparable with the mixing time result in strong variations of the atmospheric composition above 65 km. We do not know which process is responsible for the proportion of mode 1 to mode 2 particles both consisting of sulfuric acid in the upper cloud layer and haze. A very small increase of SO_2 at 50 km results in pronounced amounts of SO_2 at the cloud tops which may be transformed photochemically to sulfuric acid particles of the upper haze. In our opinion volcanic eruptions can drive these processes by both the gas injection and the enhanced convection and mixing.

Ultraviolet Contrasts. The Venusian disk is very uniform in the visible range, but its images in UV light show distinct features connected with atmospheric dynamics at the top clouds. Barker et al. (1975) proved that dark feature spectra are similar to that of the whole planet. Comparison of contrasts at 270 nm and 365 nm observed by the PV polarimeter (Esposito and Travis 1982) results in a correlation coefficient of 0.74. Despite different sources of contrasts at these wavelengths (SO_2 for 270 nm and an unknown absorber at 365 nm, see below), a common process underlying the UV contrasts at all wavelengths is shown. A steep decrease of the SO_2 mixing ratio with height leads to the assumption that both absorbers are located in the main cloud deck. Then two probable processes responsible for the dark feature formation may be considered: an increase of absorber amounts and/or a decrease of the upper haze optical thickness. Esposito and Travis (1982) checked both possibilities by modeling the measured dependence of polarization of dark and bright points on the phase angle at four wavelengths (270, 365, 550, and 935 nm). Their model consists of a pure gas layer, the upper haze of sulfuric acid drops with $r = 0.23$ µm (see the preceding paragraph) and the main cloud deck with H_2SO_4 drops of $r = 1$ µm. Single scattering albedos of the upper haze at both bright and dark points and that of the main cloud deck at bright points are assumed equal to and matching the Venusian spectrum. The gas and haze optical thicknesses and the single scattering albedo of the main cloud deck at dark points are chosen to fit the observations. The modeling shows that the upper haze optical thickness equals 0.39 and 0.27 in the UV region at bright and dark points, respectively, the main cloud deck defined as a level of $\tau = 1$ at 365 nm is located at 29 and 37 mbar, i.e., lower at dark points by 1.2 ± 0.6 km. The UV single scattering albedo is 0.982 and about 0.965 (0.975 in the upper part with $\tau \leqslant 1.25$ and 0.953 below) for bright and dark points, respectively.

In the visible range the upper haze optical thickness constitutes 0.27 and 0.19, and the unit optical thickness corresponds to 40 and 43 mbar at bright and dark points, respectively, while the single scattering albedo equals 0.999 everywhere. Formation of dark features may be described by dynamical and chemical models. The former (Fig. 113) requires vertical convergence and horizontal divergence. According to the chemical model, water vapor and/or molecular oxygen are less abundant at dark points, thus decreasing SO_2 destruction and sulfuric acid formation. In this case both a H_2SO_4 to SO_2 ratio and single scattering albedo become lower as well as the upper haze thickness.

IUE Satellite Observations (Conway et al. 1979). The UV spectrum of Venus at 208 to 218 nm with the 0.03 nm resolution obtained by this satellite shows well-defined SO_2 bands. The simple reflection model gives the SO_2 amount equal to $4 \times 10^{16}\,cm^{-2}$. The spectrum was normalized to the Pioneer Venus measurements at 240 nm (see above), then the single scattering albedo for a homogeneous model was calculated. Thus, obtained ω was used for the calculation of $f_{SO_2} = (1 \text{ to } 8) \times 10^{-7}$. The absorption takes place at a level of 3 to 26 mbar (72 to 78 km).

Analysis of Ground-Based Observations. The above measurements (Barker et al. 1975, Barker 1979, see Figs. 109 and 110), similar to the UV spectroscopy on board the Pioneer Venus orbiter, make it possible to obtain the SO_2 vertical distribution. Besides, they permit to define the nature of the absorption at longer wavelengths. Both problems were dealt with by Zasova et al. (1981).

The calculation of the SO_2 amount from the absorption band measurements entails the problem of applicability of average absorption coefficients of the bands consisting of a large number of lines (see Chap. 1.1). The simple absorption law and average absorption coefficients can be used for unsaturated lines in bands or for intervals between the lines which are smaller than the line widths. These intervals are large both for hydrids because of their small moment of inertia and for linear molecules with only one rotational degree of freedom. SO_2 does not belong in these types of molecules, and indeed, laboratory measurements (Warneck et al. 1964) revealed the regularity of the absorption coefficients for a pressure interval of 2.5 to 13 mbar and a gas layer transmission greater than 10%. It should be taken into account that

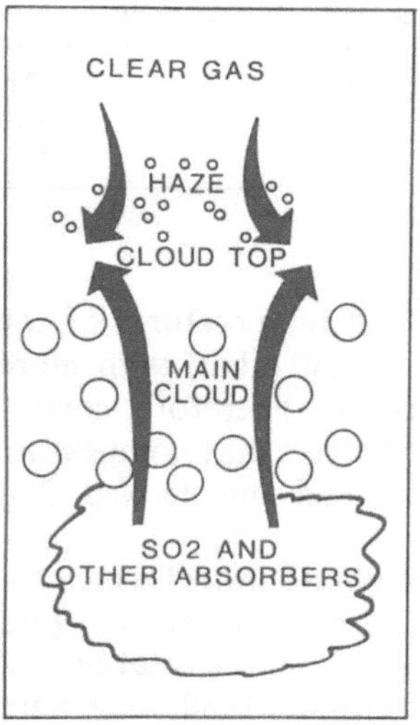

Fig. 113. A dynamical explanation of the UV contrasts (Esposito and Travis 1982)

the equivalent width of the saturated line with a collisional profile is proportional to the square root of the amount of an absorber multiplied by the unsaturated line width. Both values being proportional to pressure, the equivalent width is proportional to pressure as well. But this effect cannot occur in the measurements by Warneck et al. because in this case the line profile is a Doppler one. From this we concluded that the simple absorption law is valid for SO_2 at an absorption thickness of <2.5. This problem is important for all SO_2 spectral observations discussed above.

Saturation of the SO_2 bands was considered recently by Belton (1982). He argues that the main properties of the Venusian SO_2 bands (i.e., much larger SO_2 abundances indicated by weaker bands and the strong phase angle dependence of equivalent SO_2 abundances) which are considered by Esposito et al. (1979) and by us as resulting from a steep decrease of f_{SO_2}, may be explained by the saturation of stronger bands, while the phase angle dependence is similar to that of CO_2 bands. He concludes that the random band model with a mean line spacing of 0.15 cm^{-1} can describe planetary and laboratory SO_2 spectra well, but a more careful laboratory study of the band structure and much more complex calculations of radiative transfer for these bands should be performed. A considerably larger SO_2 mixing ratio and scale height at the 40 mbar level are expected from these calculations.

Despite the correctness of Belton's approach some arguments in favor of SO_2 mixing ratio profiles deduced by the simple absorption law can be adduced along with above mentioned. The phase angle dependence of the SO_2 bands measured by Barker (1979) is much steeper than that of CO_2 bands, thus indicating the upward decrease of f_{SO_2}. Belton's calculations indicate that the simple absorption law agrees better with the measured spectrum of Venus than the band model adopted. Laboratory measurements of the band absorp-

Table 18. Wavelengths, cross-sections, and SO_2 amounts for three bands (Barker 1979)

λ (nm)	305	313	321
$\sigma \times 10^{19}$, cm^2	7	1.5	0.24
$N \times 10^{-18}$, cm^{-2}	0.45	1.1	4

tion cross-sections were performed with pressures and temperatures of the same order of magnitude as in the Venus atmosphere (40 mbar) with more effective pressure broadening by SO_2 taken into account. Therefore, corrections of the measured cross-sections for the saturation effect may compensate those of the interpretation of atmospheric spectra, and the derived SO_2 density profiles may be a fair estimate.

For our analysis, we used a homogeneous atmospheric model which included, besides the scattering by gas and sulfuric acid droplets of $r \simeq 1$ μm, small particles of $r \simeq 0.2$ μm constituting about one-third of the total optical thickness of the upper cloud layer (Knollenberg and Hunten 1980, see Chap. 3.7). The properties of the middle and lower cloud layers with larger particles could not sufficiently affect the results of the calculations.

The SO_2 absorption cross-section is reduced by a factor of several hundreds in the 320 to 340 nm interval, so the absorption at $\lambda \gtrsim 330$ nm cannot be caused by SO_2. We assume the second absorber cross-section independent of λ for $\lambda < 330$ nm. Using the spherical albedo at 330 nm we can obtain $\omega = 0.98$. Let us assume further

$$[SO_2] = A\tau^m .$$

A is the SO_2 number density at $\tau = 1$ (68 km), m is the ratio of an extinction coefficient scale height (5 km) to that of SO_2. Calculations were made for three SO_2 bands; their wavelengths, cross-sections, and SO_2 amounts derived for the simple reflection model by Barker (1979) are shown in Table 18.

The calculations matched the observations at $A = 10^{11}$ cm^{-3} and $m = 5$. Evidently, the SO_2 profile obtained can be used in some limited altitude interval. To define it, let us calculate the SO_2 mixing ratio for each band assuming an uniformly mixed atmosphere ($m = 1$); then the measurements of a given band can be attributed to the altitude of the coincidence of the actual ($m = 5$) and uniform ($m = 1$) distributions. The calculation shows that the SO_2 profile can be applied in the optical depth range of 1 to 2.5 (68 to 64 km), the corresponding SO_2 range being 10^{11} to 10^{13} cm^{-3}.

This profile is consistent with the Pioneer Venus SO_2 measurements (see above) which covered the SO_2 range of 3×10^9 to 10^{11} cm^{-3}. Heights differ slightly; our calculations give $[SO_2] = 10^{11}$ cm^{-3} at 68 km and that of Esposito et al. (1979) at 70 km. The discrepancy is probably due to the fact that a dark region with a maximum SO_2 amount is considered by Esposito et al. and Barker's measurements are closer to the average situation at low and middle

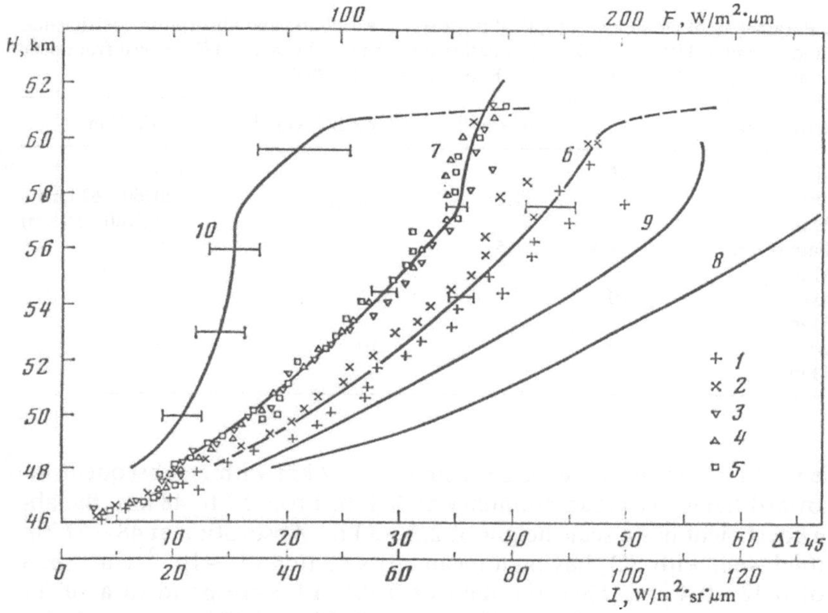

Fig. 114. Vertical profiles of the 365 nm radiation in the clouds of Venus; *1* to *5* are intensities measured at zenith angles of 45°, 315°, 228°, 133°, and 164°, respectively; *6* and *7* are mean intensities for 45° and 315°, and 133° and 228°, respectively; *8* and *9* are total fluxes from the upper and lower hemispheres; *10* is the difference between *6* and *7*, ΔI_{45} (Ekonomov et al. 1983)

latitudes. Actually the mean SO_2 number density is $2 \times 10^{10}\,cm^{-3}$ at 70 km according to Esposito and Gates (1981) and coincides with our result.

UV photometric observations aboard the V14 lander probe were performed at 365 nm with a band pass of 70 nm (Ekonomov et al. 1983). A proper choice of the detector ruled out any leaks higher than 10^{-3}. A scanning system provided measurements in eight directions with two of them used for a zero level correction (black surface was fixed at 90° and 270°) and with parachute disturbances at the zenith measurements. Technique of interpretation was the same as for other Venera photometric measurements (Chap. 3.5). Averaging of measurement points, derived downward and upward fluxes and their difference, the so-called net flux, are shown in Fig. 114. The obtained characteristics of the cloud layers are given in Table 19 with $1-g=0.2$ assumed.

Total measured optical thickness of the cloud layer, $\tau = 55$, seems to be large in comparison with that measured in the visible range ($\tau \simeq 30$) by the same probe and the same authors (Moshkin et al. 1983). Most of the solar near UV radiation is absorbed above 62 km, and one may expect a maximum of absorption near 65 km. The data on the absorbed radiation are very important for the problem of the atmospheric thermal regime. The absorption

Table 19. Parameters of cloud layers (optical thickness τ, extinction and absorption coefficients, σ and k, and energy of UV ($\lambda < 0.42$ μm) radiation absorbed in the layer, ΔF, derived from near UV photometric observations aboard V 14 (Ekonomov et al. 1983)

Atmospheric region	τ	σ, km^{-1}	$k \times 10^3$, km^{-1}	ΔF, W m^{-2}
$h > 62$ km	15			120
Upper layer, 58 – 62 km	5	1.2	20^{+20}_{-14}	20 (60 – 62 km) 2.5 (60 – 58 km)
Intermediate region, 55 – 58 km	4.5	1.5	1	0.7
Middle layer, 52 – 55 km	10	3	3	1
Lower layer, 48 – 52 km	20	5	10	3

coefficient has a very steep decrease from 60 to 57 km with a subsequent increase toward the lower cloud boundary at 48 km. From 57 to 48 km, the absorption coefficient has a scale height of 2.5 – 3 km. Absorption at 48 – 57 km is identified well, with SO_2 having a mean cross-section of $\sim 10^{-22}$ cm^{-2} in a region of 0.32 – 0.4 μm. Then the data of Table 19 correspond to a sulfur dioxide mixing ratio of 50, 20, and 10 ppm at 50, 54, and 57 km, respectively, and can supplement gas chromatographic measurements below clouds (~150 ppm at ~20 km) and the spectroscopic measurements. Interpolation between 20 ppb at 70 km and 10 ppm at 57 km results in a sulfur dioxide scale height of 1.5 km which is in fair agreement with the SO_2 band interpretation. The altitude distribution of the absorber coincides with that of the first particle mode, thus indicating that these particles contain an absorber. However, this indication is not strong enough, due to the fact that some other components may feature a steep decrease of absorption at these altitudes as well.

Other Absorbers. Ferric Chloride. Many indications that SO_2 cannot provide absorption at $\lambda > 330$ nm exist: a quantitative discrepancy between a small change of the albedo in the 320 to 340 nm region and a sharp change of the SO_2 cross-section by almost three orders of magnitude; the absence of a band structure of the spectrum at $\lambda > 330$ nm; the disappearance of the SO_2 bands at phase angles exceedings 90°, while the absorption at longer wavelengths remains almost the same.

The latter gives important information on the altitude profile of the second absorber: in contrast to SO_2, it should be mixed more uniformly in the upper cloud layer. Figure 115 exhibits albedos of Venus calculated by the author for various absorbers (aerosol of elemental sulfur S_a, gases NO_2, SCl_2, Cl_2). The mixture of SCl_2 and S_a agrees rather well with the measurements, but the required SCl_2 mixing ratio equals 10^{-7}, which exceeds our aeronomic estimate by more than one order of magnitude.

The sulfuric acid aerosol of the second mode with a sprinkling (~1%) of ferric chloride $FeCl_3$ shows the best agreement with the observations. The

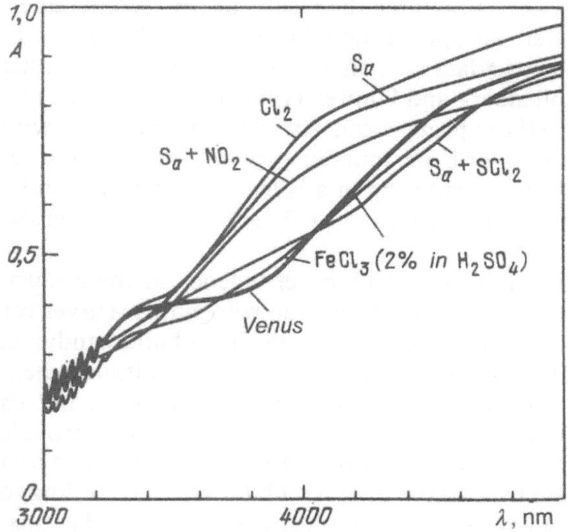

Fig. 115. Calculations for various absorbers and measured (*solid line*) albedo of Venus in 300 – 500 nm. S_a is the aerosol of elemental sulfur

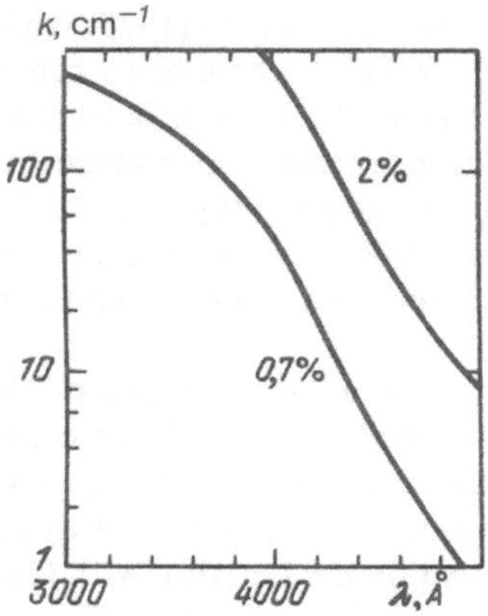

Fig. 116. Absorption coefficient of the $FeCl_3$ solution with a concentration of 0.7 and 2% in 80% sulfuric acid (Zasova et al. 1981)

wavelength dependences of the absorption coefficient of this solution for various concentrations are shown in Fig. 116. The dependence on the concentration is nonlinear and is better described by the square law.

Ferric chloride in the Venusian clouds cannot be formed by an HCl effect on iron-bearing compounds in meteoritic matter: the influx of Fe atoms from meteorites is smaller than their downward flux as a part of the cloud particle matter of the first and second modes by more than an order of magnitude. Another possibility is the evaporation on the Venusian surface of $FeCl_2$ or other components which can produce $FeCl_3$.

Chlorine. The UV absorption was investigated also by Pollack et al. (1979, 1980). The first of these papers offered SO_2 as an absorber responsible for all effects of the UV absorption. Let us examine the results of the second paper. To calculate a radiative transfer, an optical model of an aerosol component was offered, based on photometric data of Tomasko et al. (1980) concerning optical thickness, data of Knollenberg and Hunten (1980) (particle sizes), and those of Kawabata et al. (1980) (haze properties). The model is consistent with the data of the observation on board the sounder probe. However, the reflected radiation is a total effect of scattering in a layer of $\tau \simeq 5$ to 10, and a very detailed structure model does not have any sufficient advantage over the homogeneous one.

Pollack et al. (1980) obtained the wavelength dependence of the absorption coefficient of the second absorber. A comparison with similar curves for various substances has led to the conclusion that the most probable candidate is Cl_2 with a mixing ratio of 10^{-6}. As will be shown below, the altitude dependence of f_{Cl_2} actually exhibits a maximum in the upper cloud layer, but its value is far lower (2×10^{-8}). Further, the second absorber should provide almost wavelength independent absorption in the 2000 to 3200 Å region and the Cl_2 absorption changes here by a factor of several hundreds. According to our calculations (see Fig. 115) Cl_2 also does not adequately match the measured spectrum.

Sulfur as an UV absorber was proposed by Hapke and Nelson (1975) with the further development of this idea by Young (1977). According to these papers most of the cloud mass (50 – 90%) is represented by large ($r \simeq 10$ μm) sulfur particles. Photochemical calculations of atmospheric composition by Prinn (1975), who considered COS as a sulfur-bearings gas, supported this conception. In this case the formation of elemental sulfur is the first step in the transformation of COS into sulfuric acid which requires some amounts of molecular oxygen.

These hypotheses were not consistent with direct measurements from PV and V 11, 12 probes (the refractive index of sulfur exceeds very much the measured one of cloud particles). However, sulfur as the second absorber was proposed recently by Toon et al. (1982). Sulfur forms as a result of SO_2 photolysis with the subsequent photolysis of SO and is removed by the reaction with O_2. In the absence of O_2 a formation of allotropes S_n ($n \leqslant 8$) occurs with S_8 as a stable allotrope under cloud layer conditions. Allotrope exchange between gas and solid sulfur particles is assumed to satisfy the saturation condition. Sulfuric acid particles absorb sulfur particles in collisions. The latter are enriched with short allotropes, of which S_3 and S_4, having strong absorption at 300 – 500 nm (maximum at 400 nm) and at 450 – 600 nm (maximum at 530 nm), respectively, are of particular importance. A few percent of these allotropes are enough to reproduce the measured UV absorption. This model successfully explains the higher than 1.45 refractive index of particles in the upper cloud layer deduced by comparison of the PV size spectrometric and nephelometric measurements (however, the difference in m_r was ascribed by Ragent and Blamont, 1980 to small particles undetectable by the size spectrometer). It reproduces the bimodal size distribution with the mean radius of the first mode of 0.1 μm. Sulfur molecules react immediately with the free oxygen, so the appearance of sulfur particles is most probable in oxygen poor regions in the upper cloud layer which according to the authors are connected with the location of upward motions, thus confirming the correlation of sulfur and SO_2 absorption. However, in our opinion this assumption seems to be valid for low O_2 number densities in the cloud layer contrary to the adopted density in the model based on the PV gas chromatography. With large oxygen amounts in the cloud layer the oxygen poor regions may form near the cloud top by vertical divergent and horizontal convergent atmospheric motions which are inconsistent with the dynamical model of the dark features (Fig. 113). Sulfur particles inside sulfuric acid drops do not contradict any in situ measurements. Thermal relaxation of short allotropes to S_8 inside sulfuric acid drops provides the observed relatively short lifetime of the dark features (of the order of days) and the absence of absorption in the middle and lower cloud layers.

This physico-chemical model was not described in detail by Toon et al. (1982). Calculations are still in process.

However, Young (1983) shows that sulfuric acid drops cannot have sulfur cores. Collisions of sulfur and sulfuric acid particles and/or condensation of both species result in sulfuric acid drops coated by sulfur particles. The fully-coated drops cannot coalesce, and this sets the upper limit on the particle size. Assuming initial drops of $r \simeq 1$ μm with ~1% of their surface coated by sulfur to be in accord with the near UV absorption, one obtains $r_{max} \simeq 100$ μm for the fully-coated drops, which is in reasonable agreement with the measurements. Further increase of sulfur coating or partial evaporation of the liquid phase results in distortion of the particles and may lead to breakup. However, we should be cautious in applying this scheme as an explanation of the mode 3 particles due to the great difference between r_{max} obtained and the mode 3 radius.

The sulfur coating on sulfuric acid drops may be responsible for the very low refractive index in the lower cloud layer. Evaporation of sulfur near 49 km may explain the sudden return to $m = 1.44$ at this height. The atmospheric temperature is 360 K here, and the appropriate sulfur vapor volume mixing ratio, f_{S_8}, equals 3 ppm. Evaporation of the sulfuric acid aerosol at 47.5 km with $T = 368$ K results in the sulfuric acid vapor mixing ratio of ~100 ppm. One should bear in mind that these values refer to vapors and are invalid for the aerosol phase above the levels of sublimation because the fluxes, but not the mixing ratios, are equal above and below these levels. However, if elemental sulfur coats sulfuric acid drops, a ratio of sulfur to sulfuric acid in the aerosol phase should be equal to the ratio of these components in the gas phase below the level of sublimation. Therefore according to this interpretation, 20% of sulfur atoms in the aerosol phase are attributed to elemental sulfur. Photochemical models by Krasnopolsky and Parshev (1980a, b, 1981) and by Yung and De Moore (1982) give 33% and no elemental sulfur aerosol, respectively (see Chap. 4.11, 15).

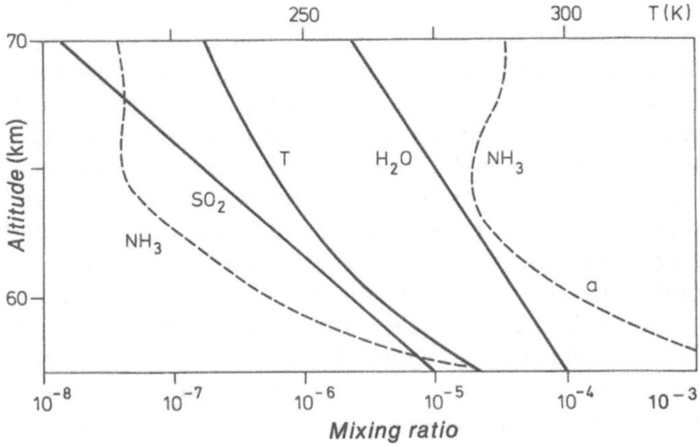

Fig. 117. Vertical profiles of SO_2 and H_2O mixing ratios and of the atmospheric temperature. *Dashed lines* are NH_3 mixing ratio profiles which yield a condensate of ammonium pyrosulfite for the temperature shown (*a*) and that lower by 15 K (Krasnopolsky 1985a)

Ammonium Pyrosulfite $(NH_4)_2S_2O_5$. Titov (1983) studied the formation of aerosol in the mixture of SO_2, NH_3, and H_2O with carbon dioxide as the dominant gas. From the stoichiometry of the reaction, he found that ammonium pyrosulfite formation occurs at $T \lesssim 285$ K:

$$2SO_2 + 2NH_3 + H_2O \rightarrow (NH_4)_2S_2O_5$$

while this mixture yields ammonium sulfite at higher temperatures

$$SO_2 + 2NH_3 + H_2O \rightarrow (NH_4)_2SO_3 .$$

Titov measured a constant equilibrium of gas near the solid phase of ammonium pyrosulfite: $K_p = P_{SO_2}^{-2} P_{NH_3}^{-2} P_{H_2O}^{-1} = 10^{(20000/T)-49}$ atm^{-5} in a range of 258 – 285 K. Then it is easy to calculate the NH_3 mixing ratio which results in $(NH_4)_2S_2O_5$ aerosol formation. This mixing ratio does not exceed the upper limit $f_{NH_3} < 3 \times 10^{-8}$ given by Kuiper (1968), except at the temperatures about 15 – 20 K lower than the mean ones (Fig. 117). Titov measured the absorption spectrum of ammonium pyrosulfite (Fig. 118) and found that its being the first mode of particles in the upper cloud layer can explain the Venusian albedo at 320 – 500 nm. However, the formation of this absorber is stimulated by low temperatures, and the Venusian clouds should have been brighter at low latitudes in the UV region and strong anticorrelation of 365 nm and 11.5 μm maps should have occurred. But this anticorrelation is inconsistent with the observations.

Nitrosylsulfuric Acid. Watson et al. (1979) and Sill (1983) draw attention to the fact that nitrogen oxides, NO_x, formed by lightnings (see Chap. 4.6) may be responsible for the appearance of nitrosylsulfuric acid cristals. Sill measured the absorption spectrum of the solution H_2SO_4 (86%) + $NOHSO_4$ (1.4%) + H_2O and found that this spectrum together with SO_2 and NO_2 explains qualitatively the UV albedo of Venus. However, the absorption coefficient of this solution is only 3 cm^{-1} at 365 nm, thus, resulting in $1 - \omega = \frac{2}{3}kr \simeq 2 \times 10^{-4}$ for the second mode particles in the upper cloud layer, i.e., lower by two orders of magnitude than that which one needs to explain the UV

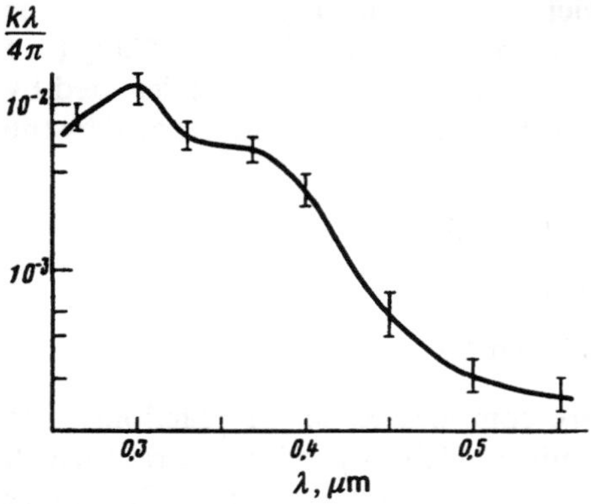

Fig. 118. Absorption coefficient of ammonium pyrosulfite (Titov 1983)

absorption. Nitrosylsulfuric acid forms over a short period of time by the equilibrium

$$NO + NO_2 + 2H_2SO_4 \rightleftarrows 2NOHSO_4 + H_2O$$

with an equilibrium constant, K_p, of $P_{H_2O}/(P_{NO}P_{NO_2})$. For the mixture of H_2SO_4 (86%) and $NOHSO_4$ (1.4%), $P_{H_2O}/(P_{NO}+P_{NO_2}) \simeq 50$ (Sill 1983). Under laboratory conditions $P_{NO} \simeq P_{NO_2}$. In and above the upper cloud layer $P_{NO} \simeq 100\,P_{NO_2}$ (see Chap. 4.15), and one can obtain the condition $[NO]/[H_2O] \simeq 0.1$ for this solution to exist in the Venusian atmosphere. Thus, f_{NO} should be ~1 ppm in the upper cloud layer, which corresponds to the upper limit derived from the absence of the absorption band at 215 nm (Owen and Sagan 1972). However, aeronomic estimates of NO production by lightning are lower by two orders of magnitude (see Chap. 4.6), and we will not consider the presence of $NOHSO_4$ even in the upper cloud layer ($f_{NO} \simeq 10$ ppm should be in the middle and lower clouds if $NOHSO_4$ is present there) and even in such a diluted form. Absorptions of NO_2 at 365 and 500 nm do not differ very much and cannot explain the UV albedo. High Venusian reflectivity at 500 nm results in the upper limit of 6 ppb to NO_2 mixing ratio (Krasnopolsky and Parshev 1979b).

Perchloric acid $HClO_4$ was also suggested as a chlorine bearing aerosol material (see von Zahn et al. 1983). This hypothesis presents the following difficulties:

1. The production rate of $HClO_4$ should be negligible. Formation and removal of ClO_x are mainly due to ozone and atomic oxygen respectively:

$$O_3 + ClO_{x-1} \rightarrow ClO_x + O_2 \qquad k_{fx}$$
$$O + ClO_x \rightarrow ClO_{x-1} + O_2 \qquad k_{rx}.$$

This results in

$$\frac{[ClO_x]}{[ClO_{x-1}]} = \frac{k_{fx}}{k_{rx}}\frac{[O_3]}{[O]}, \qquad \text{and} \qquad \frac{[ClO_2]}{[Cl]} < 10^{-2}e^{-2800/T}\left(\frac{[O_3]}{[O]}\right)^2$$

after substitution of the ratio coefficients (only the upper limit of k_{f2} is measured, and k_f, k_r are not known for $x > 2$). Therefore the ClO_2 (not ClOO!) number densities do not exceed $10\,cm^{-3}$ in all models calculated by Yung and De Moore (1982). Consider now formation of perchloric acid by the following reactions:

$$HO_2 + ClO \rightarrow HOCl + O_2; \quad k = 5 \times 10^{-12}\,cm^3/s$$

$$HOCl + h\nu \rightarrow OH + Cl; \quad J = 4 \times 10^{-4}\,s^{-1}$$

$$HOCl + O_3 \rightarrow HClO_4; \quad k < 10^{-14}\,cm^3/s.$$

Actually $HClO_4$ is produced by a more complicated set of reactions. The above approach results in the upper limit of $HClO_4$ production rate which does not exceed $3 \times 10^6\,cm^{-2}\,s^{-1}$ for the models of Yung and De Moore (1982), i.e., more than three orders of magnitude lower than that of mode 1.

2. The saturated vapor pressure of water solutions of $HClO_4$ is too large. It equals 15 mbar at 273 K for pure $HClO_4$ (data on $HClO_4$ are taken from Gmelin, 1969) and has a sharp minimum of 0.12 mbar at 273 K for $HClO_2 \cdot 2H_2O$. It corresponds to azeotropic mixture, i.e., a proportion of the components in saturated vapor is the same as in solution. For this minimum a temperature dependence of saturated vapor pressure can be approximated by $\lg P(\text{bar}) = 3.264 - (1965/T)$. Then for $HClO_4 \cdot 2H_2O$ to be present in the cloud layer water vapor mixing ratio should have been equal to 100 ppm at the level of line formation (62–65 km) and 3×10^{-3} at 50 km with those of $HClO_4$ being half of these values. The values for water vapor contradict the measurements, and for $HClO_4$ seem quite improbable. For other proportions of $HClO_4$ and H_2O the situation is even less satisfactory.

Perchloric acid does not form compounds with sulfuric acid. Therefore addition of sulfuric acid to water solution of perchloric acid results in an increase of $HClO_4$ to H_2O ratio in vapor. Amounts of $HClO_4$ comparable with those of H_2O are too large to be considered.

3. Infrared spectrum of dry perchloric acid demonstrates some absorption features, the most prominent of them being the bands at $1320\,cm^{-1}$ (νCl–O), $1040\,cm^{-1}$ (νCl–O), $740\,cm^{-1}$ (νCl–OH), $580\,cm^{-1}$ (δO–Cl–O). Spectra of water solutions of $HClO_4$ have a strong band of $\nu_3 ClO_4^-$ at $1100\,cm^{-1}$ with an absorption coefficient of $2 \times 10^3\,cm^{-1}$ (symbols ν and δ denote valent and deformation vibrations). None of these bands is present in the Venus spectra measured by Veneras 15 and 16 (Ertel et al. 1984).

3.9 Investigation of a Cloud Layer Elemental Composition by X-Ray Fluorescent Spectroscopy in the Region of 1 to 20 Å (0.6 – 13 keV)

This experiment was carried out on board the Venera 12 and 14 descent modules, its results were described by Surkov et al. (1981, 1982) and Petryanov et al. (1981a, b). The idea of the method is as follows. If a substance is irradiated by X-rays with a photon energy in excess of ionization energies of the atomic *K* shell; then *K* shell ionization occurs. This *K* shell is restored by the capture of an electron from outer shells thereby emitting a photon. Thus, the spectroscopy of this fluorescence provides information on the elemental composition of the substance being analyzed. The ^{55}Fe isotope conversion to ^{55}Mn by the *K* capture producing the Mn 5.9 keV line was used in the Venera 12, 14 instruments. This energy is sufficient enough to excite *K* lines of atoms with *Z* from 13 to 25 (from Al to Mn). To excite atoms with higher *Z*, the ^{109}Cd isotope was also used in the V 12 device. Lines of elements with $Z < 13$ have energies less than 1 keV, i.e., out of the instrument range. The fluorescent spectrum measurements were conducted with a proportional counter and a multichannel amplitude analyzer in 0.6 to 8 keV (V 12, 14) and 1 to 13 keV (V 12) ranges.

The instrument had the system of blowing an atmospheric gas through a filter to collect aerosol particles. It was this deposit that was analyzed, although the atmospheric gas and construction materials of the instrument made some contribution to the measured spectra. These spectra (Fig. 119) also have the 5.9 keV manganese peak from the source of radiation which was used for the verification of energy and intensity scales. An increase in atmospheric pressure brings about the increase of both gas scattering and the 5.9 keV peak intensity (by a factor of 1.5 for the atmospheric pressure changing from 0.1 to 1.5 bar). At the same time, the fluorescent radiation of the deposit decreases (by a factor of 2 at 2.6 keV for 1.5 bar). The total counting rate was measured by the intensimeter (Fig. 120) which made it possible to check the time dependence of the aerosol collected. The change in atmospheric pressure was taken into account in data processing to retrieve the altitude distribution of the collected aerosol (Fig. 120b).

The instruments were switched on at the 100 mbar level, operated for 9 min and was turned off at 1.5 bar. Over this time, about 1 m^3 of the atmospheric gas was blown through the filter. The fluorescent spectra have a peak near 2.5 keV apart from the 5.9 keV manganese source radiation. However, a more detailed analysis has shown that both peaks are wider than expected; the 2.6 keV peak consisting of four components: Cl 2.62 keV, S 2.31 keV, Ar 2.96 keV, and Al 1.49 keV. The 5.9 keV peak includes the radiation of Ti 4.5 keV together with the manganese line. The Al and Ti lines are attributed to the materials of the measuring cell. Surkov et al. (1981, 1982) give the follow-

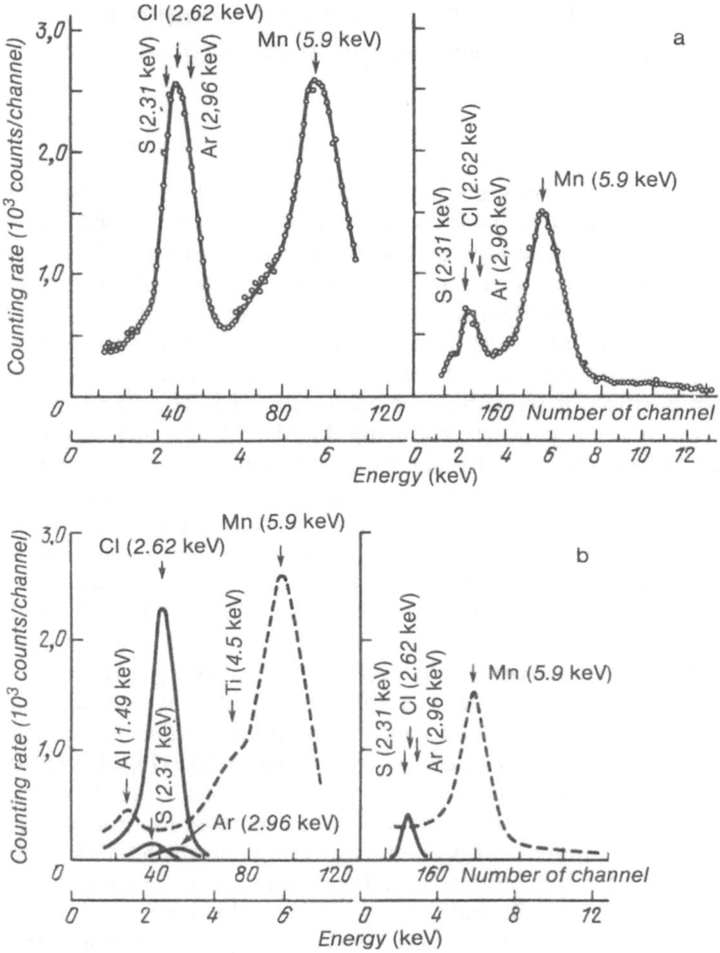

Fig. 119. a X-ray fluorescent spectra of cloud aerosol measured by V 12; **b** contributions of various components. Similar spectra measured by V 14 have peaks near 2.5 keV, i.e., slightly shifted to the left (Surkov et al. 1981, 1982)

ing (related to 1 m^3 of gas blown through) average concentrations of atmospheric and cloud layer elements (in 10^{-9} g cm^{-3}):

	V 12	V 14
Chlorine	2.1 ± 0.6	0.16 ± 0.04
Sulfur	0.17 ± 0.03	1.1 ± 0.13
Mercury	<0.05	–
Argon	200 ± 100 ppm	–

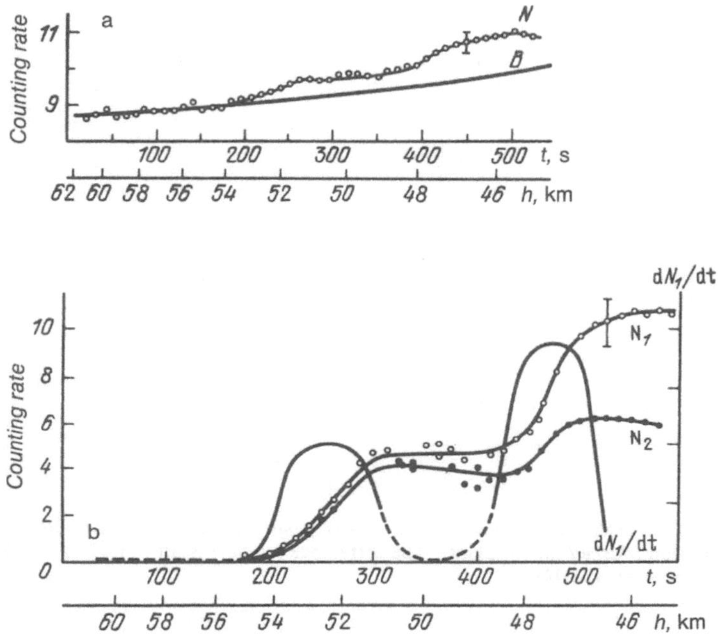

Fig. 120. a Total counting rate and contribution of atmospheric scattering vs height; **b** aerosol accumulation normalized by atmospheric volume (N_1) and mass (N_2) (Surkov et al. 1981)

Thus, sulfur- and chlorine-bearing aerosol components appear to be very variable. Surkov et al. (1981) give the curve of a counting rate and the contribution of atmospheric gases to the rate (see Fig. 120). The difference between these curves defines the quantity of the aerosol collected and its time derivative should be proportional to the aerosol density, or, to be more exact, to chlorine and sulfur quantities in the aerosol phase. This derivative indicates the presence of aerosol layers at 55 to 51 and 49 to 46.5 km. However, one should bear in mind that competing processes, aerosol accumulation and its evaporation by the atmospheric gas flow, occur in the instrument.

It should be taken into account that the V 12 counting rate at the 5.9 keV peak was higher by a factor of 5 than during the laboratory calibration of the instrument. Petryanov et al. (1981a, b) argued that the concentrations measured should be reduced to one-fifth. They give the following values averaged over the descent time:

Cl	4.3×10^{-10} g cm^{-3}
S	$<10^{-10}$ g cm^{-3}
Hg	$<10^{-10}$ g cm^{-3}
f_{Ar}	$<2 \times 10^{-4}$.

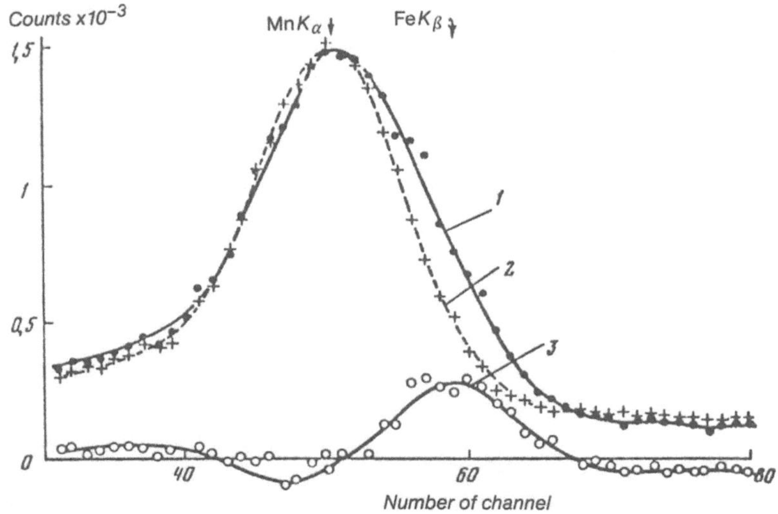

Fig. 121. The shape of the peak near 6 keV measured by V 12 (*solid line, dots*) and that obtained by laboratory calibration (*dashed line, crosses*). The difference (*circles*) is attributed to the Fe 6.9 keV line (Petryanov et al. 1981b)

They examined the 5.9 keV peak in detail and found that the Fe 6.9 keV line also contributed to the peak (Fig. 121). The intensity ratio of the chlorine and iron lines proved to be 0.84 ± 0.2. Their laboratory measurements yielded just the same ratio for the ferric chloride $FeCl_3$: 0.846 ± 0.022. $FeCl_2$ did not exceed 0.6, thus giving an additional and convincing argument for $FeCl_3$ being present in the Venusian clouds exists. The average $FeCl_3$ concentration equals $(6.4 \pm 1.2) \times 10^{-10}\,g\,cm^{-3}$. Excitation and detection of the iron line was impossible by the V 14 instrument due to the absence of the ^{109}Cd X-ray source. Preliminary analysis of the Venera 12 spectra did not reveal any features produced by this source, and it was not used by the V 14 spectrometer.

3.10 Summary of the Data on the Tropospheric and Cloud Layer Composition

Summary of the data on the tropospheric chemical composition is given in Table 20. The nitrogen mixing ratio measured by PVGC seems to be the most accurate and is given as the preferable value. Istomin et al. (1983) thoroughly analyzed the discrepancy between their results on argon and those of Hoff-

Table 20. Summary of data on the tropospheric chemical composition (in ppm)

Component	Veneras 11–14			Pioneer Venus		Spectroscopy (63 km)	Recommended values
	MS[a]	GC[b]	OS[c]	MS	GC		
N_2%	4±0.5	2.5±0.3	–	4±2	3.4	–	3.4%
Ar	100	40±10	–	70^{+50}_{-30}	67±2	–	90
Ne	7	–	–	10^{+20}_{-6}	4.3±0.7	–	7
He	–	–	–	12	–	–	12
Kr	0.04	0.7±0.3	–	0.05	<2	–	0.04
Xe	≤0.02	–	–	<0.1	–	–	≤0.02
HCl	–	–	–	<10	–	0.6	0.6
HF	–	–	–	–	–	5×10^{-3}	5×10^{-3}
CO	–	30 (42–35 km) 17 (20–0 km)	–	–	32 (42 km) 20 (22 km)	50	50 (60 km) 30 (40 km) 15 (0 km)
SO_2	–	130±35	10 (57 km) 20 (54 km) 50 (50 km)	<10 (55 km) <300 (0–20 km)	180±50	0.02 (70 km) 2 (66 km)	0.02 (70 km) H = 1.5 km 2 (66 km) 10 (57 km) 20 (54 km) 50 (50 km) 150 (<40 km)
H_2O	100±20 (0–25 km)	700±300 (58–49 km) 50^{+100}_{-30} (42–0 km)	500±200 (55–50 km) 200^{+200}_{-100} (50–40 km) 100^{+100}_{-50}	<1000	5000(42) 1350(22)	0.5–30 30 (58 km)	10 (63 km) 30 (58 km) 200 (50 km) 20 (0 km)

Table 20 (continued)

Component	Veneras 11 – 14			Pioneer Venus		Spectroscopy (63 km)	Recommended values
	MS[a]	GC[b]	OS[c]	MS	GC		
O_2	–	18 ± 4 (58 – 35 km)	(40 – 20 km) 20^{+30}_{-10} (20 – 0 km) < 30 (20 – 0 km)	< 30	44 (52 km) 16 (42 km) – (22 km)	< 0.3	< 0.3 (63 km) or $< 1.5 \times 10^{18}\ cm^{-2}$ at 60 km
S_2	–	–	$0.1^{+0.1}_{-0.08}$	–	–	–	–
SO	–	–	(20 – 0 km)	–	–	$< 2 \times 10^{-3}$	–
H_2S	–	80 ± 40 (37 – 29 km)	–	3 ± 2 (20 – 0 km)	< 2 (22 km)	< 0.1	–
H_2	–	25 ± 10 (58 – 49 km)	–	–	< 10 (22 km)	–	–
COS	–	40 ± 20 (37 – 29 km)	–	< 3 (20 – 0 km)	< 2 (22 km)	$< 10^{-2}$	–
C_2H_6	–	–	–	2	< 1 (22 km)	< 1	–
SF_6	–	0.2 (58 – 35 km)	–	–	–	–	–
CH_4	–	–	–	–	< 0.6 (22 km)	< 1	–
NH_3	–	–	–	–	–	< 0.03	–
NO	–	–	–	–	–	< 1	–
NO_2	–	–	$< 5 \times 10^{-4}$	–	–	$< 6 \times 10^{-3}$	–
Cl_2	–	–	< 0.1	< 10		< 0.7	–
Br_2	–	–	$< 10^{-4}$	–	–	$< 10^{-3}$	–

[a] MS mass spectrometry
[b] GC gas chromatography
[c] OS optical spectroscopy

Table 21. Isotopic ratios in the Venus and Earth atmospheres

Ratio	V11–14	PV	Earth
D/H	–	$(1.6 \pm 0.2) \times 10^{-2}$	1.5×10^{-4}
$^{3}He/^{4}He$	–	$<3 \times 10^{-4}$	1.4×10^{-6}
$^{13}C/^{12}C$	1.12×10^{-2}	1.12×10^{-2}	1.12×10^{-2}
$^{15}N/^{14}N$	–	$(3.66 \pm 0.75) \times 10^{-3}$	3.66×10^{-3}
$^{18}O/^{16}O$	2×10^{-3}	2×10^{-3}	2×10^{-3}
$^{22}Ne/^{20}Ne$	$(8.2 \pm 0.2) \times 10^{-2}$	$(7 \pm 2) \times 10^{-2}$	0.1 $(7.3 \pm 0.2) \times 10^{-2}$ [a]
$^{21}Ne/^{22}Ne$	<0.07	–	0.029
$^{40}Ar/^{36}Ar$	1.11 ± 0.02	1.03 ± 0.04	296
$^{38}Ar/^{36}Ar$	0.183 ± 0.005	0.18 ± 0.02	0.187

[a] Solar wind ratio

man et al. (1980), and the recommended value of 90 ppm is based first of all on VMS and PVGC data. The MS data on other noble gases are normalized to ^{36}Ar. It is quite clear now that V 11, 12 MS results on krypton were subject to contamination, and very careful calibration and measurements on board V 13, 14 confirm PVMS values. However, the problem of linearity of the MS response for such low mixing ratio still remains unsolved (Istomin et al. 1983). The helium mixing ratio is the result of a downward extrapolation from 130 km (von Zahn et al. 1980). Of all water vapor measurements we prefer the VOS results which are much less sensitive to disturbances, impurities, adsorption, etc. than in situ measurements.

The oxygen mixing ratios measured by both VGC and PVGC coincide and are consistent with other observations. However, as the coexistence of CO and O_2 in the hot Venusian atmosphere seems rather doubtful we, therefore, avoid recommending these values. The O_2 spectroscopic limit derived by Trauger and Lunine (1983) is given as a mixing ratio at 63 km and a column density at 60 km (Sze and McElroy 1975). The identification of H_2S, COS, and H_2 by V 13, 14 GC does not formally contradict the PV upper limits, because they refer to different altitudes (above 30 km and below 20 km for V 13, 14 and PV, respectively); however a steep increase of these components is rather unexpected in this altitude range.

Isotopic ratios of some elements are given in Table 21 along with those for the Earth.

The Chemical Composition of the Clouds and Hazes. The main physical properties of Venusian clouds and hazes are summarized in Table 22. Data on parameters of three modes are taken from the PV particle size spectrometric measurements with the reduction of the third mode optical thickness and mass loading by factors of 2.5 and 4, respectively, in the lower cloud layer to be in accord with the PV photometric observations. Optical thickness values are based on the PV size spectrometric and photometric results and on the Veneras nephelometric and photometric measurements. Upper haze data have

Table 22. Properties of the Venusian clouds and hazes

Region	h, km	mode[b]	r[a], μm	σ, μm	n, cm^{-3}	t, km^{-1}	m, 10^{-9} g cm^{-3}	τ/τ_{total}
Upper haze	90 – 70	1	0.13	2.16	0.1 – 100	0.01	10^{-5} – 10^{-2}	1
Upper cloud layer	70 – 57	1[c]	0.18	2.16	600	0.3	0.2	0.3
		2[c]	1.1	0.4	70	0.7	0.75	0.7
Middle cloud layer	57 – 50	1	0.15	1.9	70 – 120	0.04 – 0.15	0.015 – 0.06	0.03
		2	1.36	0.28	30 – 55	0.5 – 1	1	0.32
		3	3.7	1.26	40	2	5	0.65
Lower cloud layer	50 – 47.5	1	0.2	1.8	150	0.3	0.08	0.1
		2	1.3	0.33	70	1.5	2	0.38
		3[d]	3.6	1.29	150	2.5	8	0.52
Precloud layers	48 – 46	1	0.15	1.9	300	0.5	0.1	0.2
		2	1.25	0.5	60	0.2	1	0.8
Lower haze	48 – 30	1	0.13	1.57	45	–	–	–
		0	0.06	–	8000	0.04	10^{-2}	1

[a] r is the modal radius; σ is its variance; n is the number density; t is the extinction coefficient; m is the mass loading, M is the columnar mass
[b] Mode 1 and 3 have log normal distribution, mode 2 has Gaussian
[c] Values of n, t, m are given for the maximum at 60 km
[d] t and m are reduced by factors 2.5 and 4 to be consistent with the PV photometric measurement. The appropriate correction were made to τ/τ_{total} and M

been deduced from the V9, 10 spectroscopic and the PV polarimetric orbiter observations. The only important difference is between data on the lower haze measured by the PV size spectrometer and by the Venera photometers. The former shows this haze to consist of the mode 1 particles, while the latter indicates very small particles with $r \simeq 0.06$ μm and very large number densities of the order of 10^4 cm^{-3}. These very small particles could not be measured by the PV particle size spectrometer.

The most convincing is the identification of the mode 2 particles with sulfuric acid drops. It is these particles that produce polarization of the reflected radiation and have been identified by ground-based observations (Chap. 3.1). The properties of the mode 2 particles are stable throughout the cloud layer, so this identification proves to be valid in all parts of the clouds.

Mode 1 particles appear similar in chemical composition in the upper haze and in the upper cloud layer. This mode number density mimics that of the mode 2, which is indicative of the fact that both modes have the same origin. The upper haze refractive index value confirms this identification.

Turco et al. (1983) note that temperature profiles measured by four PV probes are consistent with the process of water vapor condensation to form detached layers at 85 – 95 km and some locations at 110 – 120 km if the water

M, g cm^{-2}	Optical thickness, (visible)								Refractive index
	PV	V9	V10	V11	V12	V13	V14	mean	
$\times 10^{-6}$	0.25	0.3	0.2	–	–	–	–	0.25	1.45 ± 0.04
$.1 \times 10^{-4}$ $).95 \times 10^{-3}$	11	7	17	7	6	3	12	8	1.48 ± 0.04
$.2 \times 10^{-4}$ $.7 \times 10^{-4}$ $\times 10^{-3}$	10	26	28	9	12	21	10	15	1.38 ± 0.02
$.2 \times 10^{-4}$ $.8 \times 10^{-4}$ $.5 \times 10^{-3}$	6	2	8	12	20	7	5	7	1.33 – 1.44
10^{-5} $.6 \times 10^{-4}$	0.3	–	–	–	–	0.2	0.6	0.3	1.48 ± 0.03
10^{-5}	0.2	<3	<3	–	–	–	–	0.2	1.7
10^{-5}	–	0.8	0.85	1.4	3	–	–	1.3	<1.8

vapor mixing ratio exceeds 1 ppm at these altitudes. With $f_{H_2O} = 5$ ppm the visible optical thicknesses of these layers are 10^{-2} and 10^{-6}, respectively. Therefore, they may contribute only a few percent to the total optical thickness of the upper haze consisting primarily of sulfuric acid. However, Mukai and Mukai (1981) find that the sulfuric acid upper haze optical thickness should be 0.15 at 3.4 μm to match the phase angle dependence of brightness (Fig. 122) measured by Martonchik and Beer (1975). The appropriate optical thickness at 940 μm is 0.75 which exceeds a typical value of ~0.2 obtained by PV polarimetry. To avoid the discrepancy, Mukai and Mukai (1981) propose absorbing particles having an imaginary refractive index of ~2 at 3.4 μm, or an absorption coefficient the aerosol material of $\sim 10^5$ cm^{-1}.

The bulk of the cloud layer consists of the mode 3 particles, therefore their identification is the most important.

The nephelometric measurements aboard four PV probes, as well as the V9 – 14 photometric and nephelometric observations, show high variability of the lower cloud layer and the stability of the middle layer. From this fact it follows that the mode 3 in these layers may be of different origin. Krasnopolsky (1985a) argues that a substantial part of the third mode in the middle cloud layer consists of sulfuric acid or products of its further conversion:

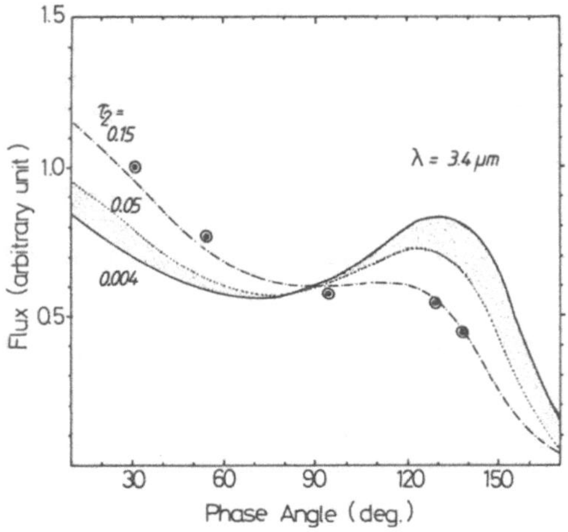

Fig. 122. Phase angle dependence of Venusian reflectivity at 3.4 μm measured by Martonchik and Beer (1975) (*circles*) and calculated for the optical thickness of the upper haze equal to 0.004, 0.05, and 0.15 at 3.4 μm or 0.02, 0.25, and 0.75, respectively at 0.94 μm (Mukai and Mukai 1981)

1. The formation of sulfuric acid entails the upper cloud layer, the atmosphere above being much dryer than the lower one. In this case the upward flux of water vapor is compensated by the downward flux of water bounded in the sulfuric acid drops (two H_2O molecules per H_2SO_4 molecule in the 85% solution):

$$\phi_{H_2O} + 2\,\phi_{H_2SO_4 \cdot H_2O} = 0\,;$$

$$\phi_{H_2O} = -K[CO_2]\,\frac{df_{H_2O}}{dz} \simeq (3 \pm 1.5) \times 10^9\,K\,(cm^{-2}\,s^{-1})\,.$$

Calculating ϕ_{H_2O}, we assume $f_{H_2O} = 10$ ppm at 63 km and 200 ppm below 55 km with a linear decrease between 55 and 63 km. The error of ϕ_{H_2O} is due to the adopted interpolation. Fluxes of the second and the third modes in the middle cloud layer calculated by the integration of their mass distribution multiplied by the Stokes-Davis velocity

$$V = \frac{2}{9}\,\frac{\varrho g}{\eta}\,r^2\left(1 + \frac{l}{r}\right) \times (1.257 + 0.4 \cdot e^{-1.1 r/l})$$

equal 3×10^{11} and $3 \times 10^{13}\,cm^{-2}\,s^{-1}$, in terms of numbers of $H_2SO_4 \cdot H_2O$ molecules for the second and the third modes, respectively (Fig. 104). Here ϱ is the density, g is the gravitational acceleration, η is the CO_2 viscosity, and l is the mean free path. With the mode 3 neglected one obtains quite an improbable value of $K = 200\,cm^2\,s^{-1}$, but if mode 3 is assumed to exist and consists of sulfuric acid droplets K equals $2 \times 10^4\,cm^2\,s^{-1}$ which is in accord with other estimates of K (see Chap. 3.13).

2. The V14 X-ray fluorescent spectroscopy results in $(1.1 \pm 0.13) \times 10^{-3}\,g\,cm^{-2}$ of sulfur and $(0.16 \pm 0.04) \times 10^{-3}\,g\,cm^{-2}$ of chlorine integrated through the clouds. These values may be regarded as more typical and representative than those of V12. Optical properties of clouds at the PV and V14 locations are quite similar (Table 22). The comparison of modal columnar masses with the measured sulfur abundance shows that sulfuric acid

and/or its products, being the main sulfur-bearing cloud material, should constitute the mode 2 and about one-half of the mode 3 columnar mass.

3. The global average CO_2 and SO_2 photolytic rates are a quarter of solar quantum intensity in ranges of $\lambda < 202$ nm and 202 – 217 nm and equal to 0.55×10^{13} and $1.8 \times 10^{13}\ \mathrm{cm}^{-2}\,\mathrm{s}^{-1}$, respectively. The values at low latitudes where PV and V measurements were carried out are twice as high as the global average values because the meridional transport time is larger than that of the mode 3 sedimentation. Quantum yield of the H_2SO_4 formation by SO_2 photolysis is defined by two competing processes (see Chap. 4.9):

$$SO_2 + O + M \rightarrow SO_3 + M\,; \qquad k_1 = 8 \times 10^{-32} e^{-1000/T}\ \mathrm{cm}^6\,\mathrm{s}^{-1}$$

$$SO + O + M \rightarrow SO_2 + M\,; \qquad k_2 = 6 \times 10^{-31}\ \mathrm{cm}^6\,\mathrm{s}^{-1}$$

and equal to

$$\gamma = \left(1 + \frac{k_2[SO]}{k_1[SO_2]}\right)^{-1} = (1 + 400[SO]/[SO_2])^{-1}.$$

The SO to SO_2 ratio at the altitude of the maximum SO_2 photolytic rate is model dependent, but only slightly, so $\gamma \simeq 0.3$ and the sulfuric acid production rate at low latitudes is $\sim 10^{13}\ \mathrm{cm}^{-2}\,\mathrm{s}^{-1}$, i.e., one-third of the mode 3 one.

So sulfuric acid and products of its further conversion should constitute a substantial part (from about one-half to unity) of the mode 3 material. For a more humid atmosphere at the middle cloud layer, this requirement is even more stringent. From this consideration it follows that K should be $(1-2) \times 10^4\ \mathrm{cm}^2\,\mathrm{s}^{-1}$.

Ferric Chloride in Clouds. Krasnopolsky (1985a) shows that some important properties of the Venusian clouds can be explained starting with an assumption that ferric chloride condensation occurs at the lower cloud boundary at 47.5 km. The Fe_2Cl_6 saturated vapor pressure equals 10^{-6} and 10^{-5} mm Hg at 349.9 K and 365.6 K, respectively; therefore its density can be written as

$$\varrho_s = 2.16 \times 10^{-10} e^{(T-368.5)/7}\ \mathrm{g\,cm}^{-3}$$

where $T_0 = 368.5$ K is the temperature at 47.5 km (Seiff et al. 1980). Hence, $\varrho_{os} = 2.16 \times 10^{-10}\ \mathrm{g\,cm}^{-3}$ is the density of Fe_2Cl_6 vapor at 47.5 km or $f_{Fe_2Cl_6} =$ 15 ppbv and 100 ppbm (by volume and mass, respectively) below the cloud layer. Then it is easy to calculate the condensate density profile by the relation

$$M = \varrho_{os} \frac{\gamma/\delta + 1/H}{V/K + \gamma/\delta + 1/H} \left[e^{-(V/K + 1/H)z} - e^{\gamma z/\delta}\right] - \frac{\phi/K}{V/K + 1/H}\left[1 - e^{-(V/K + 1/H)z}\right]$$

which is the result of integration of the equation

$$\phi = -K\left[\frac{d(M + \varrho_s)}{dz} + \frac{M + \varrho_s}{H}\right] - MV.$$

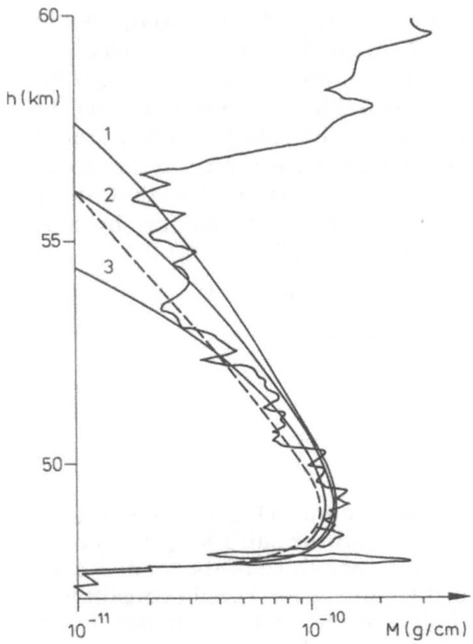

Fig. 123. Altitude profiles of the $FeCl_3$ aerosol mass loading in the clouds of Venus calculated for $K = 2 \times 10^4\ cm^2\,s^{-1}$ with $\phi = (0.5, 1,$ and $2) \times 10^{-12}\ g\,cm^{-2}\,s^{-1}$ (*curves 1, 2, 3*) and for $K = 10^4\ cm^2\,s^{-1}$ and $\phi = 0$ (*dashed curve*). The measured mass loading of the mode 1 particles corrected by a factor of 1.5 due to higher density of $FeCl_3$ than $2\ g\,cm^{-3}$ adopted by Knollenberg and Hunten (1980, Fig. 105c) is also shown (Krasnopolsky 1985a)

Here, $\gamma = -9\ K\,km^{-1}$ is the temperature gradient, $\delta = 7\ K$ is the parameter in the relation for ϱ_s, H is the scale height, V is the sedimentation velocity, and ϕ is the flux of a condensate.

Independently of ϕ, K and V values, the calculated condensate mass loading coincides fairly well with that of the mode 1 in the lower cloud layer (Fig. 123), especially with a correction for a larger density of ferric chloride ($3\ g\,cm^{-3}$ instead of $2\ g\,cm^{-3}$ adopted by Knollenberg and Hunten for all modes). The condensate mass loading profile in the middle cloud layer depends on ϕ, K, and V. To calculate it we employ the value of the effective radius of the mode 1 corresponding to a maximum contribution to mass loading, i.e., a maximum of $r^3 n(r)$. Then $r_{eff} = 0.45\ \mu m$ and $V = 0.028\ cm\,s^{-1}$ at 54 km. The best agreement of the calculated and measured profiles is for $K = 2 \times 10^4\ cm^2\,s^{-1}$ and $\phi = 10^{-12}\ g\,cm^{-2}\,s$. Almost the same coincidence for $K = 3 \times 10^4\ cm^2\,s^{-1}$ and $\phi = 2.5 \times 10^{-12}\ g\,cm^{-2}\,s$ exists, but K should be in the range of $(1-2) \times 10^4\ cm^2\,s^{-1}$, and the amount of sulfuric acid in the cloud layer should exceed the measured one by the PV size spectrometer in the case of $K > 2 \times 10^4\ cm^2\,s^{-1}$ (see above). The coincidence is not so good for $K = 10^4\ cm^2\,s^{-1}$ even for $\phi = 0$. The obtained upward flux of $FeCl_3$ may be compared with the downward flux of the mode 2 particles in the upper cloud layer. The latter is easy to calculate, and equals $10^{-10}\ g\,cm^{-2}\,s$ in the maximum at 60 km. Therefore, the $FeCl_3$ to sulfuric acid flux ratio is 1%, just the value one needs to explain the UV albedo of Venus! So the upward flux of $FeCl_3$ mode 1 particles is absorbed by sulfuric acid drops via collisions or via the condensation with $FeCl_3$ particles as cores.

The calculated columnar mass loading of $FeCl_3$ is equal to 6×10^{-5} g cm^{-2} or the chlorine abundance in the clouds is four times as small as measured by V14. According to the V11, 12 photometry (see Chap. 3.5), $1-\omega \simeq 5 \times 10^{-4}$ in the middle and lower cloud layers at 490 nm. With the mode 1 partition being ~3% here this results in $k \lesssim 500$ cm^{-1} for solid $FeCl_3$ at 490 nm. This restriction is not severe (we have no data on k of the pure $FeCl_3$).

It is evident from the mode 1 profile that these particles have a different nature in the upper cloud layer as compared to those in the middle and the lower ones. Perhaps they are similar to the upper haze particles consisting of sulfuric acid (see Chap. 3.7) or some part of them consists of elemental sulfur (see Chap. 3.9).

Zasova (personal communication) evaluates the lifetime of the $FeCl_3$ solution in concentrated sulfuric acid to be about 1 week at room temperature. This lifetime depends strongly on temperature, thus, it is much shorter at elevated temperatures with the evaporation of HCl and colorless ferric sulfate formation. The sedimentation velocity of the mode 2 particles is 0.1 cm s^{-1} at 60–57 km and the sedimentation time is about 1 mo. The mean temperature is ~0°C here, therefore the lifetime and the sedimentation time are nearly the same. Therefore, a sharp decrease of true UV absorption should occur here in accord with the V13, 14 observations (Chap. 3.8).

The $FeCl_3$ und H_2SO_4 interaction can be neglected in the gaseous phase because of its very low reaction rate in solution. Solid particles of $Fe_2(SO_4)_3$ should appear after the evaporation of sulfuric acid at the lower boundary of clouds. Profiles and main properties of the lower haze were discussed in Chap. 3.5. The aerosol mass loading is 1.5×10^{-11} g cm^{-3} at 35 km with a mean particle radius of 0.06 µm and a density of aerosol material of 2 g cm^{-3}. This value of the mass loading coincides with that of the first mode in the upper part of the middle cloud layer, thus indicating the equality of the eddy diffusion coefficients at 55–58 and 35 km. However, the formation of such small particles is still unclear.

Thermochemical calculations by Lewis (1970) gave the $FeCl_2$ mixing ratio as 10^{-10}, i.e., only by two orders of magnitude smaller than the value necessary to provide the condensation of $FeCl_3$ at 47.5 km. The difference is not considered large due to the uncertainty in input data of such calculations. The transformation of $FeCl_2$ to $FeCl_3$ with the subsequent conversion of ferric sulfate to chlorides needs special investigation.

So the presence of $FeCl_3$ in the clouds (1) agrees with the UV albedo of Venus; (2) is confirmed by the direct X-ray fluorescent measurements; (3) agrees with the altitude profiles of the mode 1 particles in the lower and middle cloud layers; (4) provides the necessary concentration of the $FeCl_3$

Table 23. The most plausible chemical composition of the cloud aerosol particles

Region	*h*, km	Mode 1 ($r \simeq 0.2$ μm)	Mode 2 ($r \simeq 1.3$ μm)	Mode 3 ($r \simeq 3.7$ μm)
Upper haze	70 – 90	H_2SO_4(~85%)	–	–
Upper layer	70 – 57	H_2SO_4(~85%), S_n	H_2SO_4(~85%) + $FeCl_3$(~1%) + S_n	–
Middle layer	57 – 50	$FeCl_3$	H_2SO_4(~85%) + $Fe_2(SO_4)_3$(~1%) + S_n	H_2SO_4(~85%)
Lower layer	50 – 48	$FeCl_3$	H_2SO_4(~95%) + $Fe_2(SO_4)_3$(~1%) + S_n	H_2SO_4(~95%)
Precloud layer	48 – 46	$FeCl_3$	H_2SO_4(~95%) + $Fe_2(SO_4)_3$(~1%)	–
Lower haze	48 – 30	$Fe_2(SO_4)_3$ ($r \simeq 0.06 - 0.1$ μm)	–	–

solution in sulfuric acid drops in the upper cloud layer; (5) explains the altitude profile of the near UV absorption coefficient; (6) is in accord with the dynamical model of dark features (Fig. 113); and (7) explains the strong correlation of contrasts at 270 and 365 nm.

The most plausible chemical composition of cloud aerosol on Venus is given in Table 23.

3.11 The Upper Atmosphere

Mass Spectrometric Measurements. A number of direct measurements in the upper atmosphere of Venus were carried out by the PV orbiter. Most of the data obtained refer to a high solar activity ($F_{10.7\,cm} \simeq 200$). The neutral atmospheric composition and its variations were studied by mass spectrometry (Niemann et al. 1980). Number densities of CO_2, CO, O, N_2, and He were measured. These number densities are almost constant on the day and night side hemispheres with steep gradients near both terminators. The data of observations were employed to simulate the neutral upper atmosphere at subsolar and antisolar points (Fig. 124) and variations of the measured components (Fig. 125). Ratio of day to night side densities ranges from one to two orders of magnitude at 150 and 200 km, respectively. Such a high density gradient appears mainly near the terminators with the expansion of day side gas to an almost perfect vacuum on the night side. The gas flow should have a velocity near the speed of sound, i.e., ~200 m s^{-1} at $T \simeq 200$ K.

Mass spectrometric measurements down to 130 km were carried out aboard the PV bus as well (von Zahn et al. 1980). Its descent was at $z_0 = 60°$ on the morning side. Number densities of CO_2, He, and CO + N_2 were measured with the separation of CO and N_2 at 140 – 153 km, an upper atmospheric model based on these observations is given in Table 24.

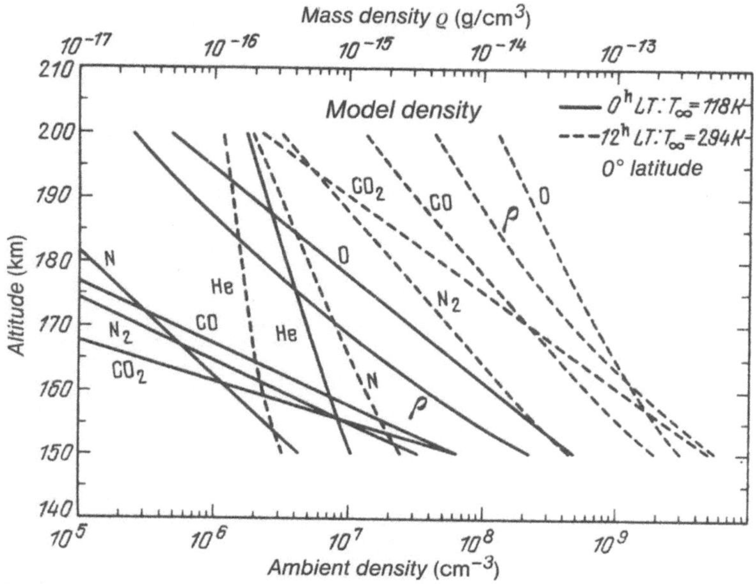

Fig. 124. Neutral composition of the subsolar and antisolar thermosphere of Venus (Niemann et al. 1980)

PV Orbiter Atmospheric Drag Measurements (Keating et al. 1980). The change of the orbital period per revolution by atmospheric drag is proportional to $\varrho H^{1/2}$, where ϱ is the atmospheric density and H is the atmospheric scale height, both values refer to the altitude of 0.5 H above the periapsis. With a number of the measured periods of revolutions having different periapsises at the same geophysical conditions, one can fit these data by the model atmosphere including the temperature and CO_2 and O number density profiles (because the height where $[CO_2] = [O]$ is inside the range of the periapsis variations, 140 – 190 km). A large set of the drag observations has been used to construct the model atmosphere, which depends only on the exospheric temperature, T_∞, and is specified by several parameters (CO_2, O, and $CO + N_2$ number densities at 115 km, the height of turbopause, parameters to set the temperature profile). Two of these parameters ($[CO_2]$ and [O] at 115 km) were deduced from the drag measurements and extrapolated to 115 km while the rest were taken from the other PV experiments.

The O and CO_2 number density profiles measured and their diurnal variations are very similar to those in Fig. 124, though the values of number densities are higher by a factor of ~ 1.5 than given in Fig. 124.

With the known model parameters every drag measurement results in the exospheric temperature which will be discussed in Chap. 3.13.

Night airglow measurements in the visible range were carried out on board the V9, 10 orbiters (Krasnopolsky et al. 1976, Krasnopolsky and Tomashova

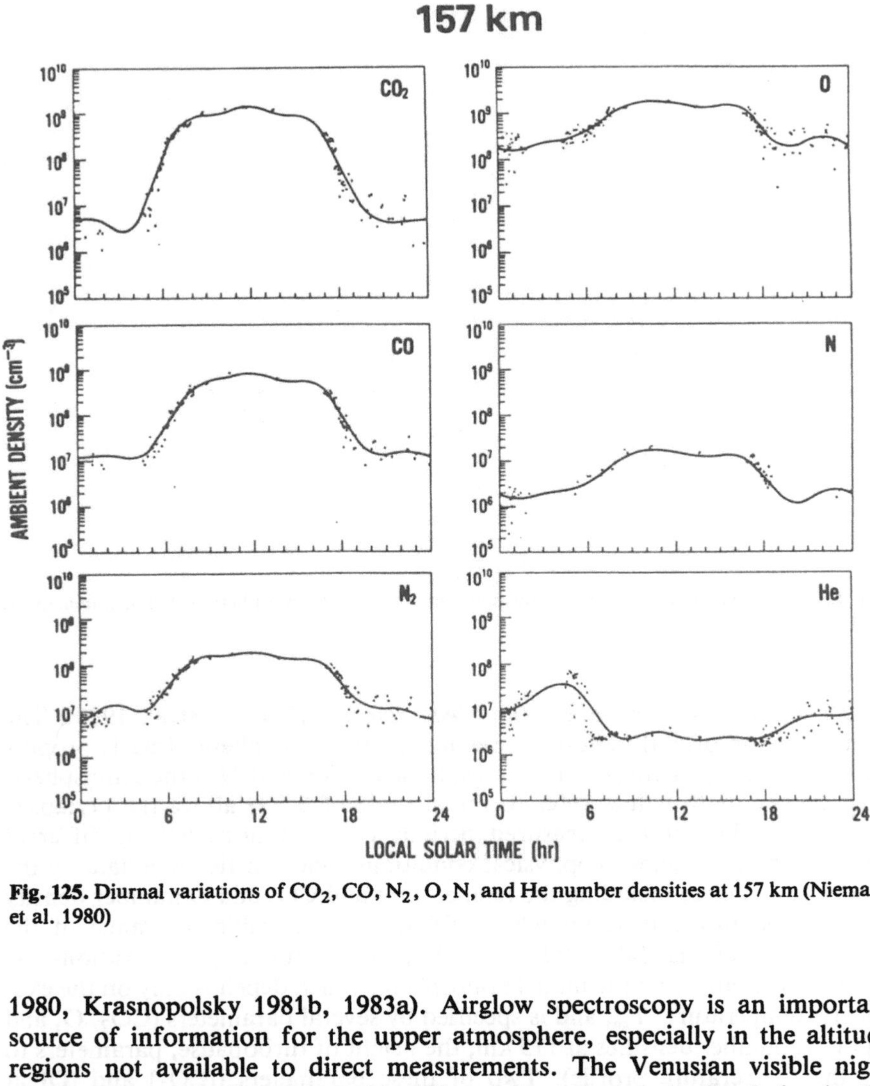

Fig. 125. Diurnal variations of CO_2, CO, N_2, O, N, and He number densities at 157 km (Niemann et al. 1980)

1980, Krasnopolsky 1981b, 1983a). Airglow spectroscopy is an important source of information for the upper atmosphere, especially in the altitude regions not available to direct measurements. The Venusian visible night airglow appears at first sight quite different from that on the Earth despite the fact that they both result from the same excitation process; three body recombination of oxygen atoms

$$O+O+M \rightarrow O_2+M .$$

The airglow spectrum (Fig. 126) represents the zeroth series of the four band systems of O_2 with mean zenith intensities of 2.7 kR for Herzberg II, 140 R for Herzberg I, 200 R for Chamberlain and 100 R for ($c^1\Sigma_u^- \rightarrow a^1\Delta_g$) band systems. Successive identification of the spectrum became possible after

Table 24. Venusian morning side model atmosphere (von Zahn et al. 1980)

h, km	CO_2, cm^{-3}	N_2, cm^{-3}	He, cm^{-3}	CO, cm^{-1}	O, cm^{-3}	ρ, g cm^{-3}	M	T, K	p, mbar	D, cm^2 s^{-1}	K, cm^2 s^{-1}
220	4.30E(4)*	6.78E(5)	1.35E(6)	1.87E(6)	3.30E(7)	1.01E(−15)	16.4	275	1.40E(−9)	2.2E(11)	5.0E(8)
210	2.43E(5)	1.86E(6)	1.56E(6)	5.12E(6)	5.89E(7)	1.92E(−15)	17.1	275	2.57E(−9)	1.2E(11)	5.0E(8)
200	1.23E(6)	5.14E(6)	1.81E(6)	1.41E(7)	1.05E(8)	3.97E(−15)	17.9	275	4.84E(−9)	6.4E(10)	5.0E(8)
190	6.32E(6)	1.42E(7)	2.10E(6)	3.87E(7)	1.89E(8)	7.95E(−15)	19.1	275	9.49E(−9)	3.2E(10)	5.0E(8)
180	3.19E(7)	3.92E(7)	2.45E(6)	1.06E(8)	3.39E(8)	1.81E(−14)	21.0	275	1.97E(−8)	1.6E(10)	5.0E(8)
170	1.58E(8)	1.09E(8)	2.90E(6)	2.93E(8)	6.12E(8)	4.66E(−14)	23.8	274	4.45E(−8)	6.8E(9)	4.3E(8)
160	7.99E(8)	3.10E(8)	3.52E(6)	8.28E(8)	1.14E(9)	1.42E(−13)	27.7	270	1.15E(−7)	2.5E(9)	2.6E(8)
150	4.92E(9)	1.07E(9)	4.83E(6)	2.78E(9)	2.51E(9)	6.05E(−13)	32.3	238	3.71E(−7)	6.1E(8)	1.4E(8)
140	4.12E(10)	4.92E(9)	8.14E(6)	1.21E(10)	6.93E(9)	3.98E(−12)	36.8	206	1.85E(−6)	9.2E(7)	5.7E(7)
130	4.10E(11)	3.02E(10)	2.02E(7)	6.58E(10)	2.39E(10)	3.50E(−11)	39.8	190	1.39E(−5)	1.0E(7)	2.0E(7)
120	4.52E(12)	2.52E(11)	9.28E(7)			3.67E(−10)		183	1.35E(−4)	9.9E(5)	6.2E(6)
110	5.26E(13)	2.60E(12)	7.30E(8)			4.17E(−9)		179	1.48E(−3)	8.7E(4)	1.9E(6)
100	6.22E(14)	2.93E(13)	7.50E(9)			4.88E(−8)		178	1.71E(−2)	7.4E(3)	5.5E(5)

* 4.30E(4) means 4.30×10^4

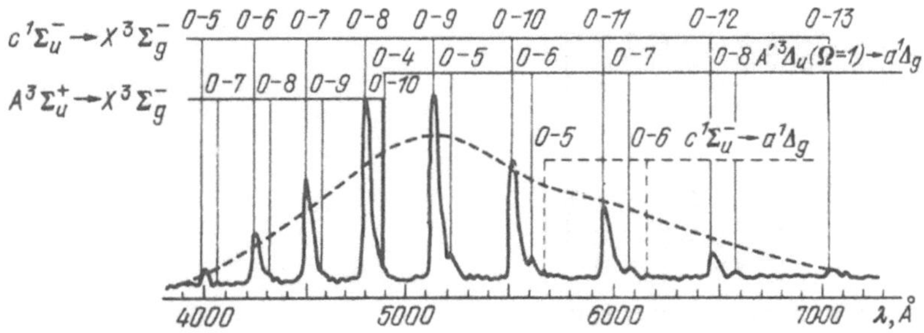

Fig. 126. Visible night airglow spectrum. Positions of four oxygen band systems are shown (Krasnopolsky and Tomashova 1980). *Dashed curve* is the instrument sensitivity

laboratory simulations by Lawrence et al. (1977), Slanger (1978), Slanger and Black (1978), and Kenner et al. (1979). More thorough studies of the Earth airglow spectra by Slanger (1979) and Slanger and Huestis (1981) indicate also the presence of Chamberlain and Herzberg II band systems of ~100 *R*.

The green oxygen line at 558 nm is absent in the Venusian night airglow with an upper limit of intensity of 5 *R*. A scheme of reactions of energy transfer, exchange, and quenching appears to be rather complicated in the oxygen airglow of terrestrial planets, and not all processes and rate coefficients in it have been investigated in the laboratory. To define these reactions and their coefficients the data on planetary airglow measurements and chemical composition of the upper atmosphere have been employed. This problem has been studied by Krasnopolsky (1981b, 1983a) and Wraight (1982). However, Krasnopolsky did not use the attractive idea of the $O_2(^5\pi)$ state with a large radius of formation, while Wraight did not consider in detail his scheme as applied to the Venusian and Martian airglow, thus, the synthesis of both approaches seems promising. Yet in every case it is impossible to deduce mean data on atmospheric composition from these airglow measurements until rate coefficients are measured in the laboratory. Therefore, only airglow variations (Krasnopolsky and Tomashova 1980) may be applied for atmospheric studies now.

The airglow forms a layer 15 km thick at 100 km with variations of a maximum in a range of ±10 km. This defines the position of the [O] maximum in the night side atmosphere. Diurnal variations (Fig. 127) have a predominant symmetric component with a maximum at 5° after midnight, thus indicating the delivery of atomic oxygen by its transport across the terminators and downward flux at an antisolar region, in accord with calculations by Izakov and Morozov (1975) and Dickinson and Ridley (1977). The same conclusion stems from latitude variations (Fig. 128). A ratio of maximum to minimum intensities at low latitudes is ~1.5, i.e., much smaller than the calculated one (calculated atomic oxygen number densities vary by a factor of 7 at 100 km, see Dickinson and Ridley 1977).

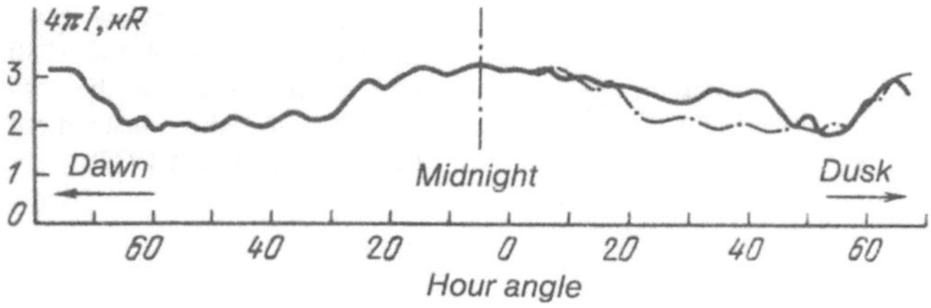

Fig. 127. Diurnal variations of the visible night airglow, *dashed line* is the symmetric continuation of the left part of the curve (Krasnopolsky and Tomashova 1980)

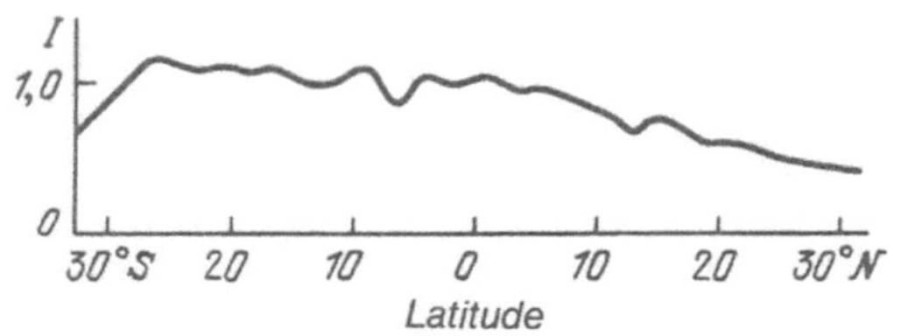

Fig. 128. Mean night airglow intensity vs latitude (Krasnopolsky and Tomashova 1980)

An asymmetric component of diurnal variations is due to the superrotation of the atmosphere. This component being small indicates the location of the upper boundary of this circulation at or below 100 km. Probably the asymmetric component is much larger in the lowest part of the airglow layer. One can see the wave-like structure of the curve in Fig. 124. It is interesting that the ratio of the wavelength to the circumference is close to the ratio of the period of the circulation (4 days) to the duration of solar day on Venus (117 days).

Oxygen airglow at 1.27 μm has been measured by Connes et al. (1979) by the same technique as on Mars (Chap. 1.3). The dependence of line intensities on their rotational excitation defines a rotational temperature as 185 ± 15 K on the night side. Day side measurements present difficulties due to an intense background radiation, therefore the day side rotational temperature was measured with large errors. The airglow intensity was determined by a comparison with the terrestrial airglow being present in the observed spectrum as small peaks in centers of strong Doppler-shifted telluric absorption lines and by a comparison with the background continuum. Both approaches result in 3.3 and 4.2 *MR* at the night and day sides, respectively. With the Venusian albedo being 0.9 at 1.27 μm, the airglow intensities are less by a factor of 2.8 and equal 1.2 and 1.5 *MR*. Actually half of the airglow photons falls on the surface, whose brightness is $1/2 \times 4\pi I_0 \times a/\pi = 2aI_0$, and the total brightness observed outside the atmosphere is $(1+2a)I_0$ where a is the surface albedo. The approximate equality of the night and day airglows indicates a small contribution of ozone dissociation to the total excitation rate. A large airglow intensity comparable with the atomic oxygen production rate by CO_2 photolysis indicates the $O_2(^1\Delta_g)$ formation as an important step in CO_2 re-

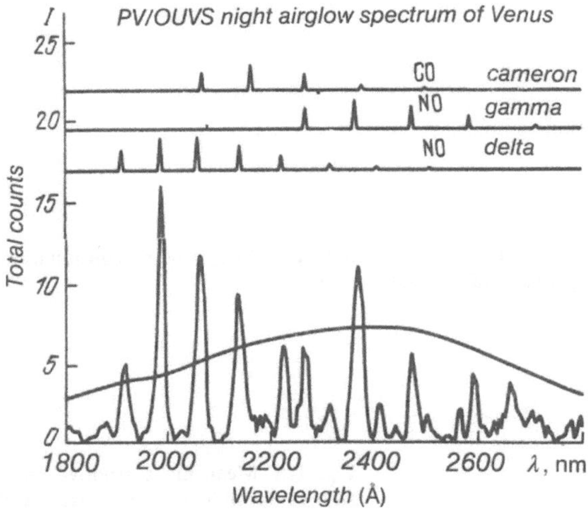

Fig. 129. Night airglow spectrum in the range of 180–280 nm. *Smooth curve* is the sensitivity of the instrument (Stewart and Barth 1979)

combination. The equality of the day and night airglow intensities shows the effective transport of atomic oxygen formed at the day side to the night side hemisphere and almost equal atomic oxygen number densities on both hemispheres at the altitude of the airglow layer.

Nitric oxide nightglow (Fig. 129) was observed by UV spectrometers aboard the PV and IUE orbiters (Stewart et al. 1979, 1980, Gerard et al. 1981, Feldman et al. 1979). The airglow is presented by the zeroth series of δ and γ band systems excited by $N + O \rightarrow NO^*$ recombination. The strongest bands of γ and δ systems are (0,1) 237 nm and (0,1) 198 nm, respectively. They were employed to study a global airglow distribution by means of the photometric mode of the PV spectrometer. Orbiter spin rotation provides monochromatic images of the planet. The planetary reflectivity and stray light intensity are very low at 198 nm due to CO_2 absorption, and the images at this wavelength are the best to study the airglow distribution. The night side mean intensity equals 0.8 *kR* at 198 nm or 2.8 and 1.6 *kR* for total δ and γ systems, respectively. An intensity maximum is located 30° after the antisolar point and equal to 5 *kR* at 198 nm. The center of the airglow layer is at 115 ± 2 km measured at the planetary limb.

Stewart et al. (1980) considered the following reactions (Table 25) to define atomic nitrogen and oxygen number densities at 80–140 km. They assumed $K = An^{-1/2} \gg D$ here (see Chap. 1.10, Chap. 3.13). The source of atomic oxygen being its downward flux, ϕ_0, its number density equals $2\phi_0 H/K$ in the region without oxygen sinks. With the altitude profile of a component formed by its downward flux and its sink in the region under consideration, its maximum number density is observed at the altitude where mixing time is equal to chemical lifetime. This rule being applied to atomic nitrogen results in $H^2/K = (k_1[O])^{-1}$. Then $M = M_p = A^2/2k_1\phi_0 H^3$,

Table 25. Reactions of N and O at 80 – 140 km (Stewart et al. 1980)

No.	Reactions	Rate coefficients ($cm^3\,s^{-1}$ or $cm^6\,s^{-1}$)
1	$O+N \rightarrow NO$	2.5×10^{-17}
2	$N+NO \rightarrow N_2+O$	2.1×10^{-11}
3	$O+N+CO_2 \rightarrow NO+CO_2$	$1.1 \times 10^{-32}(300/T)^{1/2}$
4	$O+O+CO_2 \rightarrow O_2+CO_2$	$4.7 \times 10^{-33}(300/T)^2$
5	$O+CO+CO_2 \rightarrow CO_2+CO_2$	2×10^{-37}
6	$O+O_2+CO_2 \rightarrow O_3+CO_2$	$1.4 \times 10^{-33}(300/T)^{2.5}$
7	$O+O_3 \rightarrow O_2+O_2$	$1.9 \times 10^{-11}\exp(-2300/T)$

$[O] = [O]_p = (2\phi_0/k_1 H)^{1/2}$ can be derived at the height of the maximum of [N], h_p. It is easy to calculate that both sinks of N, reactions 1 and 3, are equal with $M = 2 \times 10^{15}\,cm^{-3}$, i.e., at 95 – 100 km, while the former is by one order of magnitude more intense than the latter at the airglow layer maximum where $M \simeq 10^{14}\,cm^{-3}$. With accumulation of NO neglected, the column rate of reactions 1 and 2 should be equal. Therefore, the quantum yield of the NO airglow is ~0.5 relative to the atomic nitrogen flux. The lower K, the larger h_p and closer to 0.5 the quantum yield (0.45 for $A = 1.6 \times 10^{13}$ and 0.36 for $A = 5 \times 10^{13}$). The measured altitude of the airglow layer coincides with the calculated one with $A = 0.8 \times 10^{13}$. The mean intensity of the NO airglow of 4.5 *kR* corresponds to an atomic nitrogen flux of $10^{10}\,cm^{-2}\,s^{-1}$. According to the mass spectrometric measurements (Fig. 124) $[O]/[N] \simeq 10^2$ in the upper atmosphere, thus $\phi_0 = 10^{12}\,cm^{-2}\,s^{-1}$. We do not present a figure with calculated N and O number densities and an NO volume emission rate due to some discrepancies. The maximum atomic nitrogen number density is 7×10^8 cm^{-3} at 114 km and that of atomic oxygen is $2.6 \times 10^{11}\,cm^{-3}$ at 105 km.

This calculation results in the following interesting data. The NO airglow has a maximum at 02.00 h LT, while the mass spectrometric measurements indicate a minimum of [N] and [O] at 157 km at this time (Fig. 125). However, the temperature is near the minimum at this time as well, and the extrapolation of the mass spectrometric data from 157 km to 140 km reveals a maximum of [N] and [O] at 140 km at 22.00 h LT with values at 20.00 h and 02.00 h lower by a factor of 2. Then the airglow maximum should be observed before midnight rather that after it. Further extrapolation to ~110 km makes this problem even more complicated.

Stewart et al. note that a ratio of the O_2 Herzberg II band intensity (they assume it equal to 5 *kR*) to ϕ_0 results in a quantum yield of ~10^{-2}, which is by one order of magnitude larger than the value (10^{-3}) recommended by Lawrence. However, with a more correct value of 2.7 *kR* one obtains 2.7×10^{-3}, and a two- to three fold difference does not seem very large.

Comparison of ϕ_0 and the O_2 1.27 μm airglow intensity assumed to be 600 *kR* (though the published value is 1.2 *MR*, see above) gives a quantum

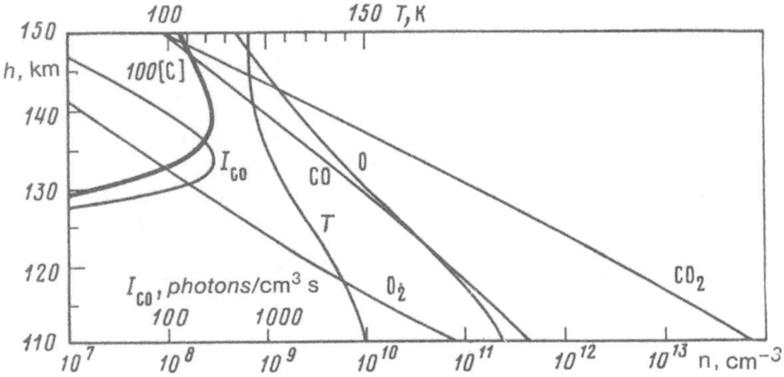

Fig. 130. Altitude profiles of CO_2, CO, O, and O_2 number densities on the night side of Venus in the range of 110 – 150 km used for calculations of atomic carbon number densities and the volume emission rate of Cameron bands of CO shown in the figure (Krasnopolsky 1982a)

yield of 1.27 μm airglow to be about unity (two in the case of 1.2 *MR*). From this, Stewart et al. concluded that all oxygen molecules formed by reactions 4 and 7 (Table 25) are in $^1\Delta_g$ state. A value of 1.2 *MR* compared with $\phi_0 = 10^{12}$ $cm^{-2} s^{-1}$ indicates, in our opinion, the inadequacy of the reaction scheme in Table 25 to describe O_2 formation, though it can be applied to the NO emission rate and N and O number density calculations. With the quantum yield of $O_2(^1\Delta_g)$ being 2, a two-step (as a minimum) process of O_2 formation should occur and every step should have a quantum yield of about unity. This problem will be discussed in detail in Chap. 4.12.

Carbon monoxide nightglow (Fig. 129) measured by the PV spectrometer (Stewart and Barth 1979) consists of several Cameron bands ($a^3\Pi - X^1\Sigma$). These bands are narrow, different from the vibrationally well-developed features of about 5 nm width in the Martian dayglow, and represent the $v' = 0$ progression only. The strongest band at 216 nm is ~100 *R*, i.e., the total band system intensity is ~250 *R*. Krasnopolsky (1982a) suggests the following reaction for its excitation:

$$C + O_2 \rightarrow CO(a^3\Pi, v' = 0) + O$$

which is slightly endothermic ($\Delta E = -0.019 \pm 0.004$ eV). He evaluates an atomic carbon downward flux as 2×10^9 $cm^{-2} s^{-1}$ on the night side, and a comparison of the calculated rate of the above process with the airglow intensity results in the yield of $a^3\Pi$ state in this reaction to be $0.7 e^{-215/T}$ with a total reaction rate coefficient of 2.6×10^{-11} $cm^3 s^{-1}$. The proposed reaction explains well the excitation of $v' = 0$ bands only. Calculations indicate the airglow layer to be 10 km thick at 135 km (Fig. 130). However, night side O_2 number densities which determine the layer altitude are very uncertain.

Night Side Emission in the Near Infrared Region. Observations of the Venus night side in the range of 1.5 – 4.0 μm with resolution of $\lambda/\Delta\lambda = 100$

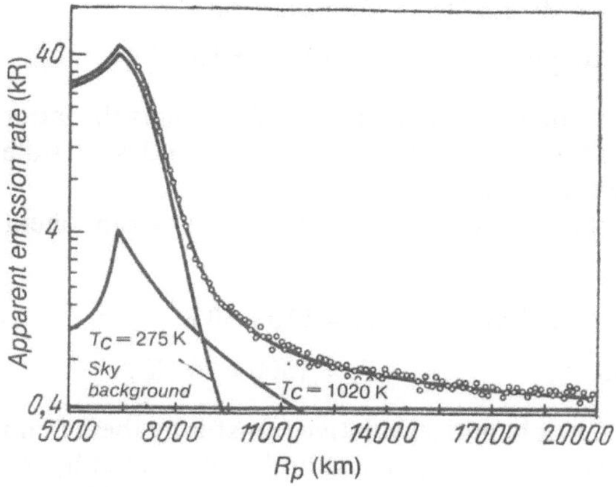

Fig. 131. Hydrogen 122 nm airglow measured by Mariner 5 is the sum of the background and two distributions with $T_c = 275$ K and $T_c = 1020$ K (Anderson 1976). R_p planetocentric distance

(Allen and Crawford 1984) reveal two emissions. The first is centered at 1.74 μm with the intensity of 250 erg s^{-1} cm^{-2} μm^{-1} sr^{-1} and is weaker by an order of magnitude at 1.69 and 1.77 μm. Its half-width is 0.03 μm and the total intensity equals 80 *MR*. The second emission is much broader and covers the range of 2.18 – 2.50 μm with the intensity of 100 erg s^{-1} cm^{-2} μm^{-1} sr^{-1} at 2.3 μm. The CO (2.0) absorption band at 2.35 μm results in attenuation by a factor of ~3 at the peaks of *R* and *P*-branches near the center of the emission. Its total intensity equals 230 *MR*. Night side images measured at 2.41 μm demonstrate a cloudlike structure. Observations on successive days reveal a retrograde rotation with the period of 5.4 ± 0.1 days.

Though the nature of the emissions and their variations have not been discussed in detail, obviously they represent thermal radiation of the lower atmosphere. The wavelengths of the emissions coincide with the highest transmission of the lower atmosphere. Being transformed to brightness temperatures, the 1.74 μm and 2.41 μm intensities give 480 K and 350 K, which correspond to the altitudes of 32 and 50 km, respectively. The rotational period of 5.4 days exhibits the zonal wind of 70 ms^{-1} typical of the middle and lower cloud layer (Chap. 3.13). Therefore the image features are of the same origin as those observed by radio occultations (Cimino 1982, Chap. 3.7) in the lower cloud layer.

Hydrogen 122 nm Airglow. The first measurements of this airglow were carried out by V4 (Kurt et al. 1968) and Mariner 5 (Barth 1968). The airglow altitude profile measured by M5 is decribed by two scale heights which differ by a factor of two with a crossover at ~3000 km altitude (Fig. 131). The hypotheses of a mixture of atomic hydrogen with molecular hydrogen or deuterium proved inadequate to explain the measurements. More reasonable is the assumption of "cold", i.e., having the atmospheric temperature, and "hot" hydrogen atoms, the latter having a large excess of energy. Using this approach Anderson (1976) explained the day side limb measurements with the following temperature and number densities at the critical level (Fig. 131):

cold component: $n_c = (2 \pm 1) \times 10^5\,\mathrm{cm}^{-3}$, $T_c = 275 \pm 50\,\mathrm{K}$

hot component: $n_c = 1.3 \times 10^3\,\mathrm{cm}^{-3}$, $T_c = 1020 \pm 100\,\mathrm{K}$.

For better agreement between calculations and the measurements the latter should be scaled downwards by a factor of 0.725. This factor includes possible calibration errors and solar intensity variations.

Night side disk measurements can be described by the two-component model as well with

cold component: $n_c = (2 \pm 1) \times 10^5\,\mathrm{cm}^{-3}$, $T_c = 150 \pm 50\,\mathrm{K}$

hot component: $n_c = 10^3\,\mathrm{cm}^{-3}$, $T_c = 1500 \pm 200\,\mathrm{K}$.

Cold atoms have a small scale height and a radiative transfer by them from day to night side is rapidly decreasing. The night side limb is presented by the hot component only. According to Anderson (1976) the most probable temperatures of the night side exosphere are in a range of 100 – 150 K. This result was confirmed by direct measurements on board the PV orbiter.

A more detailed study of hydrogen airglow was made by the V 9, 10 orbiters. In this study, a line width as well as its intensity were measured (Bertaux et al. 1978) to determine the atmospheric temperature at every point of the altitude profile. Both methods, the line width and altitude distribution, yield $T_c = 500 \pm 100\,\mathrm{K}$ and $n_c = 1.5 \times 10^4\,\mathrm{cm}^{-3}$ at $h < 1500\,\mathrm{km}$. Above 3000 km a steep increase in line width was observed up to 4500 km with a subsequent gradual decrease above 4500 km. This effect was explained by the charge exchange of exospheric hydrogen with solar wind protons near the bow shock which exhibits the increase of energy of hydrogen atoms. Hot atoms have satellite orbits and the hot hydrogen number density constitutes 10% of that of the cold component at 3500 km. Therefore, hot atoms appear outside the ionosphere. Their role in hydrogen escape is unclear.

Similar analysis was made for the Mariner 10 (Takacs et al. 1980) and V 11, 12 (Bertaux et al. 1982) observations. A summary of these data is given in Table 26. Bertaux et al. emphasize that their data coincide with the diurnal minimum of the hydrogen number density at 15.00 h LT obtained from the ion mass spectrometric measurements (see Chap. 4.4). The V 11, 12 disk observations yield the column integrated amount of hydrogen equal to $2 \times 10^{12}\,\mathrm{cm}^{-2}$ in a range of 110 – 250 km.

Dayglow Spectra in a Region of 110 – 350 nm. High quality spectra in this range were measured by a rocket spectrometer (Rottman and Moose 1973) in 1971 (Fig. 132). The spectrum reveals hydrogen, oxygen, and carbon lines, CO fourth positive band system along with two unidentified emissions at 140 and 190 nm (Table 27). In our opinion, the 190 nm emission may be attributed to the $C(^1D)$ 193 nm line, while the identification of the emission at 140 nm will be discussed below.

Only the 120 – 180 nm part of the dayglow spectra observed by the PV ultraviolet spectrometer is available now (Fig. 133). A small part of the

Table 26. Summary of the data on hydrogen airglow observations (Bertaux et al. 1982)

Space probe and year	Thermal component				Nonthermal component			
	Day side		Night side		Day side		Night side	
	n_c, cm^{-3}	T_c, K	n_c, cm^{-3}	T_c, K	n_c, cm^{-3}	T_c, K	n_c, cm^{-3}	T_c, K
M5 (1967)	$(2\pm1)\times10^5$	275 ± 50	$(2\pm1)\times10^5$	150 ± 50	1.3×10^3	1020 ± 100	10^3	1500 ± 200
M10 (1974)	$(1.5\pm1)\times10^5$	275 ± 50	$(1\pm0.5)\times10^5$	150 ± 25	$(5\pm1)\times10^2$	1250 ± 100	–	–
V9, 10 (1975)	1.5×10^4	500 ± 100	–	–	–	>1000	–	–
V11, 12 (1978)	$(4^{+3}_{-2})\times10^4$	300 ± 25	–	–	10^3	1000	–	–

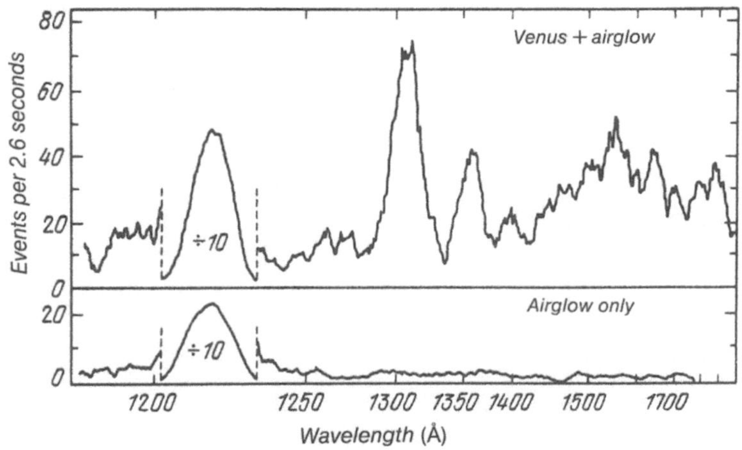

Fig. 132. Venusian UV dayglow spectrum measured by a rocket borne spectrometer (Rottman and Moos 1973). The terrestrial airglow measured during the same flight is shown in the lower part of the figure. A scale for the hydrogen line at 122 nm is reduced by a factor of 10

Table 27. Identified emissions and upper limits in the Venusian dayglow spectrum 120 – 190 nm (Rottman and Moose 1973)

λ, Å	Species	Transition	Intensity, *kR*
1193	CI	$^3P-^3D^0$	1.5 ± 1.0
1200	NI	$^4S^0-^4P$	<1.0
1216	HI	$^1S-^1P$	27 ± 3.0
1261		$^3P-^3P^0$	
	CI		1.0 ± 0.5
1278		$^3P-^3D^0$	
1302 – 1306	OI	$^3P-^3S^0$	5.5 ± 0.5
1329	CI	$^3P-^3P^0$	0.5 ± 0.5
1356	OI	$^3P-^5S^0$	2.7 ± 0.5
1463		$^1D-^1F^0$	
	CI		1.0 ± 1.0
1467		$^1D-^1P^0$	
1561	CI	$^3P-^3D^0$	2.4 ± 1.2
1657	CI	$^3P-^3P^0$	4.0 ± 1.5
	CO	$X^1\Sigma-A^1\Pi$	25 ± 5

Venusian UV spectrum (130 – 140 nm) was measured by IUE with a resolution of 0.04 nm (Durrance 1981).

Spectrophotometric Measurements on Board Mariner 10 and V11, 12 (Broadfoot et al. 1974, Bertaux et al. 1981). Both experiments are very similar in goals and techniques used. Intensities of the UV spectrum have been measured at the locations of ten emission lines: He^+ 30 nm, He 58 nm, Ne 74 nm,

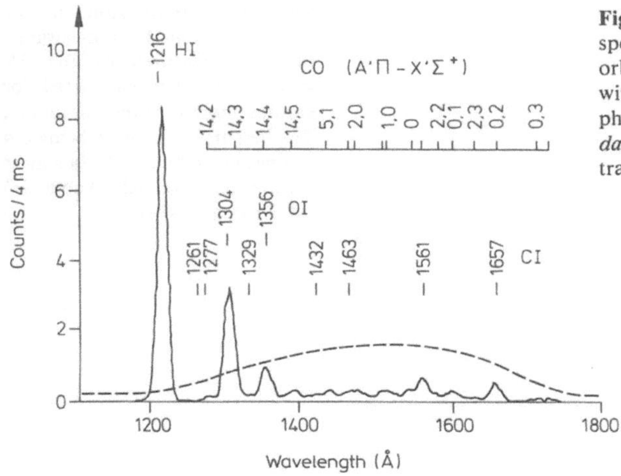

Fig. 133. Venusian UV dayglow spectrum measured by the PV orbiter (the sum of 480 spectra with incident, emission, and phase angles less than 70°). The *dashed curve* is the relative spectral sensitivity (Durrance 1981)

O^+ 83 nm, Ar 87 nm and 105 nm, H 122 nm, O 130 nm, CO 150 nm, C 166 nm, the zeroth order intensity being the integral of the spectrum. The M 10 instrument incorporated detectors (channeltrons) sensitive to radiation with $\lambda > 170$ nm, so stray light background was rather high. Therefore, of all these emissions, the data on the H 122 nm and He 58 nm emissions have been analyzed as most accurate. The V 11, 12 channeltrons featured a steeper longward decrease of sensitivity for $\lambda > 170$ nm and a better signal to background ratio. However, argon and neon emissions cannot be separated from the background. Hydrogen, helium, neutral and ionized oxygen lines have been interpreted so far.

Atomic Oxygen. A technique for interpretation of atomic oxygen airglow at 130 and 136 nm was described in Chap. 7.5. Being applied to the rocket spectrometric data it yielded $[O]/[CO_2] = 10\%$ at the height where CO_2 column amount equals 4×10^{16} cm^{-2} (Rottman and Moose 1973). This height is about 135 km with $[CO_2] \simeq 10^{11}$ cm^{-3}. The mixing ratio obtained was one order of magnitude higher than that on Mars.

Sholokhov and Burgin (1982) analyzed 130 nm airglow on the Venusian disk measured by V 11, 12 flyby spectrometers. They derived $[O] = 4 \times 10^8$ cm^{-3} at the height where $[CO_2] = 2 \times 10^{10}$ cm^{-3} or $[O]/[CO_2] = 7 \times 10^{-3}$ at 135 km which appears very low. The airglow intensity constituted 6 *kR* on the Venusian disk.

Meier et al. (1983) calculated radiative transfer on the assumption of partial frequency redistribution (PFR) which suited the problem better for a large optical thickness than the complete frequency redistribution used previously. Being applied to the PV data on 130 and 136 nm lines these calculations result in $[O]/[CO_2] = 4.5\%$ at 135 km. Peak limb intensity is 20 *kR* at 130 nm and 15 *kR* at 136 nm (Figs. 134 and 135).

Two excitation processes, resonant scattering of solar photons and photoelectron impact excitation, are dominant for the 130 nm line, while photoelec-

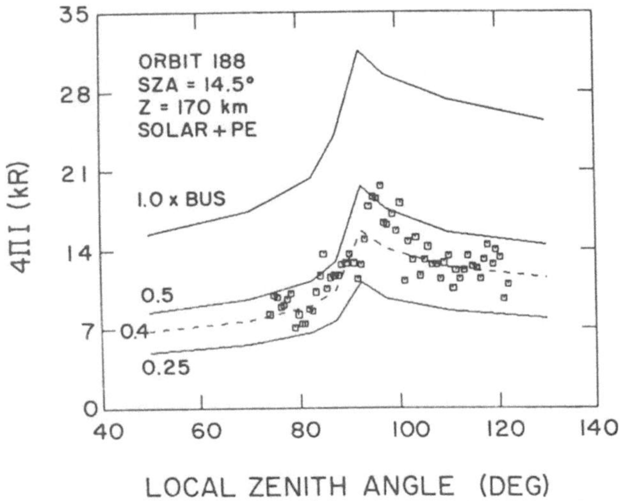

Fig. 134. Comparison of the atomic oxygen 130 nm airglow intensity measured by the PV spectrometer and calculated for various atomic oxygen amounts. The best fit is for those being 0.4 of ones given by the PV bus mass spectrometer model (Table 24) (Meier et al. 1983)

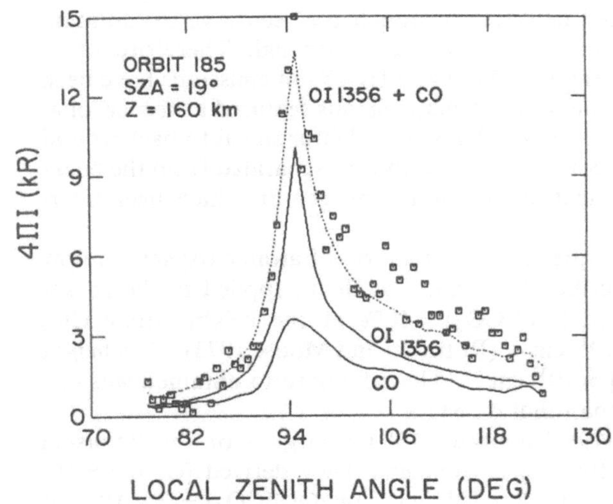

Fig. 135. Contribution to the feature at 136 nm from the atomic oxygen line and the CO (14.4) band. The latter has been derived by the appropriate scaling of the measured CO (14.5) band which is unblended (Meier et al. 1983)

tron impact on O and resonant scattering of solar 122 nm radiation by CO molecules produce the spectral feature at 136 nm which, as shown by Durrance et al. (1980), is actually the sum of the O 135.6 nm line and CO (14.4) band at 135.3 nm (Fig. 135).

Extrapolation of the PV orbiter mass spectrometric data (Fig. 124) gives $[O]/[CO_2] \simeq 10\%$ at 135 km.

Hot Oxygen. PV limb observations of 130 nm airglow above 350 km reveal a component with a scale height of 400 km (Nagy et al. 1981). Its limb intensity is 550 R at 400 km. With oxygen atoms as a source of this airglow one finds their temperature to be 6400 K. As soon as the measured intensity is smaller than for unit optical depth (875 R) a single scattering approximation becomes valid, which results in a hot oxygen number density of 4×10^4 cm^{-3} at 400 km with an emission rate factor, g, of 3.5×10^{-5} s^{-1}. One may assume that hot atoms do not scatter the "cold" radiation field. The source of hot oxygen atoms, namely, recombination of O_2^+ ions and charge exchange of hot thermal ionospheric oxygen ions with neutral atomic oxygen and hydrogen will be discussed in Chap. 4.4.

Helium 58 nm. From M 10 helium emission data, Kumar and Broadfoot (1975) obtained $f_{He} = (10 \pm 5)$ ppm at a level of $[CO_2] = 2 \times 10^{11}$ cm^{-3} and $T_c = 375 \pm 100$ K. The PV bus mass spectrometer gives $f_{He} = 70$ ppm at this height (Table 24). The difference can be partly accounted for if in helium radiation calculations T_c is taken as 275 K.

With $f_{He} = 10$ ppm Kumar and Broadfoot calculated the loss rate of helium by ionization and capture by solar wind being 2×10^5 cm^{-2} s^{-1}, an order of magnitude lower than the helium production rate by radioactive decay in rocks equal to 2×10^6 cm^{-2} s^{-1} according to V 8 γ-radiation measurements (Vinogradov et al. 1973). This discrepancy disappears with the PV data when f_{He} is taken into account.

The V 11, 12 helium line measurements result in a helium optical thickness of $\tau = 0.7$ and its number density of 4.5×10^5 cm^{-3} at a level of $[CO_2] = 2 \times 10^{10}$ cm^{-3}, i.e., ~140 km (Burgin and Sholokhov 1981). More careful interpretation (Kurt et al. 1983) gives $\tau \simeq 4$ corresponding to 2×10^{13} atom cm^{-2} and 3×10^6 cm^{-3} at 140 km, in accord with the PV mass spectrometric measurements.

CO ($A^1\Pi - X^1\Sigma$) Band System. Assuming the contribution of fluorescent scattering to be ~70% of total band system intensity measured, Rottman and Moos (1973) obtain $[CO]/[CO_2] \simeq 10\%$ at a level of $[CO_2] = 10^{11}$ cm^{-3}. The V 11, 12 observations at 150 nm yield an intensity of ~1 kR nm^{-1} (Bertaux et al. 1981), which agrees well with the rocket data.

PV and especially IUE spectra (Durrance 1981) show that solar Lyman-alpha radiation plays an important part in resonant excitation of CO (14.0) band. Fluorescent scattering of this radiation results in (14.2) 128 nm, (14.3) 132 nm, (14.4) 135 nm, and (14.5) 139 nm airglow (Fig. 136) and explains the anomalous 130/136 nm intensity ratio and the feature at 140 nm. The (14.5) band is not blended and other excitation processes (photoelectron impact of CO and CO_2, photodissociation of CO_2 and dissociative recombination of CO_2^+) should contribute only slightly to this band, therefore it may be employed to measure CO amounts. Its emission rate factor equals 5.8×10^{-9} s^{-1}. Durrance (1981) has not published any quantitative data on CO, but from Fig. 136 an estimate of the CO slant column amount can be made to give 3.5×10^{16} cm^{-2}. This amount can be referred to the level with unit CO_2 slant optical thickness at 140 nm, i.e., 1.6×10^{18} cm^{-2} (~122 km). So the column ratio of CO to CO_2 is ~2% and $[CO]/[CO_2] \sim 1.5\%$ at 122 km. Other ways

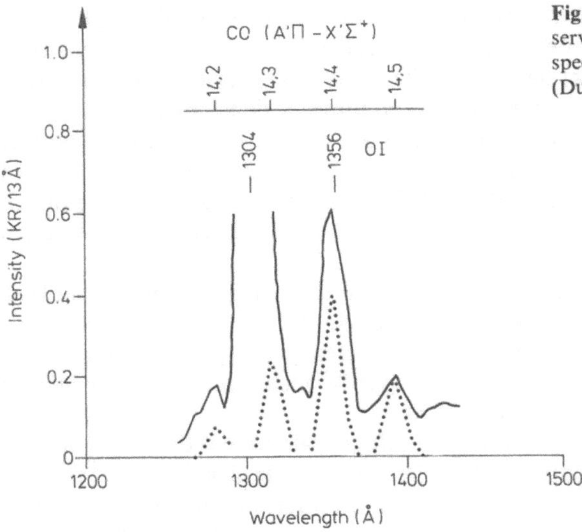

Fig. 136. Comparison of the observed spectrum and the synthetic spectrum of the CO (14, v'') bands (Durrance et al. 1980)

to measure CO amount exist: inversion of limb airglow measurements (Fig. 135) and interpretation of self-absorption (see Chap. 1.5). The presence of self-absorption stems from the fact that in the spectrum in Fig. 133 the (1.0) band is slightly less intense than the (0.1) band, while its emission rate factor is about 1.7 times larger.

Carbon Lines. Carbon lines as measured in the rocket experiment are shown in Fig. 132. They have been also measured by M10 and V11, 12. The 166 nm line is the most intense, having an intensity of 4 *kR* in the rocket spectrum (Fig. 132), 30 *kR* according to M10 observations (Broadfoot et al. 1974) and 10 *kR* measured by V11, 12 (Bertaux et al. 1981). This line intensity was calculated by Krasnopolsky (1982a) in the same way as for Mars (see Chap. 2.5) and appeared to equal 560 *R* with the contribution of resonant scattering constituting only 80 *R*. However, a comparison of the PV data (Figs. 133 and 136) results in the 166 nm line intensity of 380 *R*, which is in fair agreement with the calculations. The high value measured by M10 can be explained by a tail of the channeltron sensitivity at $\lambda > 200$ nm and a large contribution of stray light. Contrary to M10, the V11, 12 instruments had no tails of sensitivity at $\lambda > 200$ nm, but unfortunately they featured too low sensitivity at 166 nm ($\sim 3 \times 10^{-4}$ of the maximal value). Thus, in this case the stray light contribution may also be large. Since resonant scattering constitutes a small part of the total intensity, it is difficult to deduce carbon amount from the airglow zenith intensity.

O^+ 83 nm. With this line intensity of 250 *R* measured by V11, 12 Burgin and Sholokhov (1981) obtained the optical thickness of atomic oxygen ions equal to 0.4 which corresponds to $[O^+] \simeq 2.5 \times 10^4\ cm^{-3}$ at 250 km.

3.12 Ionosphere

Many and varied Venusian ionospheric problems include interaction of solar wind with planetary ionosphere and neutral atmosphere, initial and induced magnetic fields and their configuration, temperatures of plasma, etc. Their detailed consideration is beyond the scope of the present book, and we restrict ourselves to the study of altitude dependences of ionic number densities and composition up to 250 km, ionization sources, and to a lesser extent, of ionospheric variations and dynamics. This approach settles the problem of the interaction of neutral and ionic components of the upper atmosphere.

Observations before the PV orbiter were conducted by means of remote sounding, i.e., radio-occultation measurements of plasma number densities (see Chap. 1.7). First results were obtained by M5 and M10 (Fjeldbo and Eskleman 1969, Fjeldbo et al. 1975). A more extended series of observations was made by the V9, 10 orbiters (Ivanov-Kholodny et al. 1979). These measurements resulted in a day side ionospheric peak at 140 – 150 km with electron number density, $n_{e_{max}}$, dependent on the solar zenith angle, z_0; $n_{e_{max}} \sim (Ch z_0)^{-1/2}$, where $Ch z_0$ is the Chapman function equal to sec z_0 for $z_0 \lesssim 80°$. This dependence describes n_e at a small bend near 130 km as well. This was predicted by the theory of simple ionization layer and confirms photochemical origin of the ionosphere in these regions. The ionopause,

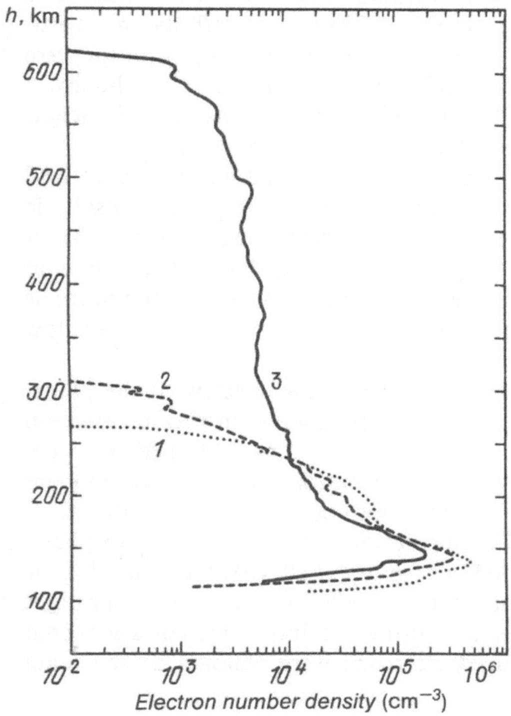

Fig. 137. Day side ionosphere of Venus measured by radio-occultations of the V9, 10 orbiters at solar zenith angles of 14°, 63°, and 83° (*curves 1, 2, 3*, respectively) (Ivanov-Kholodny et al. 1979)

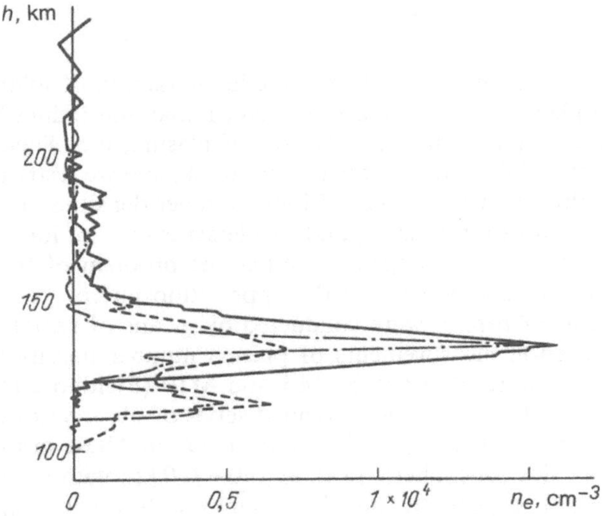

Fig. 138. Night side ionosphere of Venus measured by radio-occultation of the V9, 10 orbiters. Three typical dependences are shown (Ivanov-Kholodny et al. 1977)

which is the upper boundary of the ionosphere where a steep decrease of n_e occurs, varies from 300 km at small z_0 to 600 km near the terminator (Fig. 137).

The night side ionosphere is highly variable and in most cases has two maxima (Fig. 138). The maxima are spaced ~15 km apart with the mean altitude of the upper maximum being 137 km. The ranges of n_e variations are $(3-16)\times 10^3\ cm^{-3}$ in the upper peak and $(1.8-11)\times 10^3\ cm^{-3}$ in the lower one. Some profiles do not exhibit lower peak. A mean plasma scale height above the upper peak is 5 km, while being ~1 km in extreme cases.

According to Chap. 2, neutral atmospheric temperature can be derived from n_e profiles in the region of photochemical equilibrium, and ionospheric scale height equals twice that of the neutral atmosphere just above the ionospheric peak. Ivanov-Kholodny et al. (1977), assuming that CO_2 is the main component of the upper atmosphere, evaluated its night side temperature as ~100 K. The same evaluation for atomic oxygen yields an extremely low value of ~40 K.

The V9, 10 orbiters revealed electron fluxes in the shadow of the planet (Gringauz et al. 1977, 1979, 1981). In these experiments, integrated electron spectra in a range of 10–150 eV were measured. Data obtained were described by the two component Maxwellian distribution with a cold component having $kT \simeq 13$ eV and $n \simeq 1.5\ cm^{-3}$ and a hot component with $kT \simeq 80$ eV and $n \simeq 0.2\ cm^{-3}$. The most careful calculations of night side atmospheric ionization by these electrons yielded $n_{e_{max}} = 1.7\times 10^4\ cm^{-3}$ at the altitude where $[CO_2] = 2\times 10^9\ cm^{-3}$. This value, together with an altitude of the night side ionospheric peak of 137 km, was one of the first indications of a very cold night side atmosphere. The measured electrons were claimed as the main source of the night side ionization.

These electrons should excite nightglow emissions, the strongest of them are the green oxygen ($^1D-{}^1S$) line at 558 nm and the CO_2^+ band at 289 nm. However, 558 nm airglow was not detected by the V9, 10 airglow spectrometers with an upper limit of 4 R (Krasnopolsky 1979d, 1982c). The interpretation of this upper limit based on the PV data on night side neutral atmosphere (Figs. 124 and 130) considers three excitation processes:

$$O_2^+ + e \rightarrow O + O(^1S);$$
$$CO_2 + e \rightarrow CO + O(^1S) + e$$
$$O + e \rightarrow O(^1S) + e.$$

Quantum yield of the first reaction is ~ 0.05 and it contributes 1 R according to the radio-occultation profile of n_e and with $n_e \simeq [O_2^+]$ at the ionospheric peak taken into account. Electron energy loss per excitation of a 558 nm photon equals 220 eV in carbon dioxide and is almost independent of electron energy, E. Quantum yield of $O(^1S)$ in the third process is ~ 0.4 independent of E with $E > 15$ eV. Krasnopolsky (1982c) also considered reflection of some part of the electron flux by the neutral atmosphere and a possible magnetic field with the inclination D (the angle of the magnetic line to the horizontal plane). With ionization by the energetic electron being proportional to $\sin D$ the upper limit of 4 R for 558 nm airglow results in $D \lesssim 18°$ being in accord with the electron fluxes measured by Gringauz et al. (1979).

Krasnopolsky (1982c) notes that the mean $n_{e_{max}}$ is $0.7 \times 10^4\,cm^{-3}$ at the time of V9, 10 operation is smaller by a factor of 2.4 than that calculated by Gringauz et al. (1981). Therefore, ionization rates differ by a factor of 6, which can be compensated by the assumption of $D \simeq 10°$. A substantial part of night side ionization should be produced by plasma transport from the day side, and to fit the electron flux data and n_e profiles measured by the V9, 10 ratio-occultations extremely low values $D \simeq 2-5°$ should be adopted. More probable is the suggestion that the electron flux data of Gringauz et al. (1979) are overestimated.

The upper limit of 558 nm airglow results in the restriction on helium ion flux to be less than $10^7\,cm^{-2}\,s^{-1}$ due to the process

$$He^+ + CO_2 \rightarrow CO^+ + O(^1S) + He.$$

This flux was considered earlier as a source of night side ionization.

Pioneer Venus Measurements. The orbiter with a low lying pericenter at 150 km gives great advantages of detailed direct ionospheric measurements. These measurements have been carried out by two ion mass spectrometers (on board the orbiter and bus), a retarding potential analyzer, and an electron temperature probe. These experiments were accompanied by radio-occultation observations and by the study of solar wind-magnetosphere interaction.

A typical day side ionospheric composition measured by the orbiter ion mass spectrometer (H. Taylor et al. 1980) is shown in Fig. 139 for $z_0 = 11°$. For larger z_0 altitudes of O_2^+, NO^+, and CO_2^+ peaks increase slightly with

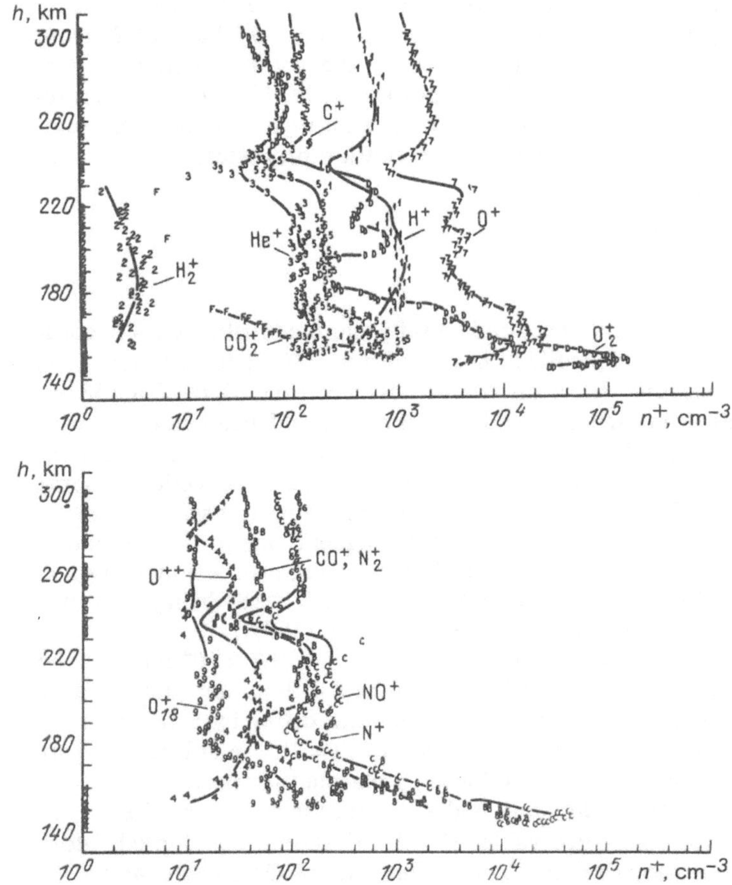

Fig. 139. Day side ionospheric composition measured by the PV orbiter ion mass spectrometer at $z_0 = 11°$ (H. Taylor et al. 1980)

their peak number densities decreasing by a factor of 2 to the terminator. Twelve kinds of ions, including double charged ion O^{++} were observed. CO^+ and N_2^+ ions were not separated. The $^{18}O^+$ and $^{16}O^+$ ratio reveals that the oxygen isotopic ratio is smaller to that on the Earth.

A variety of ionospheric phenomena was discovered on the night side of Venus. Some orbits reveal an "established" ion distribution similar to that on the day side (Fig. 140). However, sometimes the night side ionosphere disappears and is depleted over the total hemisphere (Fig. 141). In other cases this disappearing ionosphere occurs in small limited areas, the so-called ionospheric holes. The transition ionosphere is also shown in Fig. 141.

The established night and day side ionospheres differ greatly in the $[H^+]/[O^+]$ ratio equal to 3×10^{-3} at $z_0 = 70°$ and reaching a maximum of

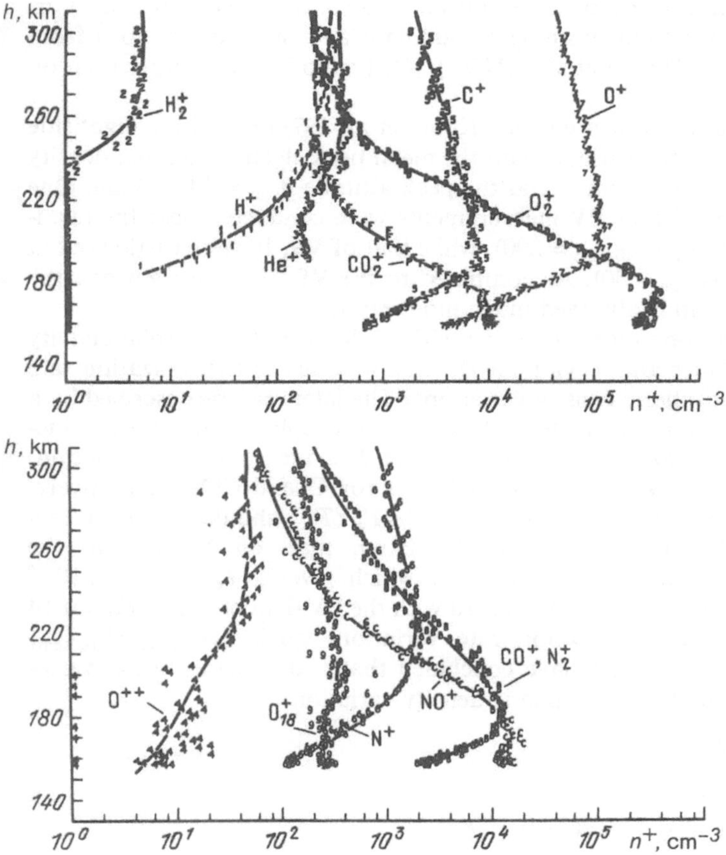

Fig. 140. The "established" night side ionospheric composition at $z_0 = 160°$ (H. Taylor et al. 1980)

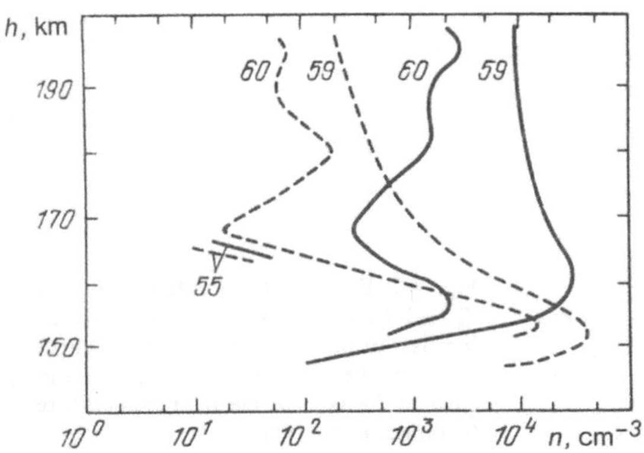

Fig. 141. Altitude profiles of O^+ (*solid lines*) and O_2^+ (*dashed lines*) number densities typical for the established, transition, and depleted night side ionosphere (orbits 59, 60 and 55, respectively) (H. Taylor et al. 1980)

~3 after midnight. Therefore, the topside ionosphere consists primarily of atomic oxygen ions from morning to midnight and of hydrogen ions from midnight to dawn. The night side $[NO^+]/[O_2^+]$ ratio is much larger than on the day side.

Radio-occultation measurements (Kliore et al. 1979) reveal the night side ionosphere to be highly variable with the mean peak electron number density equal to $(1.7 \pm 0.7) \times 10^4\,cm^{-3}$ and the peak altitude $h_{max} = 142 \pm 4$ km. One should keep in mind that PV measurements were conducted near the maximum of solar activity ($F_{10.7\,cm} \simeq 200$), while that of V9, 10 were carried out at the minimum ($F_{10.7\,cm} \simeq 80$). As compared to the V9, 10 measurements, the secondary maximum is observed much more rarely.

Correlation of ion number densities with 27 day variation of solar activity was observed (Bauer and Taylor 1981). This was due to the ionization rate and neutral atmospheric density variations. The latter can be described by a temperature variation which depends on the atmospheric thermal conductivity, k. With $k = k_0 T^s$, $T_\infty = a I^{1/s}$, where I is the solar UV radiation, and $\Delta T_\infty / T_\infty = s^{-1} \Delta I / I$; $s = 0.71$ and 1.23 for O and CO_2, respectively. Then the relative temperature variation ($\Delta T_\infty / T_\infty$) should be by a factor of $1.23/0.71 = 1.73$ larger on the Earth than on Venus. On the Earth $\Delta T_\infty / T_\infty = 1.2 \times 10^{-3} \times \Delta F_{10.7\,cm}$, which gives $\Delta T_\infty / T_\infty \simeq 7 \times 10^{-4} \times \Delta F_{10.7\,cm}$ on Venus, which is in accord with the PV data or $\Delta T_\infty / T_\infty \simeq 0.14\, \Delta F_{10.7\,cm} / F_{10.7\,cm}$. The ionization rate variation equals $\Delta F_{10.7\,cm} / F_{10.7\,cm}$, therefore Bauer and Taylor have concluded that it dominates in ion variations. However, neutral atmospheric density variations

$$\frac{dn}{n} = \frac{\Delta z}{H} \frac{dT_\infty}{T_\infty}$$

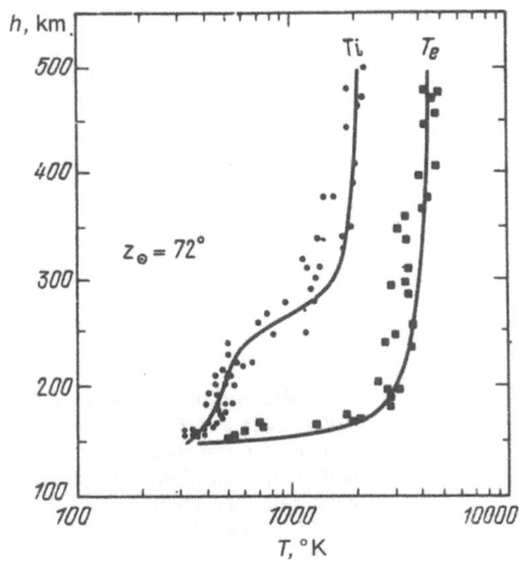

Fig. 142. Electron and ion temperatures, T_e and T_i, in the day side ionosphere of Venus. Both temperatures are almost equal in the night side ionosphere and close to the day side electron temperature (Knudsen et al. 1980)

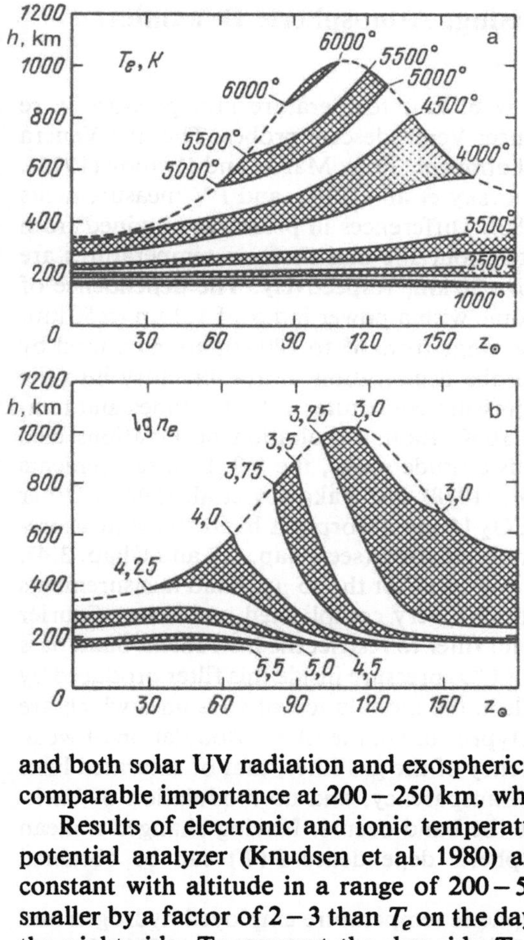

Fig. 143 a, b. Contour maps of T_e and n_e (Brace et al. 1983)

and both solar UV radiation and exospheric temperature variations can be of comparable importance at 200 – 250 km, where $\Delta z/H$ is rather large.

Results of electronic and ionic temperature measurements by a retarding potential analyzer (Knudsen et al. 1980) are given in Fig. 142. T_e is fairly constant with altitude in a range of 200 – 500 km and with local time. T_i is smaller by a factor of 2 – 3 than T_e on the day side and almost equal to that on the night side. To support the day side T_e an energy influx of 3×10^{-2} erg $cm^{-2} s^{-1}$ should be near the ionopause, and joule heating of 3×10^{-3} erg $cm^{-2} s^{-1}$ is necessary at 150 – 250 km to provide the day side T_i measured. These values are 7×10^{-2} erg $cm^{-2} s^{-1}$ and 5×10^{-3} erg $cm^{-2} s^{-1}$, respectively at nighttime, the latter being referred to 150 – 500 km. The energy influx to electron gas is about 10% of the solar wind drag energy on the sunlit hemisphere.

Altitude and solar zenith angle dependence of electron temperature and number density are presented in Fig. 143. This figure indicates also the position of the ionopause.

Electron fluxes in the range of 10 – 70 eV similar to those measured by Gringauz et al. (1979) on board the V9, 10 orbiters were observed by the PV orbiter (Intrilligator et al. 1979). No data on airglow emissions produced by these electrons have been published as yet.

3.13 Temperature, Eddy Mixing, Atmospheric Dynamics, and Lightning

Temperature. Tropospheric (below 65 km) temperature and pressure were measured by 11 Venera and 4 Pioneer Venus descent probes. The first Venera data served as a basis for a model atmosphere by Marov and Ryabov (1972), V 11, 12 data were given by Avduevsky et al. (1979), and PV measurements were described by Seiff et al. (1980). Differences in pressures obtained from V 11, 12 and PV probes constitute about 1% and those in temperatures are ~5 K and ~10 K below and above 30 km, respectively. The dependence of pressure on temperature is polytropic with a power index of 1.2 up to 50 km.

Temperature and pressure in a range from 65 to 120 km are measured by an accelerometer which gives a probe deceleration during its entry into the atmosphere. A typical scatter in pressure constitutes 1.3 – 1.7 times and temperature data are scattered by ~10 K. Radio-occultation observations also provide reliable information in this altitude range, the V 9, 10 measurements being considered by Kolosov et al. (1978) and Yakovlev et al. (1978). Other remote sounding techniques are CO_2 15 μm absorption band shape measurements or limb darkening of infrared radiation (see Chap. 1.8 and Chap. 3.4). A pressure modulated radiometer was used for the 15 μm band measurements on board the PV orbiter rather than a very complicated and heavy Fourier spectrometer. It incorporated a solid filter to restrict the instrument band pass and a CO_2 gas filter. Oscillations of CO_2 pressure inside this filter produced by a device similar to a speaker result in the modulation of CO_2 lines which are close to saturation at the mean CO_2 pressure in the filter. Modulation of weak lines is low due to a high transparency of the gas filter, that of saturated lines is low as well due to a very low transparency, and no modulation of background between lines and outside the band occurs. Having changed a mean pressure in the filter, an atmospheric dependence $P(T)$ can be obtained (Taylor et al. 1980).

The first estimates of thermospheric temperature employed hydrogen and helium airglow observations (275 K by Anderson 1976; 375 ± 100 K by Kumar and Broadfoot 1975; 500 K by Bertaux et al. 1982). Early theoretical evaluations gave the day side temperature in a range of 400 – 800 K (McElroy 1969, Marov and Ryabov 1972) depending on solar activity, and a night side temperature of 350 K at middle solar activity (Izakov and Morozov 1975). Dickinson and Ridley (1977) calculated $T = 330$ and 170 K at the day side maximum and the night side minimum, respectively. Night side temperatures derived by Anderson (1976) (150 ± 50 K) from airglow measurements and by Ivanov-Kholodny et al. (1977) and Gringauz et al. (1977) from a night side ionospheric profile (~100 K) turned out to be the best at that time.

Detailed temperature measurements above 150 km were made by the PV orbiter neutral mass spectrometer (Niemann et al. 1980) and by a PV orbiter atmospheric drag (Keating et al. 1980). These data show that thermospheric temperature at high solar activity has a day side maximum of 290 – 300 K and

Table 28. Temperature and pressure in the day side atmosphere of Venus

h, km	p, bar	T, K	h, km	p, bar	T, K
0	92.0	732	70	3.35 − 2*	231
5	66.7	692	80	4.25 − 3	197
10	47.3	654	90	3.22 − 4	170
15	32.8	617	100	1.95 − 5	174
20	22.3	579	110	1.75 − 6	183
25	14.7	537	120	1.4 − 7	185
30	9.4	495	130	1.39 − 8	190
35	5.77	453	140	1.85 − 9	206
40	3.4	416	150	3.71 − 10	238
45	1.91	385	160	1.15 − 10	270
50	1.02	350	170	4.45 − 11	274
55	0.51	301	180	1.97 − 11	275
60	0.224	263	190	9.49 − 12	275
65	0.091	243	200	4.84 − 12	275

* 3.35 − 2 means 3.35×10^{-2}

is almost constant on the day side hemisphere with a mean value of 285 K with a steep decrease near the terminator to a mean night side value of 110 − 120 K. The planetary averaged thermospheric temperature is 200 K.

Very small day-to-day variations of the day side thermospheric temperature observed by both techniques are consistent with the simple analysis by Bauer and Taylor (1981, see Chap. 3.12) which results in $\Delta T/T = 7 \times 10^{-4}$ $\Delta F_{10.7\,\mathrm{cm}}$.

The extremely low night side temperatures demonstrate negative temperature gradients, thus forming cryosphere instead of thermosphere. The nature of this cryosphere is still unexplained.

Neutral atmospheric temperatures were deduced by Bauer et al. (1979) from ion composition measurements, because the altitude profile of ionospheric plasma in the region of photochemical equilibrium is determined by neutral atmospheric temperature (see Chap. 2.1). They revealed temperatures increasing from 180 K at $z_0 = 90°$ to 275 K at $z_0 = 60°$. Mass spectrometric and accelerometric measurements by the PV bus (von Zahn et al. 1980) at the morning side with $z_0 = 60°$ as well as accelerometric and direct measurements of pressure and temperature by the PV sounder and day probes are summarized in Table 28. We will not dwell upon problems of atmospheric thermal regime. Mean net thermal emission of Venus is 157 $W\mathrm{m}^{-2}$, corresponding to an equivalent temperature of 230 K and bolometric albedo of 0.76 (Schofield and Taylor 1982).

Eddy Mixing. Determination of the eddy mixing coefficient, K, from a nonuniform aerosol and gaseous component distribution were considered in Chap. 1.10. An altitude profile of aerosol particles forming below the region discussed is defined by the competition of eddy mixing and sedimentation,

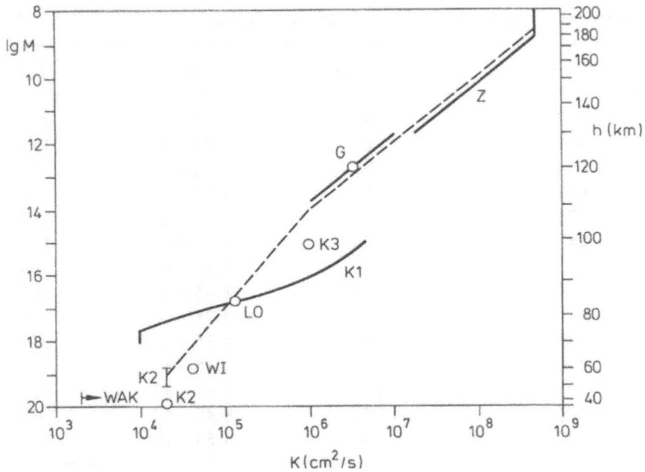

Fig. 144. Various data on the eddy mixing coefficient: *Z* von Zahn et al. (1980); *G* Gerard et al. (1981); *LO* Lane and Opstbaum (1983); *WI* Woo and Ishimaru (1981); *WAK* Woo et al. (1982); *K1* Krasnopolsky (1979c); *K2* Krasnopolsky (1985a); *K3* Krasnopolsky (1983a). The *dashed line* shows $K = 10^{13} n^{-1/2}$ cm^2 s^{-1} and $K = 4.5 \times 10^{10} n^{-1/3}$ cm^2 s^{-1} above and below 110 km, respectively, n is total number density

and K can be obtained from particle size measurements and their scale height, H_a:

$$K = \frac{VHH_a}{H - H_a}$$

where H is the atmospheric scale height and V is the sedimentation velocity calculated from the Stokes-Davis formula (Chap. 3.10). This relation being applied to the upper haze measurements by Krasnopolsky (1980, Fig. 108) and Lane and Opstbaum (1983, Chap. 3.7) results in $K(z)$ as shown in Fig. 144.

Strong arguments in favor of $K = 2 \times 10^4$ cm^2 s^{-1} at 50 – 60 km are given in Chap. 3.10, and the coincidence of the lower haze mass loading at 35 km with that of the mode 1 particles at 55 km was considered there as indicating the coincidence of K at these altitudes. Therefore, $K = 2 \times 10^4$ cm^2 s^{-1} at 35 km is also shown in Fig. 144. Analysis of small-scale temperature fluctuations observed by radio-occultations yielded the eddy mixing coefficient associated with these turbulent motions equal to 2×10^3 cm^2 s^{-1} at 45 km (Woo et al. 1982). It should be considered as a lower limit of K here. Woo and Ishimaru (1981) derived $K = 4 \times 10^4$ cm^2 s^{-1} at 60 km from an evaluation of wind shear instability.

Mass spectrometric measurements of helium, N_2, and CO_2 near the homopause by the PV bus (von Zahn et al. 1980) result in $K = 1.4 \times 10^{13} M^{-1/2}$ cm^2 s^{-1} at 130 – 170 km where M is the total number density. As we discussed in Chap. 1.10, this relation well describes eddy mixing in the Earth and

Martian atmospheres (von Zahn et al. 1980). This relation being applied to NO airglow interpretation (Chap. 3.11) Gerard (1981) obtained $K = 0.8 \times 10^{13} M^{-1/2}$ cm^2 s^{-1} from the measured altitude of the airglow layer.

Fluctuations of nightglow emissions may serve as a measure of eddy mixing at the height of the layer maximum. These fluctuations for $O_2(c^1\Sigma_u^- - X^3\Sigma_g^-)$ on Venus and $O_2(A^3\Sigma_u^+ - X^3\Sigma_g^-)$ on the Earth appear to have very similar size distribution as measured by Krasnopolsky (1983a). In both cases the excitation process is a three body association of oxygen atoms, therefore, mixing coefficients on the Earth at 95 km and on Venus at $\sim$100 km should be equal, the former being 10^6 cm^2 s^{-1}, which is accord with K obtained from the PV bus measurements.

So thermospheric eddy mixing may be given as $K = 10^{13} M^{-1/2}$ cm^2 s^{-1}. However, extrapolation of this relation to the mesosphere (110 – 60 km) turns out to be invalid. In our opinion the most reliable data on K in the mesosphere are $K = 2 \times 10^4$ cm^2 s^{-1} at 50 – 60 km based on comparison of the mode 3 mass loading in the middle cloud layer and the water vapor altitude profile, and $K = 1.3 \times 10^5$ cm^2 s^{-1} at 84 km given by both upper haze measurements. Therefore, $K = 4.5 \times 10^{10} M^{-1/3}$ cm^2 s^{-1} may be recommended in the mesosphere (Fig. 144).

Atmospheric Dynamics. Detailed consideration of atmospheric dynamics is beyond the scope of this book. The lower and upper atmosphere present two extreme cases in Venusian atmospheric dynamics. In the lower atmosphere the time of cooling by convection and radiative transfer exceeds the duration of the solar day (117 Earth days). Therefore, atmospheric circulation with poleward winds in the upper part of this circulation cell and with pole to equator winds in the lower part occurs. In the upper atmosphere the time of cooling by thermal conductivity to the mesopause with subsequent radiative cooling is much smaller than the solar day. The circulation from the subsolar to antisolar region and in the opposite direction near the mesopause forms here. Gas flows in both direction are equal, yet wind velocities are much larger in the upper part of the circulation cell due to smaller atmospheric densities. A thick atmospheric region centered at the cloud layer exists between these two extreme regions. The thermal regime of this region is rather complicated, and quite different flow conditions on its low and upper boundary should be taken into account. Here westward winds (which is *in* the direction of the rotation of the planetary body) with velocities of about 100 m s^{-1} form, though the processes underlying their formation are not quite clear. These winds are responsible for the effect of a quasi-four-day rotation of the atmosphere and cloud layer which is retrograde relative to the rotation of the other planets.

Though the four-day rotation was discovered from ground-based observations of features in UV images, a modern experimental investigation is based on Doppler shift measurements of radio signals from descent probes. For a long time, Venera probes were the only sources of information on wind velocities on Venus (Kerzhanovitch et al. 1979). Four PV probes and V11 – 14

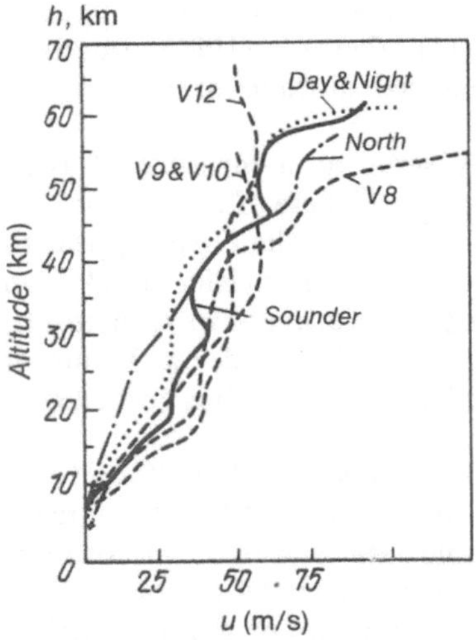

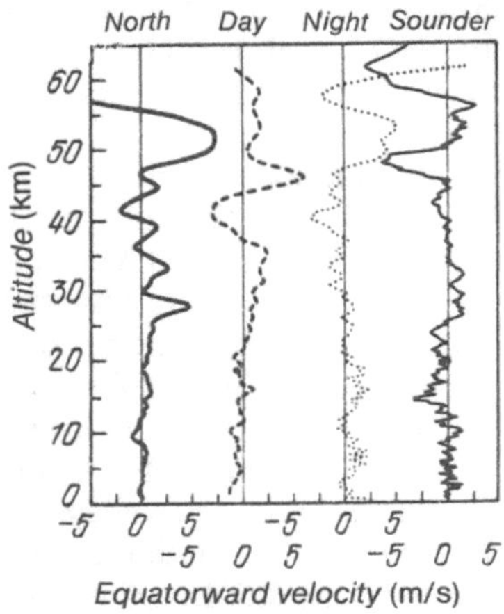

◄ Fig. 145 ▲ Fig. 146

Fig. 145. Zonal (latitudinal) component of the wind speed measured by V and PV probes

Fig. 146. Meridional component of the wind speed measured by four PV probes (Counselman et al. 1980)

measurements have provided new data in the field. Both Doppler shift observations and the determination of probe locations by means of a ground-based radio interferometer were employed to obtain wind velocity profiles from the PV probe descents (Seiff et al. 1980, Schubert et al. 1980, Counselman et al. 1980). All results agree rather well (Figs. 145 and 146). Wind velocity measurements were supplemented by the pressure and temperature determinations which provide data on stability of different parts of the atmosphere.

Wind velocities can be derived also from pressures and temperatures measured in different points in the approximation of cyclostrophic balance, when wind speed forms a centrifugal force which compensates the difference of pressure:

$$u^2 = \frac{g \Delta z}{\ln \frac{\sin \theta_1}{\sin \theta_2}}$$

where u is the wind speed, g is the gravitational acceleration, and Δz is a difference in altitudes with equal pressures at latitudes θ_1 and θ_2. Calculations (Seiff et al. 1980) based on the direct PV measurements of pressure on the day and the North probe agree well with the upper part of the curve derived from radio-interferometric measurements (Fig. 147), provide wind velocities at altitudes up to 110 km, and reveal a steep decrease of zonal circulation above 90 km. Above 115 km a circulation from subsolar to antisolar regions occurs.

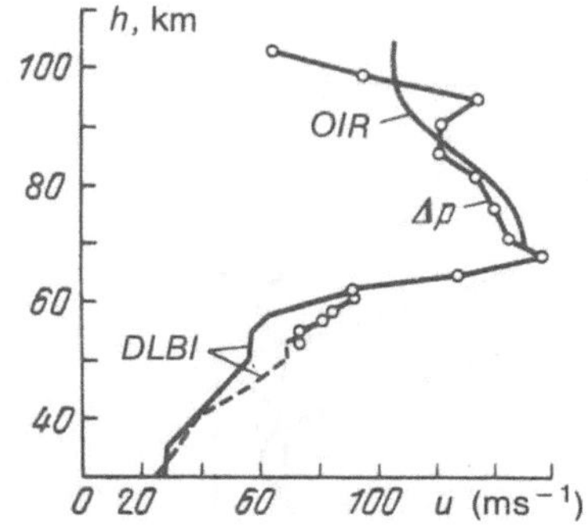

Fig. 147. Zonal wind velocities derived from the PV radio interferometric (*DLBI*), infrared radiometric (*OIR*), and pressure (Δp) measurements, the latter obtained by two probes with approximation of cyclostrophic balance (Seiff et al. 1980)

Another important technique of investigation of atmospheric dynamics is temperature profile measurements by the PV infrared radiometer orbiter (Taylor et al. 1980). These temperature profiles give a force field moving atmospheric masses. To calculate wind velocity profiles, a velocity field at the lower boundary should be known. The position of the lower boundary of infrared orbiter observations corresponds to $\tau = 1$ in the infrared region, i.e., located at 65 km. For $u = 100\ \mathrm{m\,s^{-1}}$ at 65 km a steep decrease of u should be observed at 85 km (Fig. 148). A cyclostrophic balance calculation gives the same result at high latitudes near 65°, but places this decrease at 110 km at low latitudes, ~30°. This technique results in maximal meridional wind velocities of $30\ \mathrm{m\,s^{-1}}$ poleward at 85 km and toward the equator velocities of $\sim 100\ \mathrm{m\,s^{-1}}$ at 100 – 110 km.

The ratio of day to night side thermospheric pressures near the terminator is so high that the situation can be approximated by gas expansion to vacuum, when a flow velocity equals the sound velocity, $220\ \mathrm{m\,s^{-1}}$ and $360\ \mathrm{m\,s^{-1}}$ in CO_2 and O respectively at 200 K. Though $[O] > [CO_2]$, a value of $\sim 200\ \mathrm{m\,s^{-1}}$ appears to be more suitable due to viscosity. Besides, it is difficult to maintain a great difference of temperature and pressure near the terminator with a wind velocity of $300\ \mathrm{m\,s^{-1}}$. Night airglow observations reveal this circulation by the enhancement of intensity at the antisolar region, relative to the mean night side region, by a factor of 6 for NO airglow at 115 km and by a factor of 1.5 for O_2 airglow at 100 km. Though the thermospheric temperature has a minimum at the antisolar region, the mesopause temperature measured by the 15 μm band radiometry features a small maximum here which can be attributed to the compression of a downward air flux and release of chemical energy by atomic components. The air flow across the terminator results in the enrichment of night side atmosphere in light components. This effect is the most pronounced for hydrogen and helium.

Lightning. Strong and frequent pulses of electromagnetic radiation at 10, 18, 36, and 80 kHz which might be caused by lightning were detected by V 11 – 14 (Ksanfomality 1979, 1983, Ksanfomality et al. 1983). These pulses are very much the same as those of terrestrial lightning. To preclude electriza-

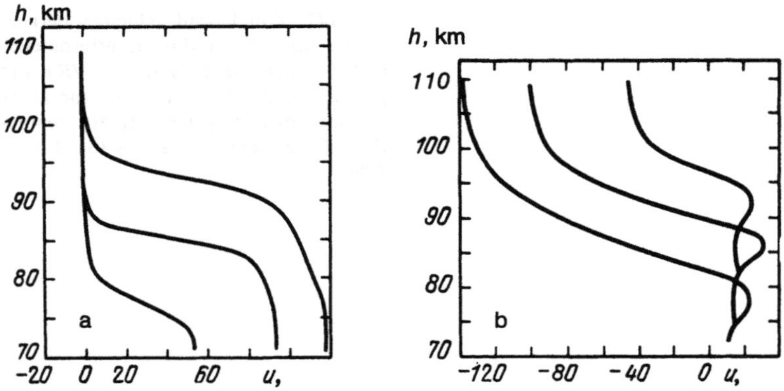

Fig. 148a, b. Altitude profiles of mean zonal (a) and meridional (b) wind a middle latitudes for various wind velocities adopted at 70 km (F. Taylor et al. 1980)

tion of probes during descent as a source of discharges, a special device to measure the discharge current was employed on board the V 13, 14 probes. A current of 55 nA was measured by both probes below 35 km; however, it proved to be due to a leak (Ksanfomality 1983). The highest frequency of the pulses has been observed below 35 km as well, but this coincidence does not indicate that the pulses are caused by microdischarges from the probe. The frequency of pulses is $\sim 30\ s^{-1}$.

A low frequency receiver with channels at 0.1, 1.73, 5.4, and 30 kHz was used on board the PV orbiter (Scarf et al. 1980, Ksanfomality et al. 1983). Some of the observed electromagnetic impulses have a source below the ionosphere and were interpreted as indicative of lightning.

In a search for lightning, Krasnopolsky (1980, 1983b) analyzed data from the V9, 10 orbiter airglow spectrometer in the visible (300 – 800 nm) range. Only one thunderstorm region (Fig. 149) having an area of $\sim 5 \times 10^4\ km^2$ and a lightning flash frequency of $\sim 100\ s^{-1}$ was detected over an area of $3.5 \times 10^7\ km^2$ on the night side of Venus in the latitude range $\pm 32°$. The pulse amplitudes were approx. 2000 times as high as the noise level and correlated with the instrument sensitivity (Fig. 149). Two locations of lightning, the cloud layer and the lowest scale height in the atmosphere, were considered. The absence of a fall off in the lightning spectra at 500 nm indicates that they occur in the cloud layer. Attenuation of light passing through the atmosphere and cloud layer is taken into account. For interpretation the light to total energy ratio is assumed the same as that on the Earth.

The mean total energy of a lightning flash on the Earth is poorly known and varies from about 10^7 to 10^{10} J. A value of 2×10^8 J recommended by Hill (1979) was used.

Figure 149 reproduces the measurements with approximate averaging because it is difficult to show all 480 points of the spectrum. The differences

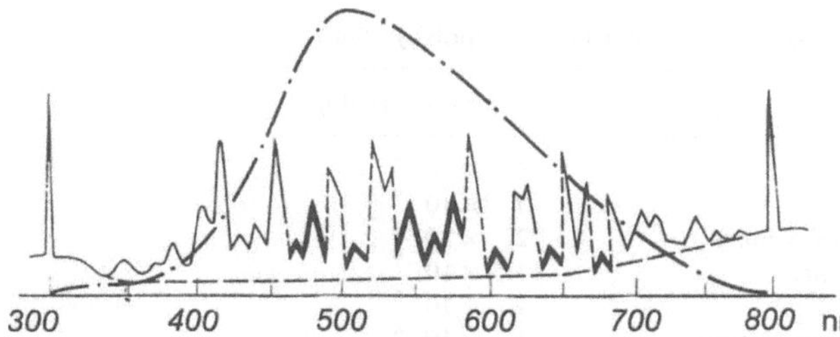

Fig. 149. One "spectrum" of the thunderstorm region. *Dashes* indicate the stray light level, *dark lines* show the reading for the gain less by a factor of 8. The night airglow spectra were measured without stray light with the gain exceeding by a factor of 10 that on the figure, for which the *dot-dash curve* indicates a reading corresponding to an energy of 1 W/A km^2 of the emission. The spectrum consists of 480 points and was obtained in 10 s; pulses at 300 and 800 nm indicate the beginning and end of the spectrum (Krasnopolsky 1980, 1983b)

in intensities between adjacent points which we refer to one flash are sometimes as high as twofold. Time intervals between successive measurements were 20 ms with the total time of the spectrum measurement of 10 s. The instrument time constant was 5 ms for weak signals, and some additional time was necessary for very short and intense light pulses to choose one of four possible gain values. All these facts should be taken into account when analyzing the absence of individual strokes in a flash. This led to some confusion in a consideration by Ksanfomality et al. (1983). The most natural explanation is that strokes are spaced at less than 20 – 30 ms, i.e., a flash consists of 10 – 20 strokes which is 3 – 5 times their number on the Earth. Based on this assumption, the data obtained are summarized in Table 29. The stroke rate in the near surface atmosphere is less than 5 s^{-1} over the entire planet if the light energy of the stroke exceeds 4×10^5 J and less than 15 s^{-1}

Table 29. Lightning and Thunderstorms on the Earth and Venus (Krasnopolsky 1983b)

Quantity	Earth	Venus
Flah duration, s	0.25	0.25
Mean number of strokes per flash	4	10 – 20
Energy of the flash in the visible range, J	5×10^6	2×10^7
Energy of the stroke in the visible range, J	1.3×10^6	1.3×10^6
Energy per unit stroke length in the visible range, J m^{-1}	500	–
Total energy of the flash, J	2×10^8	8×10^8
Thunderstorm area, km^2	10^3	5×10^4
Lightning rate within the thunderstorm area (per 10^3 km^2, s^{-1})	0.05	2
Lightning rate over the planet,	100	700
s^{-1} km^{-2} yr^{-1}	7	45
Mean energy release by lightning, erg cm^{-2} s^{-1}	0.04	1

Table 30. Illumination of the night side of Venus (Krasnopolsky 1980)

Source	Brightness (stilb)
Airglow	6×10^{-8}
Light of the stars	7×10^{-9}
Light of Earth (at phase 90° of Venus)	2×10^{-9}
Lightning (average for the planet)	1.5×10^{-7}
Threshold of detection	2×10^{-5}
Lightning (average for the storm region)	1×10^{-4}

for $(1-4) \times 10^5$ J. The light energy of a terrestrial stroke is $\sim 10^6$ J. Table 30 gives the night side atmospheric brightness produced by different sources.

A search for lightning by the PV orbiter star sensor gave negative results (Borucki 1982). The upper limit of the product of peak optical power and flash rate obtained depends on the assumed stroke duration, Δt. With $\Delta t \simeq 2$ ms, which is typical for Earth lightning, it equals 4.5×10^9 W km^{-2} yr^{-1}. This results in a flash rate less than 7 km^{-2} yr^{-1} for a peak optical power of 0.65×10^9 W according to Table 29. The upper limit obtained is less by a factor of 6 than the value derived by Krasnopolsky. However, both values have a great statistical uncertainty (one thunderstorm region was observed by Krasnopolsky and none by Borucki). Borucki calculated also the lower limit of the flash rate being 0.08 cm^{-2} yr^{-1} based on electric field detector measurements (Scarf et al. 1980).

Evidence for Vulcanism. Topographic features similar to terrestrial volcanic land forms were discovered by radar studies on Venus (Masursky et al. 1980). A thorough study of Whistler propagation demonstrates these features being a source of most of the Whistlers observed (Scarf and Russell 1983).

Actually 65% of the observed Whistlers originate from 5% of the studied area or 0.64% of the Venusian surface which coincides with the possible locations of volcanos. This can be explained by electrical discharges which accompany eruptions. Long-term variations of SO_2 at 40 mbar and upper haze optical thickness may be caused by volcanic eruptions as well (see Chap. 3.8). Sulfur dioxide in the lower atmosphere (Prinn 1984a, b) is out of equilibrium with the composition of rocks measured by V13, V14 (Surkov et al. 1983b) and need a source at the surface which may be volcanic eruptions.

4 Photochemistry of the Venusian Atmosphere

4.1 Day Side Ionosphere

Neutral atmospheric composition determines unambiguously the lower ionospheric composition in daytime with a minor influence of solar wind and magnetic field in this ionospheric region. Although strictly speaking, inverse assertion is not correct, because different models of neutral atmosphere giving a similar ionospheric composition can be proposed, this approach is usually used in practice. Comparison with results of measurement calculations of ionospheric composition with known neutral composition makes it possible to check our knowledge of processes forming the ionosphere and the reliability of the experimental data. On the other hand, the investigation of some ionized components gives clues for the study of some neutral components which are not easily accessible by direct measurements.

We will not dwell upon calculations of the Venusian ionosphere made prior to PV. Calculations based on the PV orbiter data were performed, e.g., by Nagy et al. (1980), Izakov et al. (1981), and Fox and Dalgarno (1981).

Nagy et al. explore neutral atmospheric composition measured by the PV orbiter mass spectrometer (see Chap. 3.11). Nitric oxide number densities are not measured, but taken from calculations by Rusch and Cravens (1979, see Chap. 4.3). Atomic hydrogen number densities were derived by Brinton et al. (1980) from data on [H^+]; therefore calculations of [H^+] by Nagy et al. are the simple inversion of this work.

Data on solar UV radiation at enhanced solar activity, ionization cross-sections with dissociative ionization of N_2 and CO_2 giving CO^+, O^+, C^+, N^+ are reviewed by Torr et al. (1979), Schunk and Nagy (1980), and Fox and Dalgarno (1979b). Table 31 represents a list of chemical reactions used. Ionic diffusion was introduced in the calculations. The results are shown in Fig. 150. The calculated and measured ionic number densities differ by a factor of $2-3$. One should bear in mind that both neutral atmospheric and ionospheric composition measurements by various instruments (the mass spectrometer and the orbiter drag measurements of neutral composition, the ion mass spectrometer, the Langmuir probe, the retarding potential analyzer, and radio-occultation measurements of ionospheric plasma) have almost the same scatter in data. Therefore, with errors of data on the reaction rate coefficient taken into account, this difference proves to be not very large, thus confirming the adequacy of the scheme of processes adopted.

Table 31. Ion-neutral reaction rates (Nagy et al. 1980)

	Reaction	Rate constant, $cm^3 s^{-1}$
*n*1	$CO_2^+ + O \rightarrow O^+ + CO_2$	9.6×10^{-11}
*n*2	$CO_2^+ + O \rightarrow O_2^+ + CO$	1.64×10^{-10}
*n*3	$CO_2^+ + NO \rightarrow NO^+ + CO_2$	1.2×10^{-10}
*n*4	$CO^+ + O \rightarrow O^+ + CO$	1.4×10^{-10}
*n*5	$CO^+ + CO_2 \rightarrow CO_2^+ + CO$	1.0×10^{-9}
*n*6	$CO^+ + NO \rightarrow NO^+ + CO$	3.3×10^{-10}
*n*7	$O_2^+ + NO \rightarrow NO^+ + O_2$	4.5×10^{-10}
*n*8	$O_2^+ + N \rightarrow NO^+ + O$	1.8×10^{-10}
*n*9	$N_2^+ + CO_2 \rightarrow CO_2^+ + N_2$	9.0×10^{-10}
*n*10	$N_2^+ + NO \rightarrow NO^+ + N_2$	3.3×10^{-10}
*n*11	$N_2^+ + O \rightarrow NO^+ + N$	1.4×10^{-10}
*n*12	$C^+ + CO_2 \rightarrow CO^+ + CO$	1.9×10^{-9}
*n*13	$O^+ + CO_2 \rightarrow O_2^+ + CO$	9.4×10^{-10}
*n*14	$O^+ + N_2 \rightarrow NO + N$	1.2×10^{-12}
*n*15	$O^+ + H \rightarrow H^+ + O$	$2.5 \times 10^{-11} T_n^{1/2}$
*n*16	$N^+ + CO_2 \rightarrow CO_2^+ + N$	1.3×10^{-9}
*n*17	$N^+ + CO_2 \rightarrow NO^+ + CO$	2×10^{-11}
*n*18	$N^+ + CO \rightarrow CO^+ + N$	5×10^{-10}
*n*19	$N^+ + NO \rightarrow NO^+ + N$	9×10^{-10}
*n*20	$He^+ + CO_2 \rightarrow CO^+ + O + He$	8.7×10^{-10}
*n*21	$He^+ + CO_2 \rightarrow CO_2^+ + He$	1.2×10^{-10}
*n*22	$He^+ + CO_2 \rightarrow O^+ + CO + He$	1.0×10^{-10}
*n*23	$He^+ + CO \rightarrow$ products	1.68×10^{-9}
*n*24	$He^+ + CO \rightarrow C^+ + He + O$	1.7×10^{-9}
*n*25	$He^+ + NO \rightarrow N^+ + He + O$	1.5×10^{-9}
*n*26	$He^+ + N_2 \rightarrow N^+ + He + N$	8×10^{-10}
*n*27	$He^+ + N_2 \rightarrow N_2^+ + He$	6.4×10^{-10}
*n*28	$H^+ + CO_2 \rightarrow CHO^+ + O$	3×10^{-9}
*n*29	$H^+ + O \rightarrow O^+ + H$	$2.2 \times 10^{-11} \times T_i^{1/2}$

Izakov et al. (1981) considered the daytime ionosphere below 220 km in the approximation of photochemical equilibrium. In their model the neutral atmospheric composition was almost the same as in the paper by Nagy et al. (1980) except that number densities of O_2, NO, C, H, H_2 were free parameters. Two additional components, CO_2H^+ and HCO^+ and their reactions

$$CO_2^+ + H_2 \rightarrow CO_2H^+ + H \quad k = 1.4 \times 10^{-9}\,cm^3 s^{-1}$$
$$CO_2^+ + H \rightarrow HCO^+ + O \quad 5 \times 10^{-10}$$
$$O^+ + H_2 \rightarrow OH^+ + H \quad 1.6 \times 10^{-9}$$
$$CO^+ + H_2 \rightarrow HCO^+ + H \quad 1.4 \times 10^{-9}$$
$$CO_2H^+ + e \rightarrow CO_2 + H \quad 3.5 \times 10^{-7}$$
$$HCO^+ + e \rightarrow CO + H \quad 3 \times 10^{-7}$$

were introduced. These ions were not detected, and their calculated number densities equal $10-100\,cm^{-3}$. The third reaction produces OH^+ ions, which

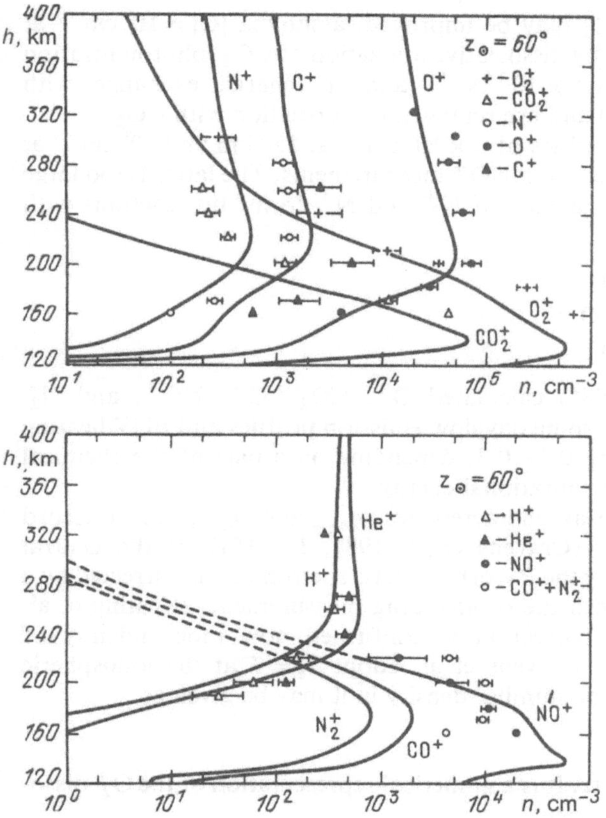

Fig. 150. Calculated (*lines*) and measured (*symbols*) day side ionospheric composition (Nagy et al. 1980)

were also not measured. A great number of ionic reactions with molecular oxygen included turns out to be unimportant for $f_{O_2} < 10^{-3}$ at 100 km. Charge exchange reaction of C^+ and CO used by Izakov et al. (1981) does not proceed for thermal components and should be eliminated.

The agreement of calculated and measured ion concentrations is slightly better than in the above mentioned paper. To fit the observed H^+, the neutral hydrogen number density should be $[H] \simeq 5 \times 10^4\,cm^{-3}$. The same procedure for H_2^+ ions with reactions

$$H_2 + h\nu \rightarrow H_2^+ + e$$
$$H_2^+ + O \rightarrow OH^+ + H$$
$$H_2^+ + e \rightarrow H + H$$

taken into account results in $[H_2] \simeq 10^5\,cm^{-3}$ at 200 km. Such a high value of H_2 can reduce $[O^+]$ and n_e at $h \sim 200$ km near the secondary ionospheric maximum.

The coincidence of $[C^+]$ may be improved, assuming $[C] \simeq 10^6\,cm^{-3}$ at 200 km. Carbon ions form by dissociative ionization of CO_2, photoionization of neutral carbon ($J = 1.5 \times 10^{-6}\,s^{-1}$) and its charge exchange with O_2^+ ($k = 3 \times 10^{-10}\,cm^3\,s^{-1}$) and are removed via a reaction with CO_2.

Nitric oxide number densities of $6 \times 10^6\,cm^{-3}$ at 150 km and $10^6\,cm^{-3}$ at 200 km should be introduced to fit NO^+ measurements. The latter is too large to be a likely value. A great part of N_2^+ and N^+ forms by reactions with excited oxygen ions

$$O^+(^2D) + N_2 \rightarrow N_2^+ + O$$
$$O^+(^2P) + N_2 \rightarrow N^+ + NO$$
$$O^+(^2P) + N \rightarrow N^+ + O\,.$$

Fox and Dalgarno (1981) calculated O_2^+, CO_2^+, O^+, NO^+, and N_2^+ number densities as well as some dayglow emission profiles and EUV heating efficiencies, the latter being 0.2–0.3, depending on a part of the chemical reaction energy released on vibrational energy.

Day side ionospheric peak characteristics, $n_{e_{max}}$ and h_{max}, are in accord with the Chapman theory (Cravens et al. 1981, Fig. 151), if the neutral number density measured by the PV orbiter mass spetrometer is corrected by a factor of 1.5, in accord with the orbiter drag measurements (Keating et al. 1980). The need of this correction is confirmed now independently of ionospheric measurements. Cravens et al. adopt $T_e = T$ at the ionospheric maximum (140 km). Electron number density in it may be given as

$$n_{e_{max}} = 520\,P^{1/2}\,T_e^{0.275}$$

where P is the ionization rate. It is a numerical representation of the O_2^+ dissociative recombination coefficient.

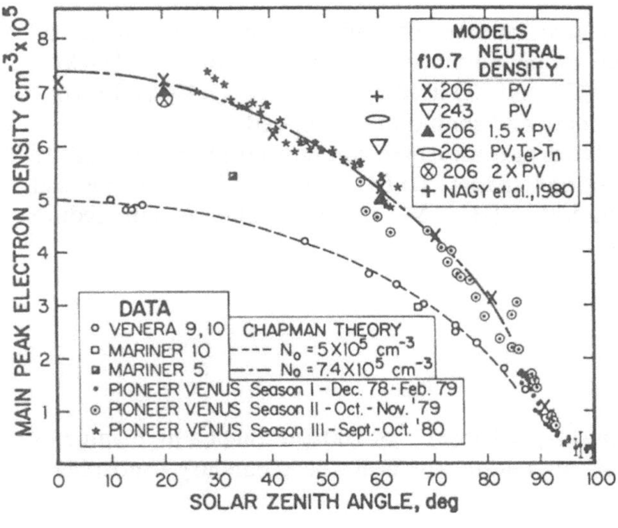

Fig. 151. Electron number densities at the ionospheric peak vs SZA, Both V9, 10 and PV data are in accord with the Chapman theory for low and high solar activity, respectively (Cravens et al. 1981)

Fox and Victor (1981) studied the altitude behavior of double ionized oxygen O^{++}. Its source is photoionization of O^+ ($\lambda < 35$ nm, $I = 0.9 \times 10^{-7}$ s^{-1} at the Venusian orbit) and double photoionization of neutral oxygen with a threshold of $\lambda < 25$ nm and a ratio of O^{++}/O^+ yields equals 10% in this spectral region. Loss mechanisms of O^{++} are reactions with molecular and atomic species having rate coefficients of 1.5×10^{-9} and 1.5×10^{-10} $cm^3\,s^{-1}$, respectively.

4.2 Nighttime Ionosphere

Two main problems are raised concerning the night side ionosphere: sources of ionization and its great variability (established, depleted, disappearing ionospheres, and ionospheric holes).

Maintenance of the Night Side Ionosphere. Four sources are usually related to the night side ionization:

1. day side plasma flow across the terminator;
2. superthermal electron fluxes;
3. photoionization by UV emissions of interplanetary gas;
4. meteorite ionization.

Let us consider the first source. Day side plasma number density is $\sim 10^5$ cm^{-3} at 150–200 km, and with a flow velocity of ~ 1 km s^{-1} it forms at the night side a downward flux of $\sim 10^8$ $cm^{-2}\,s^{-1}$. The ion flow at different altitudes and locations was measured by the PV retarding potential analyzer (Knudsen et al. 1980). Directions and velocities of the flow were derived from the instrument output variations caused by the orbiter rotation. The velocity projection in a direction away from the sun was measured to be ~ 2 km s^{-1} at 300 km and may be as high as 8 km s^{-1} near the ionopause. Then the mean night side downward flux of O^+ ions dominating in the upper ionosphere is

$$\phi_{O^+} = [O^+]\, V \frac{2\pi r \Delta r}{2\pi r^2} \simeq 1.5 \times 10^8\ cm^{-2}\,s^{-1}$$

where r is the radius of the planet and $\Delta r \simeq 500$ km is the thickness of the ionosphere near the terminator. The product $[O^+]\, V \simeq 2 \times 10^9$ $cm^{-2}\,s^{-1}$ is almost independent of height near the terminator. Ionospheric data on the night side give integrated ionization rates as 3×10^9, 5×10^8, and 7×10^7 $cm^{-2}\,s^{-1}$ at $z_0 = 90$, 95, and $\geqslant 110°$, respectively, with a hemispheric mean value of 2×10^8 $cm^{-2}\,s^{-1}$. Downward diffusion calculations fit the observed altitude and electron number density at the ionospheric peak with $\phi_{O^+} = 6 \times 10^7$ $cm^{-2}\,s^{-1}$ (Fig. 152).

One can estimate plasma loss during its transport to the night side by the time of this transport, $\tau \simeq r/V \simeq 1$ h. O^+ recombination is radiative and has a very low rate coefficient, 5×10^{-12} $cm^3\,s^{-1}$. Of greater importance are the

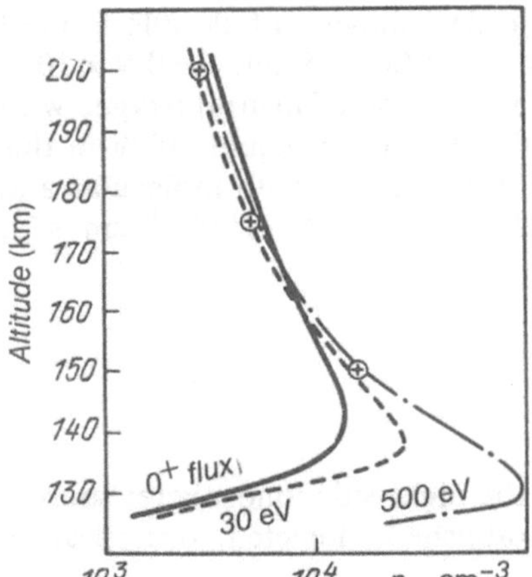

Fig. 152. Calculated electron number densities in the night side ionosphere for different sources of ionization: O^+ flux equal to $6 \times 10^7\,cm^{-2}\,s^{-1}$, and electron fluxes of 4×10^8 and $10^8\,cm^{-2}\,s^{-1}$ with energies of 30 and 500 eV, respectively (Brace et al. 1979)

reactions with the molecular species, H_2, N_2, and CO_2, their concentrations should be $3 \times 10^5\,cm^{-3}$ to produce a lifetime of ~ 1 h. For CO_2 and N_2 they correspond to 165 km at the night side, which is below the bulk ionospheric flow. However, the predawn bulge of H_2 may prevent O^+ propagation. If this effect is discovered it can verify the role of the mechanism considered.

Superthermal Electron Fluxes. As we discussed in Chap. 3.12, these fluxes were discovered by the V 9, 10 orbiters and claimed as the main source of the night side ionization. Similar fluxes were measured by PV, and their role as well as that of ionic flow was analyzed by Spenner et al. (1981). Their approach is based on PV electronic flux, night side electron number density, and ionic composition measurements. Correlation of electron fluxes with n_e and a ratio of O^+ O_2^+ which is different for both mechanisms have been taken into account.

Though the electron energy loss processes are well known (despite the absence of some data on cross-sections), their exact descriptions are rather complicated. For $E \gtrsim 30$ eV molecular bounds become negligible and molecular cross-section is equal to the sum of atomic ones. At $E \lesssim 30$ eV elastic scattering of electrons dominates with $\sigma_e \simeq 10^{-15}\,cm^2$ for CO_2. In this energy range it is almost isotropic. At $E > 50-100$ eV forward elastic scattering becomes pronounced, therefore this process may be neglected here in calculations of mean path and ionization and excitation of atmospheric species. In this range the electron mean path in CO_2 is given by

$$10^{15} + 10^{17}(E^{1.6} - 0.1^{1.6})\,cm^{-2}$$

(Krasnopolsky 1979d), with E expressed in KeV. Spenner et al. (1981) use equivalent optical thickness of the atmosphere,

$$\tau = \sin^{-1} D \cos^{-1} \theta \sum_k n_k H_k [\sigma_a(\sigma_a + \sigma_e)]^{1/2}$$

where D is the magnetic inclination, i.e., the angle of a magnetic field line to a horizontal plane, θ is the mean angle of the electron velocity to the magnetic field line, n_k and H_k are the number density and the scale height of the component k, σ_a is the absorption cross-section. With equal probabilities of forward and backward scattering at $E < 100$ eV the ionization rate is given by

$$q_k = n_k \int_0^\infty \frac{\sigma_{ik}(\varepsilon) I_0(\varepsilon) e^{-\tau} \cdot d\varepsilon}{1 + [\sigma_a(\sigma_a + \sigma_e)]^{1/2}}$$

where $I_0(E)$ is the electron flux outside the atmosphere. The altitude of ionospheric maximum and the O^+/O_2^+ ratio in it depend on D. To fit the observations the measured electron fluxes should contribute ~25% to the total ionization rate, the rest being caused by the day side plasma flow.

Measurements of electron spectra made in the range 15 – 45 eV by the PV orbiter gave practically the same results as V9, 10. At higher energies a linear extrapolation in semilogarithmic scale to 70 eV were used in calculation with no electrons having $E > 70$ eV adopted. It may be readily shown (Krasnopolsky 1982c) that electrons with $E > 70$ eV should increase the total ionization rate by factors of 1.5 and 5 for the Boltzmann energy distribution and for the distribution similar to that measured by the V9, 10 orbiters, respectively. Therefore, we feel that the part of the ionization rate caused by the electron flux may be larger, 25 – 50%. Mean $n_{e_{max}} = 0.8 \times 10^4$ and $1.7 \times 10^4\,cm^{-3}$ under conditions of the V9, 10 and PV observations, i.e., at low and high solar activity, respectively. Therefore, with the total ionization rate being $2 \times 10^8\,cm^{-2}\,s^{-1}$ for PV observations that for V9, 10 should be $5 \times 10^7\,cm^{-2}\,s^{-1}$. $[O^+]$ near the terminator were smaller by a factor of 3 at the time of the V9, 10 measurements than for PV, so the contribution of electron fluxes should be $\sim 10^7\,cm^{-2}\,s^{-1}$ for the V9, 10 observations, or 5×10^8 eV $cm^{-2}\,s^{-1}$. To be in accord with the measurements (Gringauz et al. 1979), D should be less than 5°, and in the absence of the magnetic field the ionization rate measured by Gringauz et al. exceeds the total one by a factor of 6. Correlation of electron fluxes and $n_{e_{max}}$ measured by V9, 10 is hardly observed and, thus cannot support the idea of electron fluxes playing a main part in maintenance of the night side ionosphere.

Some important aspects of ionospheric behavior were explained by the two-dimensional model of the Venusian ionosphere (Cravens et al. 1983). Initial data of this model are very similar to those of Nagy et al. (1980). The altitude h and the solar zenith angle z_0 were chosen as coordinates in this model, ionic horizontal velocities being additional parameters. To fit the observations, the latter are given by

$$V(h, z_0) = \sin z_0 \times [1 + 0.012(h - 200\,\text{km})]\ \text{km s}^{-1}$$

and the ionopause height near the terminator assumed to be 270, 500, and 876 km for different variants of calculations. Two assumptions concerning molecular hydrogen were made, namely, $[H_2] \simeq 10^5\,cm^{-3}$ and no H_2. Pre-

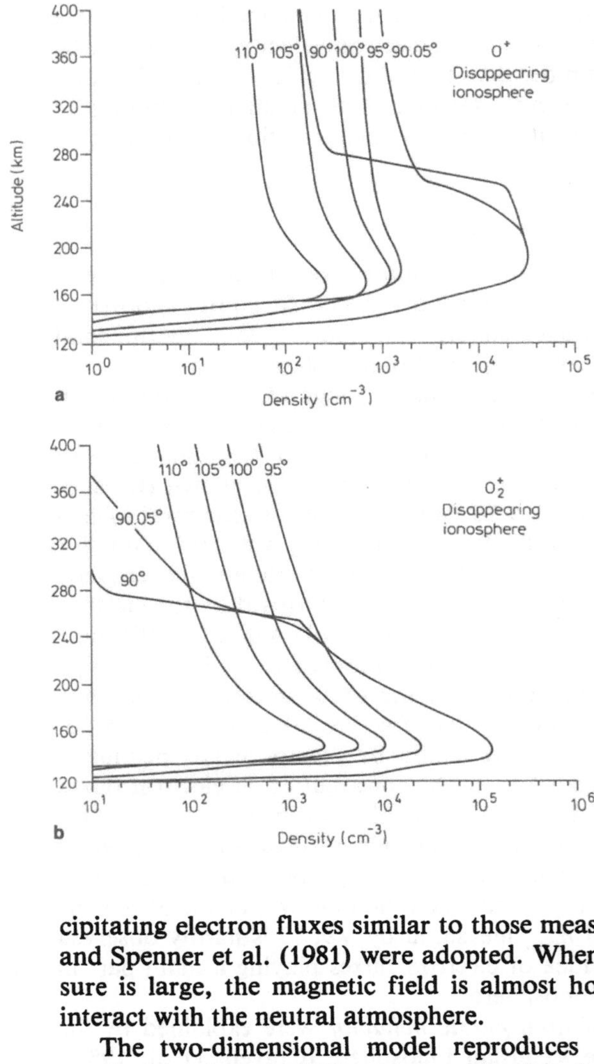

Fig. 153a, b. Calculated O^+ (**a**) and O_2^+ (**b**) number densities as a function of solar zenith angle; the assumed terminator ionopause height is 270 km (Cravens et al. 1983)

cipitating electron fluxes similar to those measured by Gringauz et al. (1979) and Spenner et al. (1981) were adopted. When the solar wind dynamic pressure is large, the magnetic field is almost horizontal and electrons do not interact with the neutral atmosphere.

The two-dimensional model reproduces slightly smaller day side ion number densities (~50% at some altitudes) than the one-dimensional one. It is due to the transport of some part of the day side ionospheric plasma to the night side. However, this difference is not very important.

Calculations with the low terminator ionopause (270 km) are shown in Fig. 153. In this case the plasma flow is very small, the height of the ionopause increases rapidly at $z_0 > 90°$, ionospheric number densities show a gradual decrease with z_0. This is the case of the disappearing ionosphere.

O_2^+ peak number densities versus z_0 are shown in Fig. 154 for the low, middle, and high ionopause as well as those of NO^+ for the high ionopause. The same for n_e at 200 km is given in Fig. 155. The results of measurements

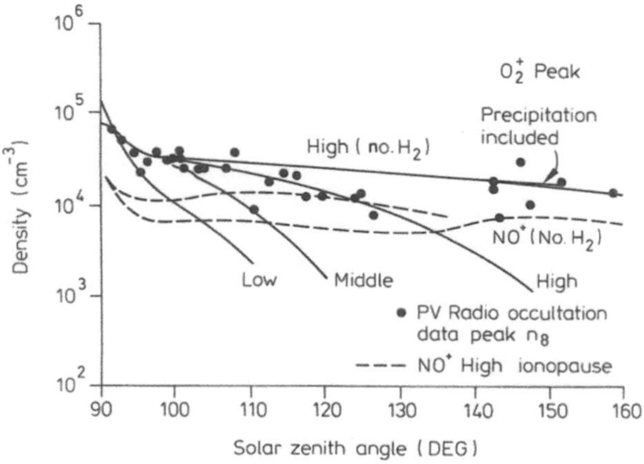

Fig. 154. Calculated peak O_2^+ density variations for the three assumed ionopause heights at the terminator; $n_{e_{max}}$ measured by the radio-occultation technique are shown for comparison (Cravens et al. 1983)

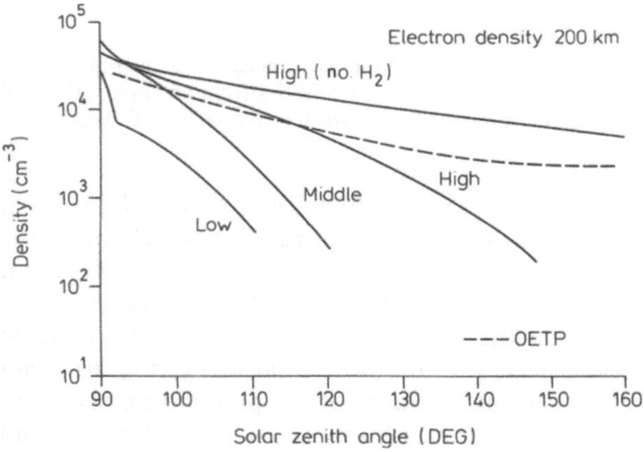

Fig. 155. Calculated electron density variations at 200 km for the three assumed ionopause heights at the terminator; values measured by the PV Langmuir probe are shown for comparison (*dashed line*) (Cravens et al. 1983)

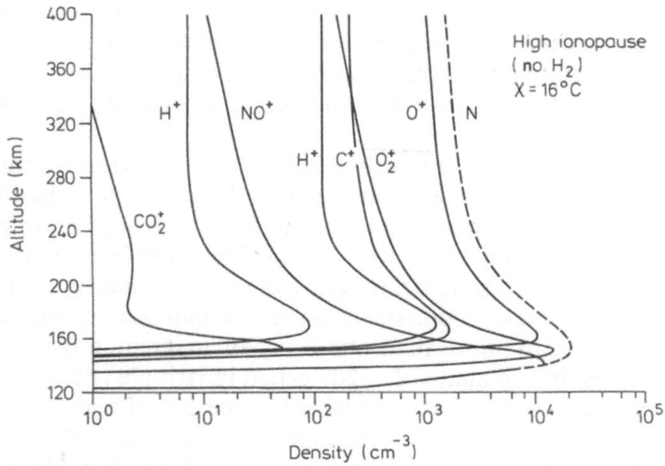

Fig. 156. Calculated ion density profiles for $z_0 = 160°$, the terminator ionopause height of 876 km, and $[H_2] = 0$ (Cravens et al. 1983)

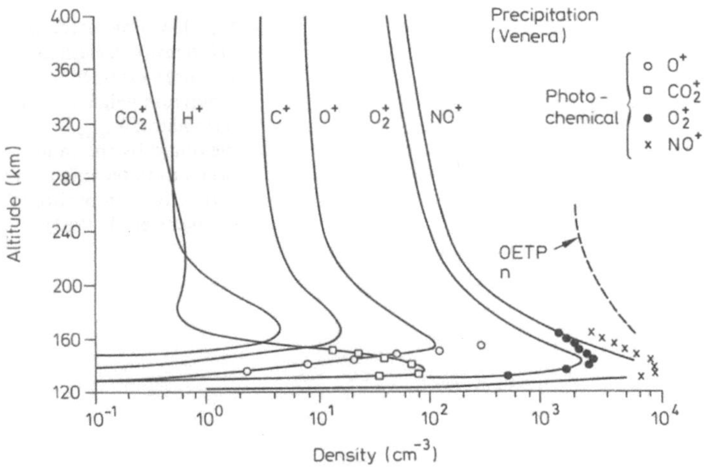

Fig. 157. Ion density profiles calculated assuming electron impact ionization as the only source. The electron flux used is that measured by V9, 10. Electron densities from an empirical model (Brace et al. 1983) based on PV measurements are also shown for comparison (Cravens et al. 1983)

are shown in both figures. They reveal that the high ionopause with no H_2 in the upper atmosphere gives the best fit to observations. Figure 156 shows the calculated ionospheric composition for $z_0 = 160°$. The measured O^+, O_2^+, and NO^+ are very nearly the same as the calculated values. The calculated ionospheric peak is located at 150–155 km. In situ PV measurements also indicate the peak near 150 km on some orbits, with radio-occultations giving the mean peak altitude of 142 km. The same calculations for the precipitating electron flux only is in Fig. 157. The main difference is low O^+ number densities and the fact that $[NO^+] > [O_2^+]$ throughout the ionosphere. The peak is near 135 km, too low even for the radio-occultation measurements. With both sources being considered, the electron fluxes make, in general, a small contribution only. However, these fluxes may be important in the case of the low ionopause.

Both sources of ionization are reduced by high solar wind pressure: the plasma flow by the lower ionopause and the electron fluxes by the induced horizontal magnetic field. The disappearing ionosphere occurs from this. In it electron fluxes are much more important, while being almost negligible in the established ionosphere. So the disappearing ionosphere is a planetary scale phenomenon. Local ionospheric depressions, so-called ionospheric holes, have plasma densities lower by more than one order of magnitude than in adjacent regions, localized in two zones of ~20° extent in latitude and ~40° in longitude. The zones are centered at ~01.00 h LT and 30° N and 12° S. The usual dimension of the holes is ~1000 km. The magnetic field is enhanced in them and almost vertical. No complete theory for their explanation has yet been developed.

Direct PV measurements provide no information on the ionosphere below the main peak. Hence in recent years no interesting results on this subject have become available. Krasnopolsky (1979d) considered two sources of ionization in this region, photoionization by He 58 nm and He^+ 30 nm emission of interplanetary helium with intensities of $10\,R$ and $\sim 1\,R$, respectively, and meteorite ionization. The maximum of photoionization should occur at the level of $[CO_2] \simeq 10^{11}\,cm^{-3}$ (~ 130 km) and yield $n_e \simeq 3\times 10^3\,cm^{-3}$ in the absence of other sources.

The average meteorite flux results in a total ionization rate of $2\times 10^5\,cm^{-2}\,s^{-1}$ with a maximum near 120 km ($[CO_2] \simeq 10^{12}\,cm^{-3}$). Its high variability gives an explanation of the changes in the lower ionospheric peak. It produces predominantly atomic ions with a low ionization potential, i.e., Mg^+, Fe^+, Si^+, etc. Their recombination is a result of the following reactions:

$$A^+ + CO_2 \rightarrow ACO_2^+ + h\nu\,, \qquad k_1 \simeq 3\times 10^{-17}\,cm^3\,s^{-1}$$
$$A^+ + CO_2 + CO_2 \rightarrow ACO_2^+ + CO_2\,, \qquad k_2 \simeq 2\times 10^{-29}\,cm^6\,s^{-1}$$
$$ACO_2^+ + CO_2 \rightarrow A^+ + CO_2 + CO_2\,, \qquad k_3 \simeq 10^{-14}\,cm^3\,s^{-1}$$
$$ACO_2^+ + e \rightarrow A + CO_2\,, \qquad k_4 \simeq 10^{-6}\,cm^3\,s^{-1}\,.$$

k_3 and k_2 have been measured for A = Na, and k_1 and k_2 are estimates. With other sources of ionization being absent the meteorite ionization forms a peak at 120 km with $n_{e_{max}} \simeq 5\times 10^3\,cm^{-3}$, which is in accord with measurements.

Reviews of data on the Venusian ionosphere where given by Brace et al. (1983) and Nagy et al. (1983a).

4.3 Metastable Species in the Venusian Ionosphere. Nitric Oxide, Atomic Nitrogen, and Atomic Carbon

Metastable species were ignored in the papers discussed in the previous two paragraphs. The role of these species was considered by Fox (1982a), who supplemented the list of ionospheric chemical reactions used by Fox and Dalgarno (1979b, 1981) for calculations of Martian and Venusian ionospheric composition (Table 8) by 47 reactions describing the chemistry of $O^+(^2D)$, $O^+(^2P)$, $N(^4S)$, $N(^2D)$, $N(^2P)$, NO, N^+, C^+, O^{++} (Table 32). Solar fluxes were used for high solar activity with $z_0 = 18°$.

Calculated number densities of atomic oxygen ions are shown in Fig. 158. Being multiplied by appropriate transition probabilities they result in volume emission rates (Table 33). Atomic nitrogen and nitric oxide number densities are given in Fig. 159. The model of Fox (1982a) is in fair agreement with measurements for both O^+ and N as well as for N^+ and CO^+.

Atomic nitrogen and nitric oxide number densities were calculated by Fox (1982a) on the assumption of diffusive equilibrium at the upper boundary. However, atomic nitrogen produced at the day side is transported to the night

Table 32. Reactions of N, NO, C, and metastable species (Fox 1982a)

Reaction number	Reaction	Rate constant
(R19)	$O^+ + N_2 \rightarrow NO^+ + N$	1.2×10^{-12}
(R20)	$N^+ + O \rightarrow N + O^+$	2.2×10^{-12}
(R21)	$N^+ + CO \rightarrow N + CO^+$	4.0×10^{-10}
(R22)	$N^+ + CO \rightarrow NO^+ C$	5.4×10^{-11}
(R23)	$N^+ + CO_2 \rightarrow CO^+ + NO$	2.5×10^{-10}
(R24)	$CO^+ + O \rightarrow CO^+ + O^+$	1.4×10^{-10}
(R25)	$CO^+ + NO \rightarrow CO + NO^+$	3.3×10^{-10}
(R26)	$CO^+ + e \rightarrow C + O$	$6.8 \times 10^{-7} (293/T_e)^{0.58}$
(R27)	$O_2^+ + N \rightarrow NO^+ + O$	1.2×10^{-10}
(R28)	$N_2^+ + O \rightarrow N_2 + O^+$	$7.4 \times 10^{-12} (T_i/300)^{-0.44}$
(R29)	$He^+ + CO \rightarrow C^+ + O + He$	1.4×10^{-9}
(R30)	$He^+ + CO_2 \rightarrow CO^+ + O + He$	7.9×10^{-10}
(R31)	$\rightarrow CO_2^+ + He$	1.1×10^{-10}
(R32)	$\rightarrow O^+ + CO + He$	9×10^{-11}
(R33)	$He^+ + N_2 \rightarrow N^+ + N + He$	8.4×10^{-10}
(R34)	$O^+(^2D) + N_2 \rightarrow N_2^+ + O$	8×10^{-10}
(R36)	$O^+(^2D) + e \rightarrow O^+(^4S) + e$	$6.6 \times 10^{-8} (300/T_e)^{0.5}$
(R37)	$O^+(^2D) + O \rightarrow O^+(^4S) + O$	1×10^{-11}
(R38)	$O^+(^2D) + N \rightarrow O + N^+$	1.5×10^{-10}
(R39)	$O^+(^2D, ^2P) + NO \rightarrow NO^+ + O$	1.2×10^{-9}
(R40)	$O^+(^2D, ^2P) + CO \rightarrow CO^+ + O$	1.3×10^{-9}
(R41)	$O^+(^2D) + CO_2 \rightarrow CO + O_2$	8×10^{-10}
(R42)	$\rightarrow CO_2^+ + O$	
(R43)	$\rightarrow CO^+ + O_2$	
(R44)	$O^+(^2P) + N_2 \rightarrow NO^+ + N(^2D)$	1.5×10^{-11}
(R45)	$O^+(^2P) + N_2 \rightarrow N_2^+ + O$	1.35×10^{-10}
(R46)	$O^+(^2P) + N_2 \rightarrow N^+ + NO$	1.3×10^{-10}
(R47)	$O^+(^2P) + e \rightarrow O^+(^4S) + e$	$3.2 \times 10^{-8} (300/T_e)^{0.5}$
(R48)	$O^+(^2P) + O \rightarrow O^+(^2D) + O$	5.2×10^{-10}
(R49)	$O^+(^2P) + e \rightarrow O^+(^2D) + e$	$1.5 \times 10^{-7} (300/T_e)^{0.3}$
(R50)	$N(^2D) + O^+ \rightarrow N^+ + O$	1.3×10^{-10}
(R51)	$N(^2D) + O_2^+ \rightarrow N^+ + O_2$	4×10^{-10}
(R52)	$N(^2D) + CO_2^+ \rightarrow N^+ + CO_2$	2×10^{-10}
(R53)	$\rightarrow NO^+ + CO$	
(R54)	$\rightarrow CO^+ + NO$	
(R55)	$N(^2D) + e \rightarrow N(^4S) + e$	$2.7 \times 10^{-13} T_e$
(R56)	$N(^2D) + CO_2 \rightarrow NO + CO$	6.8×10^{-13}
(R57)	$N(^2D) + O \rightarrow N(^4S) + O$	2×10^{-12}
(R58)	$N(^2D) + NO \rightarrow N_2 + O$	6.3×10^{-11}
(R59)	$N(^2D) + CO \rightarrow N(^4S) + CO$	2.5×10^{-12}
(R60)	$N(^2D) + N_2 \rightarrow N(^4S) + N_2$	1.7×10^{-14}
(R61)	$N(^2P) + CO_2 \rightarrow N(^4S) + CO_2$	1.25×10^{-13}
(R62)	$N(^2P) + CO \rightarrow N(^4S) + CO$	8×10^{-14}
(R63)	$N(^2P) + NO \rightarrow N_2 + O$	3×10^{-11}
(R64)*	$N(^2P) + O \rightarrow N(^2D) + O$	1×10^{-11}
(R65)	$N(^4S) + CO_2 \rightarrow NO + CO$	1.7×10^{-16}
(R66)	$N(^4S) + NO \rightarrow N_2 + O$	3.4×10^{-11}

* This reaction may proceed according to $N(^2P) + O \rightarrow NO^+ + e$, but the net effect will be the same since NO^+ is lost by dissociative recombination: $NO^+ + e \rightarrow N(^2D) + O$

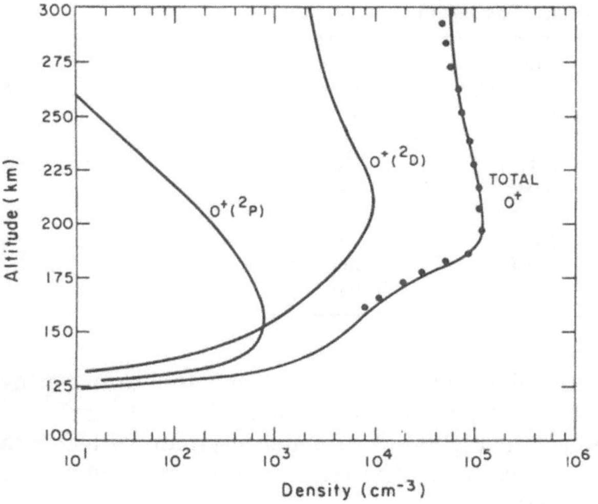

Fig. 158. Calculated atomic oxygen number density profiles; the *circles* are O^+ densities measured (Fox 1982a)

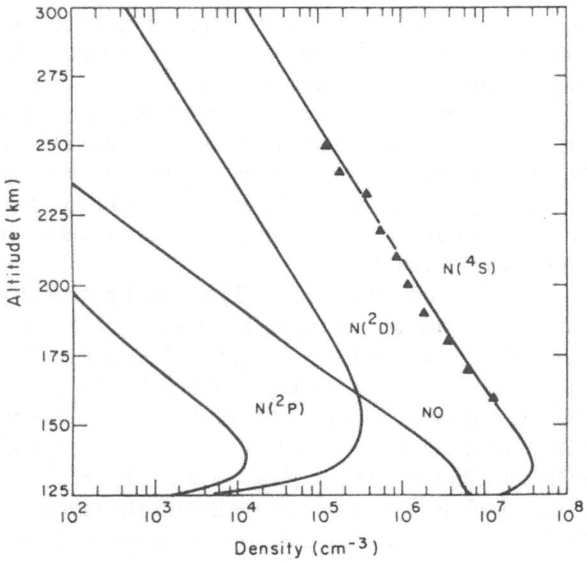

Fig. 159. Calculated atomic nitrogen and nitric oxide density profiles (Fox 1982a). The *triangles* are the total atomic nitrogen densities measured by Kasprzek et al. (1980)

side and is the source of the nitric oxide nightglow which needs the downward mean flux of N to be $10^{10}\,cm^{-2}\,s^{-1}$ (see Chap. 3.11). N and NO number densities and N flow to the night side were calculated by Krasnopolsky (1983b) with the usual N and NO chemistry being involved.

The mean day side N and NO (the latter forms as a result of the conversion of $N(^2D)$ in a reaction with CO_2) production rates are $7.2 \times 10^9\,cm^{-2}\,s^{-1}$ and $4 \times 10^9\,cm^{-2}\,s^{-1}$, respectively, at high solar activity. A major sink of both components being the reaction of N and NO, the excess of N equals 3.2×10^9 $cm^{-2}\,s^{-1}$ and is a subject of the downward transport to the lower atmosphere

Table 33. O^+ and N dayglow of Venus (Fox 1982a)

λ, nm	Transition	Transition probability, s^{-1}	Zenith intensity, *R*
732	$O^+(^2P \rightarrow ^2D)$	0.17	720
247	$O^+(^2P \rightarrow ^4S)$	0.048	200
373	$O^+(^2D \rightarrow ^4S)$	5×10^{-3}	4
520	$N(^2D \rightarrow ^4S)$	10^{-5}	16
1040	$N(^2P \rightarrow ^2D)$	0.079	2500
347	$N(^2P \rightarrow ^4S)$	0.005	160

or of the horizontal flow to the night side. It is easy to show the former being negligible as compared with the latter, and the flow of $3.2 \times 10^9\ cm^{-2}\ s^{-1}$ nitrogen atoms should produce the nightglow of δ and γ systems of NO with intensities of 1 and 0.5 *kR*, respectively. The latter values are smaller by a factor of 2.5 than measured ones.

Similar calculations being applied to the Martian atmosphere result in NO number densities less by a factor of 3 than the measured ones (see Chap. 2.2). Therefore, an attempt to explain the large NO nightglow intensity on Venus by physical quenching of $N(^2D)$ rather than by its conversion to NO leads to a pronounced discrepancy with the nitric oxide number densities measured by the Vikings in the Martian atmosphere. So we should fix the discrepancy between the N and NO theoretical predictions and the Martian NO measurements, on the one hand, and the Venusian NO nightglow measurements, on the other hand, both being two- to threefold.

The lower boundary of the horizontal flow is at ~135 km and practically coincides with the maximum of the atomic nitrogen production rate. This flow forms a flux $\phi = 2[N]_0 H_N V(2\pi R)/(2\pi R^2) = 2[N]_0 H_N V/R$, where $[N]_0$ is the N number density at 135 km, H_N is the scale height of N, $V \simeq 200\ m\ s^{-1}$ is the flow velocity, and *R* is the radius of Venus. The coefficient 2 is due to the fact that the flow boundary coincides with the N formation maximum and only half the atoms appear below 135 km. Hence, $[N]_0 = 3.2 \times 10^7\ cm^{-3}$ and $[N] = 2 \times 10^7\ cm^{-3}$ at 150 km, which is in a fair agreement with the measurements (see Fig. 124). N, $N(^2D)$, and NO are formed in a layer about 20 km thick, so the NO formation rate at 135 km is $P_{NO} = 4 \times 10^9\ cm^{-2}\ s^{-1}/2 \times 10^6\ cm = 2 \times 10^3\ cm^{-3}\ s^{-1}$ and $[NO] = P_{NO}/k_{66}\ [N] = 1.8 \times 10^6\ cm^{-3}$. The photochemical lifetime of NO is much shorter than the mixing time here.

Atomic carbon number densities were calculated by Krasnopolsky (1982a) and Fox (1982b). The list of chemical reactions employed by Krasnopolsky is shown in Table 10. That in the paper of Fox (Table 32) was supplemented by C^+ chemistry which is of minor importance for neutral carbon number densities, but of interest to compare with the results of $[C^+]$ measurements.

[C] depends strongly on assumed molecular oxygen number densities. We took them from our model calculations (see below, Chap. 4.11), and Fox

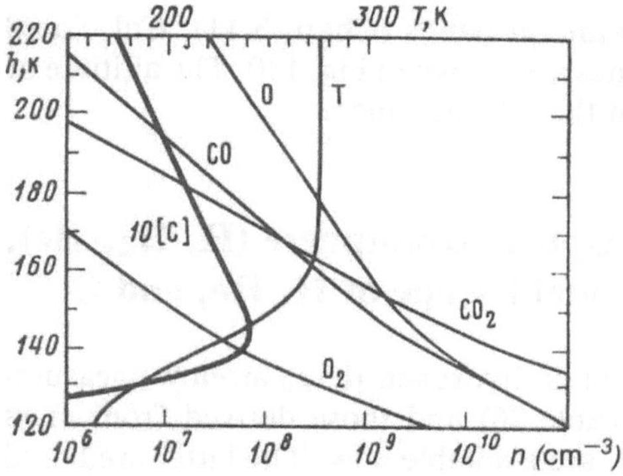

Fig. 160. Calculated atomic carbon number densities based on the profiles of CO_2, CO, O, and O_2 number densities shown (Krasnopolsky 1982a)

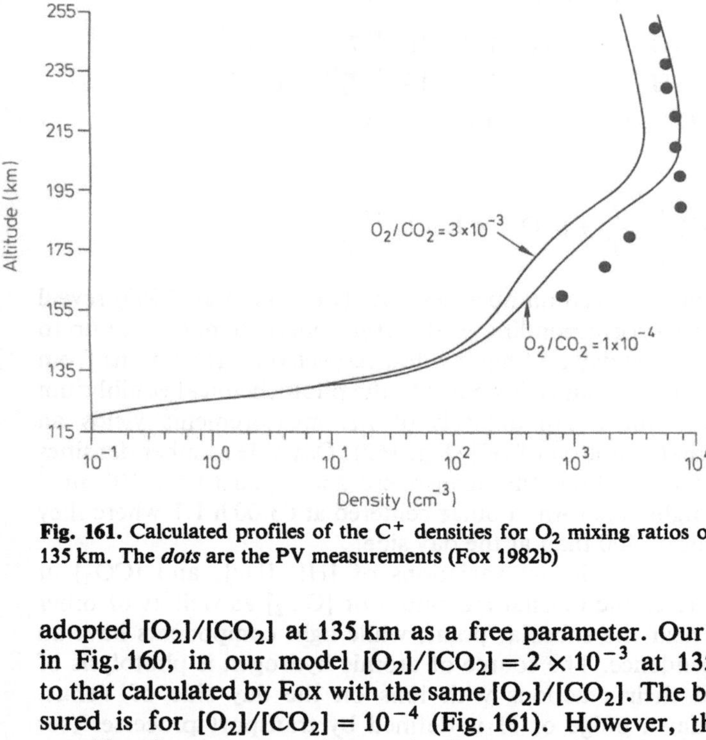

Fig. 161. Calculated profiles of the C^+ densities for O_2 mixing ratios of 3×10^{-3} and 10^{-4} at 135 km. The *dots* are the PV measurements (Fox 1982b)

adopted $[O_2]/[CO_2]$ at 135 km as a free parameter. Our result is presented in Fig. 160, in our model $[O_2]/[CO_2] = 2 \times 10^{-3}$ at 135 km. It is similar to that calculated by Fox with the same $[O_2]/[CO_2]$. The best fit to $[C^+]$ measured is for $[O_2]/[CO_2] = 10^{-4}$ (Fig. 161). However, this fact cannot be considered as strong evidence in favor of this value. The assumption of $[O_2]/[CO_2] = 2 \times 10^{-3}$ results in $[C^+]$ reduced by a factor of 2 only which is of the order of experimental errors. Having relatively large amounts in the upper atmosphere, carbon atoms are transported by the gas flow to the night side. The appropriate flux is $\phi_c \simeq 3 \times 10^9\ \mathrm{cm^{-2}\,s^{-1}}$ (Krasnopolsky 1982a). The mean transfer time is $t = R/2V \simeq 1.5 \times 10^4$ s equal to the chemical time at 145 km on the night side ($[O_2] \simeq 3 \times 10^6\ \mathrm{cm^{-3}}$). Therefore, some part of C is removed, and a mean night side downward flux is $\phi_c = -2 \times 10^9\ \mathrm{cm^{-2}\,s^{-1}}$.

It forms an airglow layer of CO Cameron bands (Chap. 3.11). Calculated night side [C] and volume emission rates are shown in Fig. 130. The altitude of the airglow layer depends strongly on the $[O_2]$ assumed.

4.4 Light Components of the Upper Atmosphere (H, H_2, He). "Hot" Atoms and Nonthermal Escape of H, He, and O

Atomic Hydrogen. Two kinds of atomic hydrogen data, airglow measurements summarized in Chap. 3.11 (Table 26) and those derived from mass spectrometric measurements of $[H^+]$, are available now. The latter are based on the photochemical equilibrium of H^+ determined by the reactions:

$$H + O^+ \rightarrow H^+ + O\,, \qquad k_1 = 2.5 \times 10^{-11}\, T^{1/2}\, cm^3\, s^{-1}$$
$$H^+ + O \rightarrow O^+ + H\,, \qquad k_2 = 2.2 \times 10^{-11}\, T_i^{1/2}\, cm^3\, s^{-1}$$
$$H^+ + CO_2 \rightarrow HCO^+ + O\,, \qquad k_3 = 3 \times 10^{-9}\, cm^3\, s^{-1}$$

resulting in

$$[H] = \frac{[H^+]}{[O^+]}\left([O]\sqrt{\frac{T_i}{T}} + [CO_2] \times 120\, T_i^{-1/2}\right).$$

The obtained neutral hydrogen number densities (Brinton et al. 1980) reveal an altitude distribution corresponding to the atmospheric temperature up to 200 km and 170 km at the day and night sides, respectively. Deviations from this distribution above these altitudes point to the photochemical equilibrium of H^+ being invalid there. An analysis of the measurements yields an extremely large diurnal variation of [H] (Fig. 162). Day side number densities are almost constant throughout the hemisphere and equal to $5 \times 10^4\, cm^{-3}$ while those on the night side form a bulge centered at 03.00 h LT where they are larger by a factor of 400 than at the day side.

Figure 163 reproduces diurnal variations of [H], [He], and $[CO_2]$ at 157 km. It demonstrates the diurnal maximum of $[CO_2]$ as well as of other heavy components near the subsolar point; while light components have an inverse diurnal dependence. The source of atomic hydrogen is photolysis of H_2O and HCl and their reactions with ions on the day side of Venus. Therefore, the predawn bulge of H is defined by transport processes (see below). Actually winds in the upper atmosphere directed to the antisolar region lead to the enrichment of the night side atmosphere by light components. This is the main reason for large night side concentrations of H, He, and other atomic components. Even before the PV measurements, Mayr et al. (1978) and Hartle et al. (1978) calculated diurnal variations of minor atmospheric components by the three-dimensional model of circulation. They obtained the night side H and He enrichment by a factor of 5, being even larger in the absence of the inverse exospheric flow. However, Kumar et al. (1978)

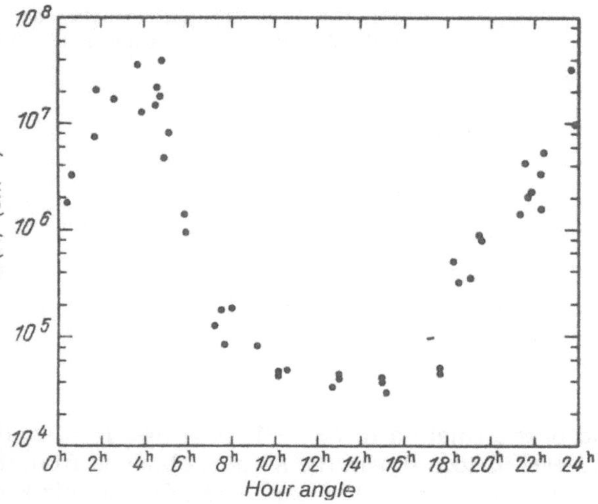

Fig. 162. Diurnal variations of atomic hydrogen number density at 165 km (Brinton et al. 1980)

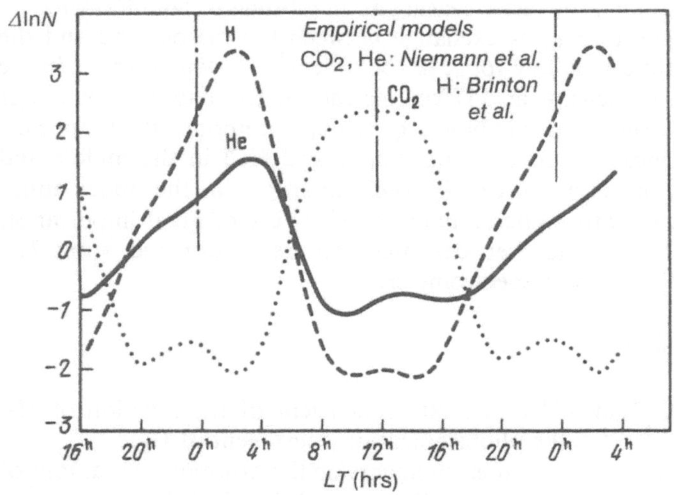

Fig. 163. Diurnal variations of CO_2, He, and H number densities at 157 km (Mayr et al. 1980)

showed this flow to be small and obtained the night side excess of hydrogen by a factor of 100 – 1000. Now this estimate is confirmed by experimental results.

The shift of the night side bulge of H and He from midnight to 03.00 h LT needs westward winds with velocities of $\sim 50\ \mathrm{m\,s^{-1}}$. A scheme of night side winds which takes into account westward winds, the flow across the terminator, and the influence of Coriolis forces is shown in Fig. 164. This problem was considered in more detail by Mayr et al. (1980). They find a

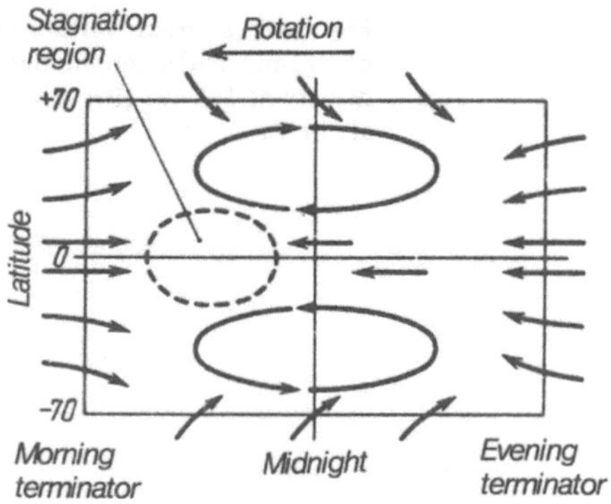

Fig. 164. Thermospheric circulation on the night side of Venus (Niemann et al. 1980)

period of the upper atmospheric rotation to be 5 – 10 days and $K = 3 \times 10^7$ $cm^2 s^{-1}$ at the turbopause level in accord with the K value obtained from the PV bus mass spectrometric measurements (see Chap. 3.13). One should keep in mind that the rotation of the upper atmosphere is not a continuation of that of the lower atmosphere, the latter ceasing at an altitude of 80 – 100 km.

Mayr et al. argue that mass exchange between the mesosphere and the thermosphere should be too low for $K \ll 3 \times 10^7\,cm^2 s^{-1}$, which results in the nighttime peak in O density at 157 km instead of the day side one. For $K \gg 3 \times 10^7\,cm^2 s^{-1}$ the night side bulges of the light components disappear.

Molecular hydrogen forms by water vapor and HCl in the middle and lower atmosphere and is a source of atomic hydrogen in the upper atmosphere. Therefore, the atmospheric amounts of H_2 are of great importance: similarly to [H] the H_2 number densities can be determined from H_2^+ measured by the PV ion mass spectrometer:

$$[H_2] = \frac{k}{I}\,[H_2^+]\,[O]$$

where $k = 1.5 \times 10^{-9}\,cm^3 s^{-1}$ is the rate coefficient of the reaction of H_2^+ with O. $I = 8 \times 10^{-8}\,s^{-1}$ is the photoionization coefficient of H_2.

The interpretation of the PV measurements on the assumption that ions of a mass number 2 were H_2^+ was given by Kumar et al. (1981). $[H_2^+] \simeq 5\,cm^{-3}$ at 250 – 300 km corresponds to $[H_2] \simeq 10^5\,cm^{-3}$ at the base of the exosphere at 200 km. Such large H_2 concentrations are important to suppress the formation of the F_2 region by the reaction

$$H_2 + O^+ \rightarrow OH^+ + H$$

with the subsequent recombination of OH^+. These reactions as well as the reaction of H_2 and CO_2^+ in the lower ionosphere are the source of atomic hydrogen.

Continuity equations for H, H_2, and H^+, H_2^+, O^+, OH^+, O_2^+, and CO_2^+ were solved in the altitude range of 120 – 300 km to reproduce the measured

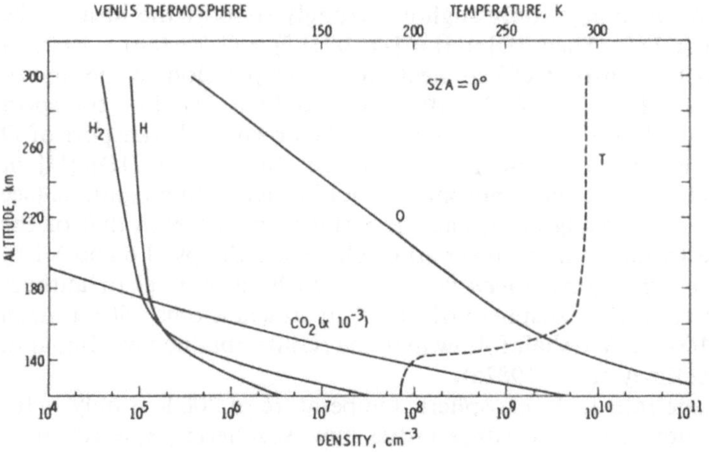

Fig. 165. Model distributions of H, H_2, CO_2, and O number densities adequate to the measured ionospheric composition with the assumption of the ion mass 2 being H_2^+ (Kumar et al. 1981)

[H], [H_2] $\simeq 10^5\,cm^{-3}$. The H_2 mixing ratio at 120 km and its vertical flow at 300 km served as fitting parameters. The results for $f_{H_2} = 10$ ppm and $\phi_{H_2} = 10^8\,cm^{-2}\,s^{-1}$ are shown in Fig. 165. If $\phi_{H_2} = 10^7\,cm^{-2}\,s^{-1}$, then f_{H_2} should equal 5 ppm to fit the observations.

Total production of H as a result of H_2 dissociation is $3.6 \times 10^8\,cm^{-2}\,s^{-1}$ with $2 \times 10^8\,cm^{-2}\,s^{-1}$ being above 250 km. Half of this value, $10^8\,cm^{-2}\,s^{-1}$, forms the hydrogen nonthermal escape flux, and the rest, $2.3 \times 10^8\,cm^{-2}\,s^{-1}$, is transported to the night side.

A value of $f_{H_2} = 10$ ppm in the lower atmosphere must be considered rather large and is rather difficult to explain. Thermochemical equilibrium near the planetary surface gives $K_p = 6$ at 740 K for the reaction

$$CO + H_2O \rightleftarrows CO_2 + H_2 .$$

Hence, $f_{H_2} = 6 f_{CO} f_{H_2O} \simeq 10^{-8}$. The Venera 13, 14 gas chromatography confirms, however, the high f_{H_2} value obtained (Table 16).

Another explanation of ion mass 2 measurements which treats them as D^+ with D/H $\simeq 10^{-2}$ in the upper atmosphere seems more plausible. This was proposed by McElroy et al. (1982a) and was strongly supported by the finding of D/H = 1.6×10^{-2} from a mass spectrometric analysis of the lower atmosphere (Donahue et al. 1982, see Chap. 3.6). Furthermore, two-dimensional modeling of the night side ionosphere fits the observations much better with low H_2 number densities, thus, confirming the idea of D^+ ions.

To identify the mass 2 ion, Hartle and Taylor (1983) studied the altitude dependence of the ratio of this ion to the H^+ number density. This ratio is proportional to e^{-z/H_H} and $[O^+]^{-1} e^{-z/H_H}$ for D^+ and H_2^+, respectively. Measurements in the predawn bulge at 155 – 165 km, where a steep change of [O^+]

occurs and the ratio changes only slightly, strongly support the idea of the mass 2 ion being D^+. Then $[D]/[H] = [D^+]/[H^+] = (1.7 \pm 0.6) \times 10^{-2}$ at 155 – 160 km. With transport effects neglected, extrapolation to the homopause yields $[D]/[H] = (2.2 \pm 0.6) \times 10^{-2}$. Corrections for ion transport appear to be small. However, in our opinion, the downward transport of D and H is very important in the predawn bulge. In this case [D]/[H] at 155 – 160 km and in the lower atmosphere should be equal in magnitude (see the last section of this paragraph), and the result coincides with that of the neutral mass spectrometer in the lower atmosphere (see Chaps. 3.6 and 3.10).

Nonthermal escape of hydrogen was discussed by a number of authors after the discovery in the exosphere of "hot" hydrogen atoms with a mean temperature of 1000 K. We shall follow here the results obtained by Kumar et al. (1978) and McElroy et al. (1982a).

A Jeans' escape rate at the exospheric temperature of 300 K is only $\sim 10^5$ $cm^{-2} s^{-1}$ and is not consistent with a rather low exospheric $[H] \simeq 10^5 cm^{-3}$ and with f_{H_2O} and f_{HCl} of the order of 1 ppm near the cloud top. For the hot hydrogen the Jeans' escape rate is $1.5 \times 10^6 cm^{-2} s^{-1}$. However, a mean temperature is not a good parameter to calculate the nonthermal escape rate which should be even higher. The total amount of the exospheric hot hydrogen was measured to be $10^{11} cm^{-2}$. Its sink is due to both escape and thermal collisions with other particles, atomic oxygen being predominant at the exobase. Ten elastic collisions are needed to reduce the excessive energy by a factor of e. A mean free time is $2H_H/V_H = 10^3$ s, where H_H is the hydrogen scale height and $V_H = 4.5 km s^{-1}$ is the thermal velocity of hydrogen at 1000 K. Therefore, the time of thermalization is 10^4 s and the source of hot hydrogen ($10^{11} cm^{-2}$) is of the order of $10^7 cm^{-2} s^{-1}$.

McElroy et al. (1982a) argue that the elastic collisions of thermal H atoms with fresh hot oxygen atoms,

$$H + O^* \rightarrow H^* + O^*$$

formed by dissociative recombination of O_2^+ and CO_2^+ ions, is a main path of the nonthermal hydrogen escape. $O(^3P)$ and $O(^1D)$ formed by recombinations of O_2^+ have velocities of $5.6 km s^{-1}$. The probability for H atoms to get the escape velocity, $10.2 km s^{-1}$, through the collisions with these hot oxygen atoms is 6% and 15% for $T = 0$ and 300 K, respectively.

According to the definition of the exobase, $\sigma N_0 = 1$, where $\sigma \simeq 3 \times 10^{-15}$ cm^{-2} is the collisional cross-section and $N_0 = 3 \times 10^{14} cm^{-2}$ is the column density of atomic oxygen at the exobase. Its ionization is $J = 1.2 \times 10^{-6} s^{-1}$ at high solar activity, giving $8 \times 10^8 cm^{-2} s^{-1}$ of fresh hot oxygen atoms as a result of the reaction

$$O^+ + CO_2 \rightarrow O_2^+ + CO$$

and subsequent recombination. Two percent of their first collisions occur with H atoms, and their corresponding day side escape rate is $10^6 cm^{-2} s^{-1}$.

Almost the same production rate of O^*, $6 \times 10^8\,cm^{-2}\,s^{-1}$ occurs on the night side. However, Knudsen et al. (1980) evaluate the mean night side total recombination rate as $2 \times 10^8\,cm^{-2}\,s^{-1}$ and only a part of it occurs above the exobase which is at 150 km on the night side. Much higher night side hydrogen number densities (by a factor of 100) result in 50% collisions of O^* with H, and 10% of the H^* atoms formed by these collisions have a speed sufficient to escape. Only half the H^* atoms which have velocities in excess of the escape one and move upward determine the night side escape rate equal to $1.5 \times 10^7\,cm^{-2}\,s^{-1}$. So the planetary average escape rate due to the reaction (1) is $8 \times 10^6\,cm^{-2}\,s^{-1}$. Charge exchange between solar wind protons and H atoms provides the rate of $2 \times 10^6\,cm^{-2}\,s^{-1}$. Ionospheric charge exchange reactions at $T_i = 1000-1500\,K$

$$H^+ + O \rightarrow H^* + O^+$$

$$H^+ + H \rightarrow H^* + H^+$$

give negligible values of the escape rate on the day side and $\sim 10^6\,cm^{-2}\,s^{-1}$ on the night side.

The contribution of the reactions

$$O^+ + H_2 \rightarrow OH^+ + H^*\,, \qquad \Delta E \simeq 0.6\,eV$$

$$OH^+ + e \rightarrow O + H^*\,, \qquad \Delta E \simeq 8\,eV$$

depends on the assumed molecular hydrogen number densities. For $[H_2] \simeq 10^5\,cm^{-3}$ (see above) the escape rate due to these reactions is equal to $10^8\,cm^{-2}\,s^{-1}$. McElroy et al. (1982a) emphasized the difficulties raised by the interpretation of ion mass 2 being H_2^+ and suggested it to be D^+. They took a value of $[H_2] \simeq 10^3\,cm^{-3}$ from the paper of Sze and McElroy (1975), thus giving an escape rate of $10^6\,cm^{-2}\,s^{-1}$. So, the total planetary average escape rate is $10^7\,cm^{-2}\,s^{-1}$.

The lifetime of the Venusian H_2O associated with the hydrogen escape equals the ratio of the present water abundance to the escape rate and constitutes 10^9 yr. The probability of the deuterium atom escaping after the collision with the fresh hot oxygen atom is much lower then for the hydrogen atom, and the escape of D is negligible. If the escape of H is proportional to the water vapor amount, then $D/H \simeq 10^{-2}$ implies an initial ratio, D/H, of 5×10^{-5} and an initial water vapor amount of $\sim 800\,g\,cm^{-2}$. The former coincides with values derived for Jupiter and the Earth. The latter now exceeds the observed amount by two orders of magnitude.

Further study of the problem of hydrogen loss on Venus will be considered in Chap. 4.16.

Oxygen Escape. Hot Atoms. The PV ultraviolet spectrometric observations revealed hot oxygen atoms having a number density of $4 \times 10^4\,cm^{-3}$ at 400 km and a scale height of 400 km which corresponds to $T = 6400\,K$ (see Chap. 3.11). The explanation of this phenomenon was given by Nagy et al. (1981) and McElroy et al. (1982b); the latter treat oxygen escape as well.

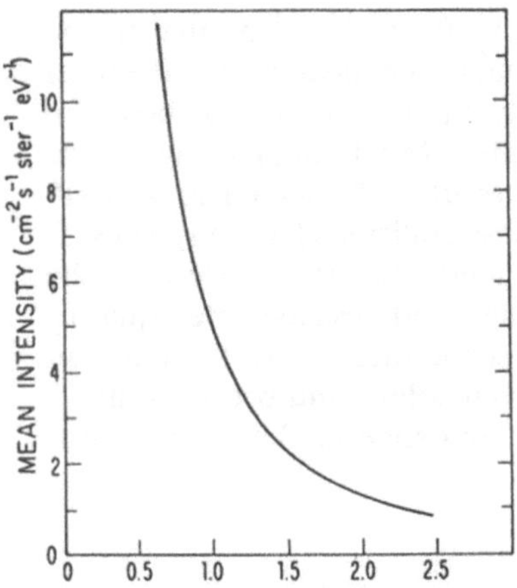

Fig. 166

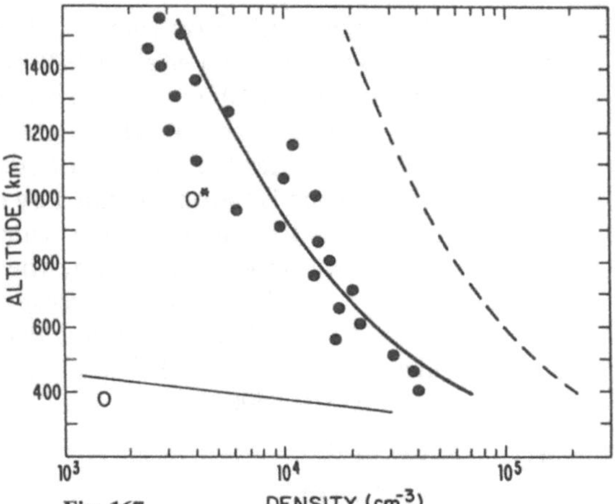

Fig. 167

Fig. 166. Energy spectrum of secondary hot oxygen atoms. The mean intensity is based on a primary source of monoenergetic O* at 2.4 eV with an intensity of 1 $cm^{-2}\,s^{-1}\,sr^{-1}$ (McElroy et al. 1982b)

Fig. 167. Altitude profiles of hot oxygen atoms calculated by McElroy et al. (1982b, *thick line*) and Nagy et al. (1981, *dashed line*); *dots* are the PV measurements. Distribution of thermal oxygen atoms is shown by the *thin line*

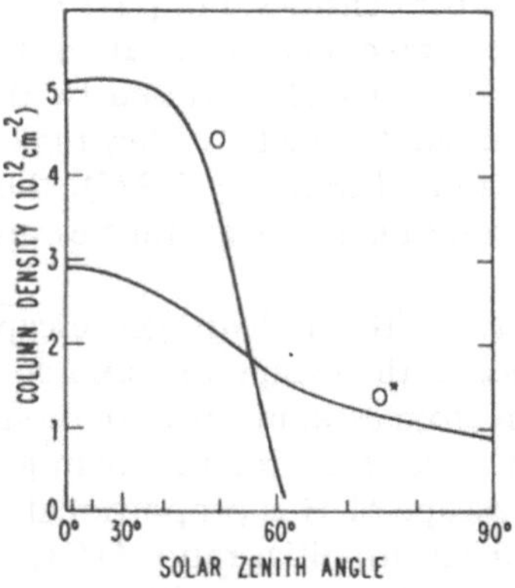

Fig. 168. Column densities of neutral oxygen above the Venusian plasmapause as a function of the solar zenith angle. The abscissa has been scaled in accord with the fractional surface area. Thermal oxygen O and hot atoms O* contribute equally to the total amount of neutral oxygen above the plasmapause approx. 8×10^{30} atoms (McElroy et al. 1982b)

As in the previous section, dissociative recombination of O_2^+ is considered as a main source of hot oxygen, giving atoms with a speed of up to $V^* = 5.6$ km s^{-1}. Their daytime production rate is $2J/\sigma$ (see previous section), their residence time is $t = (2\,V^* \cos\varphi)/g$ (φ is the angle of the velocity to the vertical and g is the gravity acceleration). The column density is $(2JV^*)/(3\,\sigma g)$ after integration over the sphere and equal to 6×10^{11} cm^{-2} for $J = 1.7 \times 10^{-6}$ s^{-1} and $\sigma = 1.2 \times 10^{-15}$ cm^2 (note that these values differ from those in the previous section).

Secondary O* number densities produced by collisions of O* and O were calculated assuming the energy independent cross-section and isotropic scattering. Figure 166 depicts a spectrum of O*. Analysis of ballistic trajectories of these atoms with the isotropic velocity distribution in a collisionless atmosphere results in [O*] (Fig. 167) which is in a good agreement with the observations.

Now with the location of the plasmapause known it is easy to calculate column densities of thermal and hot oxygen above the plasmapause as a function of the solar zenith angle (Fig. 168). These oxygen atoms are subject of photoionization with $J = 1.2 \times 10^{-6}\,s^{-1}$ and of electron impact ionization in the plasma sheet with $J_e = 2.2 \times 10^{-6}\,s^{-1}$. Ions formed are removed by solar wind. Thus, a mean oxygen escape rate is $6 \times 10^6\,cm^{-2}\,s^{-1}$ with a column density of O* equal to $2 \times 10^{12}\,cm^{-2}$.

The escape of O is regulated by solar UV radiation and solar wind and that of H by mixing, chemical reactions, and supply of H to the exosphere. The calculated escape rates are in accord with the H_2O stoichiometric ratio. Oxygen escape regulates the hydrogen escape keeping this ratio. An increase of oxygen escape enhances CO production and reduces sulfuric acid formation. The latter increases the H_2O supply to the mesosphere with subsequent increase in hydrogen abundance in the exosphere.

Similar to oxygen hot carbon and nitrogen atoms number densities have been calculated and observed by the PV ultraviolet spectrometer (Nagy et al. 1983b).

Helium Escape (Prather and McElroy 1983). Photoionization and electron impact ionization of helium atoms above the plasmapause with the rates of $3.4 \times 10^{-7}\,s^{-1}$ and $6.4 \times 10^{-7}\,s^{-1}$, respectively, result in the escape of ions which are swept off by the solar wind. The escape rate is $10^6\,cm^{-2}\,s^{-1}$ based on the mass spectrometric measurements of helium (Chap. 3.11). Extrapolation of these measurements to the lower atmosphere gives $f_{He} \simeq 12$ ppm, i.e., 1.8×10^{22} cm^{-2}, thus a lifetime of helium equals 6×10^8 years. Therefore, the present helium is not primordial but formed by current decay of uranium and thorium with the rate close to that on Earth. Similar analysis based on the M 10 helium emission data (Kumar and Broadfoot 1975) was considered in Chap. 3.11.

Measurements of ^{40}Ar formed by decay of ^{40}K result in the source of ^{40}Ar on Venus by a factor of 4 less than on Earth though the potassium amounts in rocks are comparable on both planets (see Chap. 4.7). This situation can be explained by different outgassing efficiencies of oxygen and helium.

Formation of the Night Side Bulge of Light Components. We will confine ourselves to the case of atomic hydrogen. Its planetary mean flux from the day side to the night side is

$$\phi_H = n_0 V \cdot \frac{2\pi r H_H}{2\pi r^2} = n_0 V \frac{H_H}{r} \simeq 10^8\,cm^{-2}\,s^{-1}$$

where $n_0 \simeq 10^5\,cm^{-3}$ is the hydrogen number density at the lower boundary of a flow from the day to the night side, H_H is the hydrogen scale height, $V \simeq 0.5$

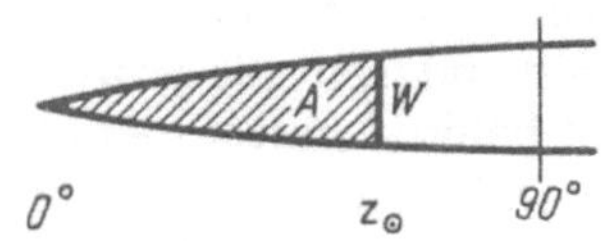

Fig. 169. Spherical triangle centered in subsolar point to calculate the gas flow across the terminator (see text, Kumar et al. 1978). Ionization of these oxygen atoms forms their escape flux

km s^{-1} is the velocity of the flow, and r is the radius of Venus. ϕ_H is positive at the day side and negative at the night side. The location of the lower boundary of the flow was discussed by Kumar et al. (1978). They considered a spherical triangle originated at the subsolar point (Fig. 169). The total upward flux of a component in it is

$$\phi A = \frac{n_0 D}{H_a} \cdot A = \frac{n_0 A b}{n_{a0} H_a}$$

where $D = b/n_{a0}$ is the diffusion coefficient, n_{a0} is the atmospheric number density at the lower boundary of the flow, $b = 10^{19}$ cm^{-1} s^{-1}, H_a is the atmospheric scale height, and A is the triangle area. The horizontal flux at z_0 formed by this upward flux is $n_0 H_i V w$, where w is the triangle base. Taking into account $A/W = r \operatorname{tg} z_0/2$, one obtains

$$n_{a0} = \frac{br}{H_i H_a V} \operatorname{tg} \frac{z_0}{2}$$

where $z_0 \simeq 90°$, $n_{a0} \simeq 2 \times 10^{10}$ cm^{-3} for the atomic hydrogen flow and 3×10^{11} cm^{-3} for the atomic oxygen flow correspond to the altitudes of the lower boundaries of 145 and 130 km, respectively.

The mesospheric mixing ratio of the chemically bound hydrogen is $f_H = f_{HCl} + 2 f_{H_2O} \simeq 2.6$ ppm (the values for the cloud tops are assumed). It was assumed to be constant up to the calculated lower boundary of the flow, 145 km, due to the large upward flux (we will see in Chap. 4.11 that f_H at the cloud tops is not adequate at the lower thermosphere due to formation of sulfuric acid above the cloud tops). Above 145 km [H] is almost constant due to its large scale height (Fig. 170). At the night side the downward flux forms almost height independent hydrogen number densities down to some level below homopause where they coincide with $f_H \simeq 2.6$ ppm typical of the mesosphere. The night to day hydrogen number density ratio in the upper atmosphere equates the atmospheric number density ratio at this level to that at 145 km. We can calculate the atmospheric density of this level with the known $\phi_H = -10^8$ cm^{-2} s^{-1}:

$$\phi_H \simeq -\frac{nK}{H}, \quad K \simeq 10^{13} n_a^{-1/2} \text{cm}^2 \text{s}^{-1}, \quad n = n_a f_H,$$

therefore,

$$n_a \simeq \left(-\frac{\phi_H H}{f_H \times 10^{13}} \right)^2 \simeq 6 \times 10^{12} \text{cm}^{-3}$$

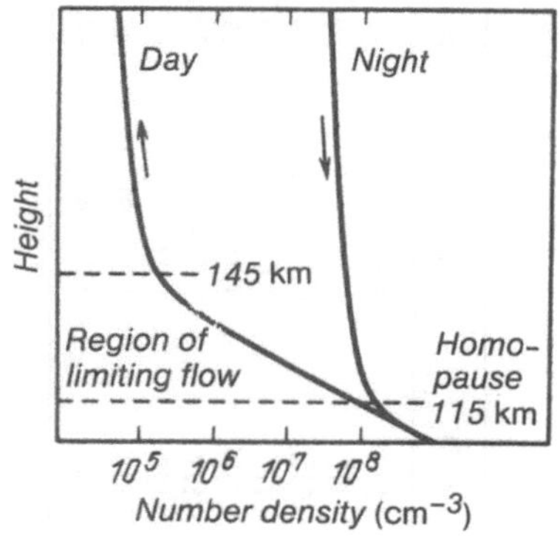

Fig. 170. Altitude profiles of atomic hydrogen number densities on the day and night sides of Venus (Kumar et al. 1978)

or 117 km. Then $[H]_{night}/[H]_{day} \simeq 6 \times 10^{12}/2 \times 10^{10} = 300$. The predicted night side hydrogen bulge was discovered by PV observation. Other light components (H_2, He) form the night side bulge in the same way. It is necessary to distinguish between the day side upward flux of atmospheric constituents caused by the transport from the day side to the night side of the planet, and the escape flux of some components caused by ionospheric reactions.

For low values of K atomic oxygen should also form the night side bulge. However, with $K \simeq 10^{13}\, n^{-1/2}\, cm^2\, s^{-1}$ and very low n at the night side the downward transport of atomic oxygen prevents the formation of bulge.

4.5 Thermospheric Models

Experimental data on the upper atmospheric temperature, pressure, mixing and chemical composition were discussed in Chaps. 3.11 and 3.13 and serve now as a basis for a thermospheric model. Thermospheric characteristics are nearly constant throughout the day and night sides, but a great difference exists between them. The following data for the day side thermosphere are now available: temperature and pressure profiles up to 120 km derived from deceleration of descent probes, temperature and density data from the PV orbiter drag, temperature and CO_2, N_2, CO, and He number density profiles down to 130 km measured by the PV bus neutral mass spectrometer, and CO_2, N_2, CO, O, N, He number density profiles down to 150 km measured by the PV orbiter neutral mass spectrometer. Though these data provide a fair basis for modeling they are not complete and represent some inconsistencies. Therefore, a simple averaging of the data cannot form the thermospheric model.

The model based on the PV bus data was described in Chap. 3.11 (Table 24). The next attempt was made by Krasnopolsky (1982b), who noted that all photochemical calculations of the mesospheric composition (see below) result

Table 34. Daytime atmosphere of Venus at high solar activity. Number densities are in cm^{-3}, M is the total number density, μ is the mean molecular mass, molecular and eddy diffusion coefficients are $D = 10^{17}\,T^{0.75}/M\,cm^2\,s^{-1}$ and $K = 10^{13}\,M^{-1/2}\,cm^2\,s^{-1}$, respectively (Krasnopolsky 1982b)

h, km	T, K	CO_2	O	CO	N_2	O_2	He	N	C	NO	M	μ
110	170	5.15+13[a]	3.0+11	5.0+11	2.06+12	1.0+11	1.18+9	–	–	–	5.45+13	43.1
120	165	3.56+12	1.07+11	1.15+11	1.95+11	4.79+9	1.15+8	1.0+7	–	–	3.98+12	42.0
130	187	2.38+11	2.1+10	1.82+10	2.19+10	4.06+8	1.85+7	3+7	4+4	1.8+6	2.99+11	39.9
140	223	2.23+10	5.25+9	3.63+9	3.39+9	5.25+7	6.13+6	2.8+7	5+6	1.8+6	3.46+10	36.5
150	253	3.01+9	2.05+9	1.01+9	7.95+8	1.12+7	3.53+6	2+7	5.9+6	4.5+5	6.9+9	31.4
160	270	5.05+8	1.05+9	3.31+8	2.63+8	3.39+6	2.66+6	1.2+7	4+6	1.6+5	2.16+9	25.9
170	275	9.79+7	5.62+8	1.16+8	9.25+7	1.00+6	2.2+6	7.1+6	2.5+6	6+4	8.8+8	22.0
180	275	1.9+7	3.0+8	4.1+7	3.25+7	2.99+5	1.85+6	4.2+6	1.7+6	2.5+4	4+8	19.6
190	275	3.67+6	1.55+8	1.44+7	1.15+7	8.83+4	1.58+6	2.5+6	1.1+6	9+3	1.9+8	18.0
200	275	7.12+5	8.32+7	5.08+6	4.04+6	2.62+4	1.35+6	1.5+6	7.2+5	3+3	9.66+7	17.1

[a] $5.15+13 = 5.15 \times 10^{13}$

Table 35. Venusian night side model (Massie et al. 1983)

Altitude, km	CO_2	CO	O	N_2	He	N	T, °K	ϱ, g/cm^3	μ
178	1.8 (3)	4.0 (4)	1.2 (7)	8.0 (4)	4.8 (6)	1.5 (5)	115	3.6 (−16)	12.7
170	3.8 (4)	2.8 (5)	3.7 (7)	5.6 (5)	6.4 (6)	4.0 (5)	115	1.1 (−15)	14.5
160	1.6 (6)	3.2 (6)	1.5 (8)	6.3 (6)	9.2 (6)	1.4 (6)	115	4.7 (−15)	16.3
150	6.7 (7)	3.5 (7)	6.2 (8)	7.0 (7)	1.3 (7)	4.8 (6)	115	2.6 (−14)	19.7
140	2.4 (9)	3.7 (8)	2.5 (9)	7.4 (8)	2.0 (7)	1.7 (7)	119	3.0 (−13)	29.2
130	8.2 (10)	3.8 (9)	1.6 (10)	8.8 (9)	3.9 (7)	8.9 (7)	122	7.0 (−12)	38.2
120	2.5 (12)	3.1 (10)	1.3 (11)	1.3 (11)	1.8 (8)	6.4 (8)	129	1.9 (−10)	41.7
110	5.7 (13)	2.2 (11)	2.4 (11)	2.4 (12)	1.3 (9)	1.0 (8)	148	4.3 (−9)	43.2
100	9.2 (14)	7.0 (11)	1.9 (11)	3.4 (13)	8.6 (9)	–	161	6.9 (−8)	43.4

Densities in cm^{-3}; read 1.8 (3) as 10^3; values apply to the midnight sector in almost the same number densities of CO, O, and O_2 at a mesopause level of 110 km. $[CO_2] = 5 \times 10^{13}\,cm^{-3}$, $[O] = 3 \times 10^{11}\,cm^{-3}$, $[CO] = 5 \times 10^{11}\,cm^{-3}$, and $[O_2] = 10^{11}\,cm^{-3}$ at this height are recommended. O, CO, and O_2 mixing ratios at the mesopause greatly exceed those in the troposphere, hence, these components are formed by CO_2 photolysis for which production rates of CO and O are equal. Formation of O_2 by O occurs below 105 km and that of CO_2 by CO and O_2 below 95 km. Therefore, $[CO] \simeq [O] + 2[O_2]$ between 90 km and homopause, and with $[O] \gg [O_2]$ above 110 km this means $[CO] \simeq [O]$. Due to diffusive separation, $[O] > [CO]$ above the homopause. Chemical reactions of CO_2, CO, O, O_2, H, N_2 can be neglected in the thermosphere except for photolysis of CO_2 and O_2. A bulk of ionospheric reactions results in dissociation of CO_2; these reactions being in equilibrium below 200 km (Chap. 2.5). This facilitates their consideration as a part of the CO_2 photolytic rate.

N, NO, and C are important by-products of ionospheric reactions which were discussed in Chap. 4.3 and are beyond the scope here. Calculations were made with the molecular and eddy diffusion coefficients equal to $D = 10^{17}\,T^{0.75}/n_a\,cm^2\,s^{-1}$ and $K = A\,n_a^{-1/2}\,cm^2\,s^{-1}$, respectively, with A as a parameter of calculations to fit the PV orbiter neutral mass spectrometric data. The results with $A = 10^{13}$ are given in Table 34. This model represents rather well all the experimental data discussed. Data on N, NO, and C number densities (Chap. 4.3) are included in it as well.

A day side thermospheric model was calculated also by Massie et al. (1983), who took into account a detailed scheme of ionospheric reactions. Fluxes at the upper boundary, 178 km, were specified by the upward extrapolation of chemical production and loss at 170 – 178 km. Number densities were specified at 178 km as well, unfortunately, without a description of how they were received. Number densities of N_2, CO, O_2, and O above 120 km appear to be essentially the same as in our model, those of CO_2 exceed ours by a factor of 1.6 typical of the PV orbiter drag and radio occultation measurements. We do not reproduce data of their day side model due to violation of the condition $[CO] \simeq [O] + 2[O_2]$ at 100 – 110 km. The night side thermospheric model is given in Table 35. Using the technique described in Chap. 4.4 and the data of Table 34, one can calculate fluxes of O, N, CO, and CO_2 from the day side to the night side of the planet which appear to be 4×10^{11}, 3×10^9, 4×10^{11}, and $10^{13}\,cm^{-2}\,s^{-1}$, respectively. The flux of nitrogen atoms coincides with that obtained in Chap. 4.3 and results in the NO nightglow intensity of 1.5 *kR*. The fluxes of atomic oxygen and carbon monoxide are equal, therefore, the thermospheric transport of O and CO from the day side to the night side does not break the equilibrium between [CO] and $[O] + 2[O_2]$.

4.6 Lightning and Lower Atmospheric Chemistry. Nitric Oxide in the Lower Atmosphere

Now we will consider the chemical composition of the middle and lower atmosphere. It is better to discuss first the lower atmospheric composition, because the results obtained can be used as boundary conditions to calculate the middle atmospheric composition. Solar UV radiation cannot reach the lower atmosphere, and consequently photochemical processes are of secondary importance in it. Therefore, lightning is left to be a main source of chemically active components here.

Possible Mechanisms of Lightning Generation on Venus. Here we will follow an approach used by Williams et al. (1983), who proceeded on the assumption that the following processes have analogy with the Earth:

1. separation of charge by collisions of large and small particles. The former falling through a cloud with a downward directed external electric field are polarized with leading and trailing edges being positive and negative, respectively. Collisions of small particles with the leading edge of a large particle provide small particles positively charged, while the large particle becomes negatively charged with a subsequent downward transport of the negative charge;

2. a capture of atmospheric negative ions by the leading edge of the falling particle. If the particle velocity exceeds ion velocities, then positive ions are not captured by the trailing edge. The final charge of the particle is $q = -0.515\ Er^2$, where E is the electric field strength and r is the particle radius. The maximum field strength produced by this process is

$$E_{\max} = \frac{8\pi}{3}\,\Omega V t r \varrho / p$$

where Ω is the ion production rate, V is the particle fall velocity, t is the time, ϱ is the particle material density, p is the precipitation rate;

3. large potential differences across the contacts of supercooled water with ice with subsequent freezing of the supercooled water.

Other possible processes of electrification are of minor importance. Contributions of these processes being calculated for terrestrial lightning result in that only the third and, perhaps, the second mechanism can produce charge separation and electric field strength sufficient to cause breakdown. Electric field strength required for a breakdown is approximately proportional to atmospheric pressure with the latter being of the order of several bars. In the case of Venus a field strength high enough to produce lightning in the cloud layer ($\sim 3 \times 10^6\ V\,m^{-1}$) may be generated in 3 h by particles with $r \simeq 20\ \mu m$ and $n \simeq 50\ cm^{-3}$. This number density of such large particles is far in excess of the measured one (see Fig. 104). Even allowing for the fact that Venusian clouds have been studied at only ten points (V9 – 14 and four PV probes), while the ratio of the total planetary area to total thunderstorm area at the low

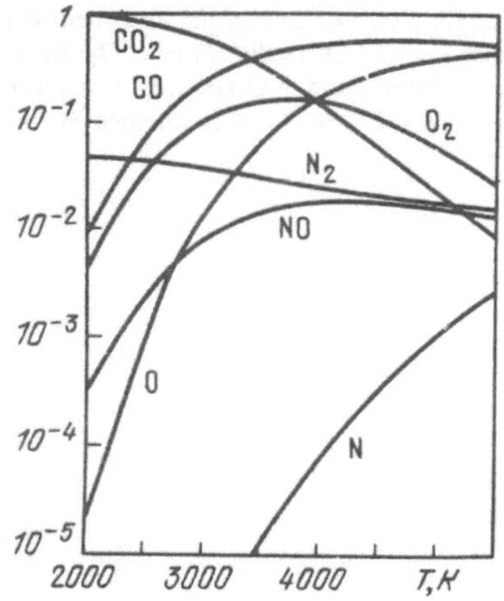

Fig. 171. Thermochemical equilibrium composition of the mixture of CO_2 (96%) and N_2 (4%) at the initial pressure of 0.7 bar as a function of temperature (Chameides et al. 1979)

and middle latitudes is 700 (Table 29), this number density appears to be too large. So the theoretical consideration does not support the idea of lightning in the Venusian clouds. It is still more difficult to explain lightning occurrence in the lower atmosphere, e.g., as a result of volcanic eruptions, due to extremely large electric field strength required for a breakdown there and the cleanliness of the lower atmosphere.

Williams et al. (1983) do not find direct unambiguous experimental evidence of lightning on Venus. A search by the PV star sensor gives the upper limit of a mean lightning rate as 30 $km^{-2} yr^{-1}$, the measurements by the V9 spectrometer (Krasnopolsky 1980, 1983b) which result in 45 $km^{-2} yr^{-1}$ arouse some doubts because of a very long duration of pulses ($\sim$0.25 s), signals of sterics detected by V11 – V14 probes (Ksanfomaliti et al. 1983) may be attributed to electrostatic discharges of the spacecraft during descent, Whistlers measured by the PV orbiter (Scarf et al. 1980) may be produced by some other phenomenon. Taking all these facts into account we will consider the results of Table 29 as the upper limit to study chemical impact lightnings on Venus.

The chemical impact of lightning was calculated by Chameides et al. (1979) and Bar-Nun (1980). Both papers follow an idea suggested by Zeldovich and Raser (1966) that the composition of an atmospheric mixture after a lightning stroke occurrence is conditioned by thermochemical equilibrium at a certain temperature T_0. The assumption is made that at the moment of a stroke a sufficiently high temperature is established, chemical reactions are very fast, and with gas cooling and hot stroke column expanding, the thermochemical equilibrium is maintained there down to T_0, the temperature at which the thermochemical equilibrium time is equal to that of gas cooling. With further decrease in temperature, chemical reactions stop and the atmospheric mixture composition is stabilized at T_0. Figure 171 shows the thermochemical equilibrium of a mixture of $CO_2 + N_2$ (4%) with the initial tempera-

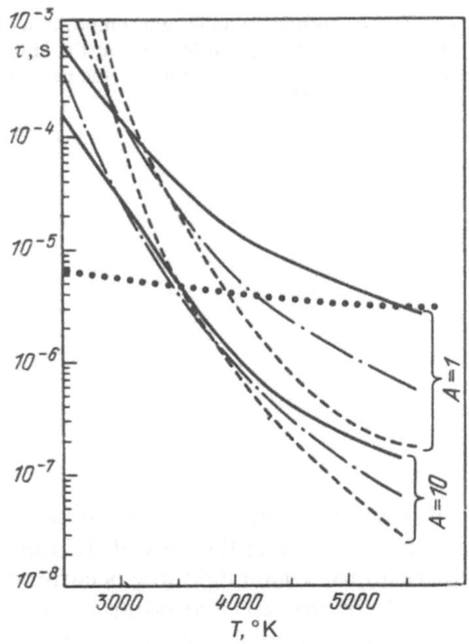

Fig. 172. Cooling time (*dots*) and chemical lifetimes of CO (*solid lines*), O_2 (*dot-dashed lines*), and NO (*dashed lines*) for $A = \varrho_0/\varrho = 1$ and 10 (Chameides et al. 1979)

ture and pressure of 310 K and 0.7 bar, respectively, which corresponds to 55 km. A technique for thermochemical equilibrium calculations will be considered in Chap. 4.7. Data presented in Fig. 171 are obtained for $A = \varrho_0/\varrho = 1$ where ϱ_0 and ϱ are the initial and final air densities. For terrestrial lightning the cooling time is

$$\tau(T) = T_s/\frac{dT_s}{dt} = \frac{5.7 \times 10^{-5} E_0^{1/2}}{\varrho_0^{1/2} T_s}$$

where the index s defines a surface of a bow shock formed by a stroke and $E_0 \simeq 10^5$ J m^{-1} is the lightning energy. Chemical equilibrium times are calculated from some known reaction rate coefficients. When $A = 1-10$, τ vs T dependencies are shown in Fig. 172. T_0 is in a range of 3500 – 5000 K, where yields of CO as well as those of NO, P_{CO} and P_{NO}, are almost temperature independent. One should account for $P_{CO} = P_{NO} + 2P_{O_2} + P_O$. The bow shock surface surrounds

$$M = 1.55 \times 10^{21} E_0/T_0 \text{ molecules m}^{-1}$$

and the yields of CO, O_2, NO, and O are equal to 1.5×10^{17}, 5×10^{16}, 5×10^{15}, and 5×10^{16} molecules J^{-1}, respectively.

Zeldovich and Raser (1966) gave $T_0 = 2000$ K for the NO formation in the air ($N_2 + O_2$). According to Bar-Nun (1980), a radius of the bow shock is 6 cm at this temperature, and the thermochemical equilibrium of $CO_2 + N_2$ (4%)

results in $f_{NO} = 2.8 \times 10^{-3}$ at 2000 K. Then the NO yield is 6×10^{15} molecules J^{-1} at 0.7 bar with $E_0 = 10^5$ J m^{-1}. Both values of P_{NO} coincide despite the large difference in the T_0 obtained. The bow shock radius and the quantity of gas in it increase with decreasing temperature, thus compensating for the effect of lowering f_{NO}. Laboratory discharge measurements (Levine et al. 1982) confirm those estimates and give $P_{NO} = (3.7 \pm 0.7) \times 10^{15}$ molecules J^{-1}. Then a mean energy release by lightning ~1 erg cm^{-2} s (Table 29) corresponds to a mean NO production rate of 5×10^8 cm^{-2} s^{-1}. Production rates of CO, O_2, O should be even larger: 1.2×10^{10}, 5×10^9, and 2×10^9 cm^{-2} s^{-1}, respectively. From Table 29 it is evident that these values are higher by three orders of magnitude in the thunderstorm region than for global mean conditions.

The production rates obtained can be compared with CO_2 and SO_2 photolytic rates and the total cloud matter flux caused by particle sedimentation, all these values being of the order of 10^{13} cm^{-2} s^{-1} greatly exceed that of lightning. However, the active component production rate of lightning is comparable to sedimentation fluxes of the first and the second modes being 10^9 and 3×10^{11} cm^{-2} s^{-1}, respectively, for a molecular mass equal to ~100. In thunderstorm regions, chemical production by lightning is even larger than photochemical production. Yet thunderstorms do not last more than ~10^4 s, and accumulated f_{CO}, f_{O_2}, and f_{NO} are 10^{-8}, 3×10^{-9}, and 3×10^{-10}, respectively.

If the lower lying atmosphere is considered as a sink of CO and O_2 produced, their number densities equal $n_i \simeq \phi_i H/K$ or 1.5×10^{12} and 5×10^{11} cm^{-3}, respectively, at ~50 km, i.e., the excess mixing ratios, Δf_{CO} and Δf_{O_2} are 5×10^{-8} and 1.5×10^{-8}, respectively.

Nitric Oxide in the Lower Atmosphere. The effect of lightning is the most important in NO production which otherwise is very low and occurs in the upper atmosphere. NO molecules formed by N atoms in the reaction

$$N(^2D) + CO_2 \rightarrow NO + CO$$
$$N + O \rightarrow NO + h\nu$$
$$N + O + CO_2 \rightarrow NO + CO_2$$

disappear with the help of the same N atoms in the reaction

$$N + NO \rightarrow N_2 + O\,.$$

Only when mixing time, $\tau_m = H^2/K$, is smaller than photochemical time, $\tau_c = (k[N])^{-1}$ (k is the rate coefficient of the last reaction), then the NO production by N atoms exceeds its removal by N atoms and forms a downward flux equal to $\phi_{NO} = -K^2/kH^3 = -5 \times 10^5$ cm^{-2} s^{-1}, i.e., three orders of magnitude less than the NO production by lightning, the latter being the main source of nitric oxide in the lower atmosphere.

NO vanishes at 80 to 90 km as a result of predissociation in the bands of 191 nm (0, 0) and 183 nm (1, 0) of the δ-system. This fact combined with the data of Nicolet and Cieslik (1980) on the predissociation rate and the solar radiation absorption in CO_2 gives $J_{NO} = 3.8 \times 10^{-6} e^{-2.8 \times 10^{-23} [CO_2] H \sec z_0} + 4.8 \times 10^{-6} e^{-4 \times 10^{-22} [CO_2] H \sec z_0}$.

N atoms formed double the NO sink, and the calculated NO mixing ratio is 4×10^{-8} at 50 km (Krasnopolsky 1983b) in the absence of other means of NO removal. If the NO sink occurs also in the atmosphere below clouds, then its mixing ratio equals 8×10^{-10}. Formation of other compounds of odd nitrogen, NO_x and HNO_x, will be discussed in Chap. 4.15.

4.7 Lower Atmospheric and Surface Rock Compositions (0 – 50 km)

To construct an aeronomic model of atmospheric composition means to calculate vertical profiles of atmospheric constituents using data on amounts of some parent components. It necessitates, in turn, the knowledge of reactions between components and their rate coefficients as well as the atmospheric temperature and eddy mixing coefficient. Unfortunately, in most cases rate coefficients of lower atmospheric reactions are not available.

The atmosphere below 50 km is not transparent for radiation with $\lambda < 400$ nm, due to the absorption of SO_2 and cloud particles in the upper layer. Therefore, photochemical processes may be significant only for constituents with dissociation energies less than 3 eV and for species chemically related to them. Optical measurements on board the Venera (Fig. 96) and PV probes demonstrate clearly the absorption of 400 – 500 nm radiation in the lower atmosphere. This absorption is due to the photolysis of S_3, and sulfur atoms formed have energies up to 0.6 eV. However, this energy is too low to allow reactions of S atoms with H_2O, HCl, CO_2 to proceed as suggested by Prinn (1979).

Even prior to the Venera 11, 12 and PV missions Prinn (1978) showed the possibility of SO_2 being the predominant sulfur-bearing constituent. He considered subsequent transformations of the H_2SO_4 downward flux:

$$
\begin{aligned}
&H_2SO_4 && \rightarrow H_2O + SO_3 \\
&SO_3 + CO && \rightarrow SO_2 + CO_2 \\
&SO_2 + CO && \rightarrow SO + CO_2 \\
&SO + CO && \rightarrow S + CO_2 \\
&S + CO + M && \rightarrow COS + M \\
&S + COS && \rightarrow CO + S_2 .
\end{aligned}
$$

He gave the chemical lifetime of SO_x compounds as

$$\tau_{SO_x} = 2 f_{SO_x}^{1/3} / 4.5 \times 10^{-12} e^{-24400/T} [M] f_{CO}^{0.76}$$

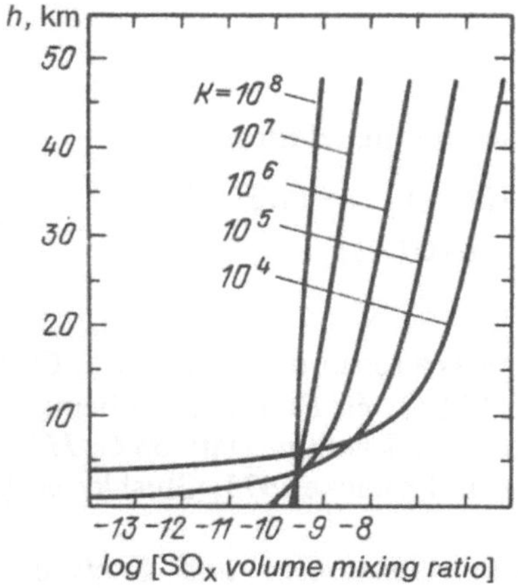

Fig. 173. Altitude profiles of the SO_2 mixing ratio for various eddy mixing coefficients (Prinn 1978)

based on investigations of the mixture SO_2 and CO at high temperatures by Bauer et al. (1971). Then, in the absence of SO_x sources, the continuity equation is

$$-\frac{d}{dz}\left\{K\left(\frac{d[SO_x]}{dz}+\frac{[SO_x]}{H}\right)\right\}=-\frac{[SO_x]}{\tau_{SO_x}} \quad \text{or}$$

$$\frac{d^2 f_{SO_x}}{dz^2}-\frac{1}{H}\frac{df_{SO_x}}{dz}-\frac{f_{SO_x}}{K\tau_{SO_x}}=0\,.$$

Solutions of this equation at different K and for a sulfuric acid downward flux of $2\times10^{12}\,\mathrm{cm^{-2}\,s^{-1}}$ at the upper boundary (50 km) are depicted in Fig. 173. A chemically passive surface ($\phi_{SO_x}=0$) was assumed for $K>10^5\,\mathrm{cm^2\,s^{-1}}$ and the active one ($f_{SO_x}=0$) for $K\leqslant10^5\,\mathrm{cm^2\,s^{-1}}$.

Thermochemical Equilibrium. According to Table 20 the compounds of O, C, N, S, H, Cl, F atoms along with noble gases exist in the Venusian atmosphere. Given known amounts of these atoms and reaction rate coefficients the atmospheric chemical composition can be calculated. However, as mentioned above, for the most part rate coefficients are not available, and an assumption of thermochemical equilibrium which is usually valid at high temperatures and pressures may be of help here. It depends on the thermodynamic properties of the components under consideration and needs neither chemical kinetics to be known nor a complicated numerical technique be applied.

For an arbitrary equilibrium reaction

$$\alpha_1A_1+\alpha_2A_2+\ldots+\beta_1B_1+\beta_2B_2+\ldots=\alpha_1'A_1'+\alpha_2'A_2'+\ldots\beta_1'B_1'+\beta_2'B_2'\ldots$$

where A_i, A_i' are gases and B_i, B_i' are present as solids and liquids with equilibrium saturated vapors in the system, the constant of equilibrium is

$$K_p(T) = \frac{P^{\alpha'_1}(A'_1)P^{\alpha'_2}(A'_2)\ldots}{P^{\alpha_1}(A_1)P^{\alpha_2}(A_2)\ldots}$$

where P is expressed in atmospheres. It is defined by the relation

$$\begin{aligned} RT\ln K_p(T) = -\Delta_r G(T) &= \alpha_1 G(A_1) + \alpha_2 G(A_2) + \ldots + \beta_1 G(B_1) \\ &+ \beta_2 G(B_2) + \ldots - \alpha'_1\, G(A'_1) - \alpha'_2\, G(A'_2) \\ &- \ldots \beta'_1\, G(B'_1) - \beta'_2\, G(B'_2) - \ldots \end{aligned}$$

where $\Delta_r G(T)$ is the change in the Gibbs' energy as a result of the reaction, G is the Gibbs's energy given by $G(T) = H(T) - TS(T)$, $H(T)$ and $S(T)$ being the enthalpy and entropy of the substance under consideration. Data on G, H, S are given by numerous handbooks (e.g., Barin and Knacke 1973, Glushko et al. 1978).

Thermochemical Equilibrium Between the Atmosphere and Surface Minerals. This problem will be discussed only briefly and is considered in more detail by Lewis (1970), Florensky et al. (1978), Lewis and Kreimendahl (1980), Barsukov et al. (1980), Khodakovsky et al. (1979), Khodakovsky (1982), von Zahn et al. (1983), and Prinn (1984a, b).

The assumption of thermochemical equilibrium between atmospheric species and rocks makes it possible to calculate the atmospheric composition with known minerals. Such calculations gave estimates of the atmospheric composition performed under the assumption for the identity of the mineralogical composition on the Earth and Venus. This approach resulted in the explanation of the measured amounts of HCl and HF due to the following equilibria (Lewis 1970):

$$\begin{aligned} &2\,NaAl + CaCl_2Si_2O_8 + SiO_2 + H_2O \rightleftarrows 2\,NaAlSiO_4 + CaSiO_3 + 2\,HCl \\ &2\,NaCl + Al_2SiO_5 + 3\,SiO_2 + H_2O \quad \rightleftarrows 2\,NaAlSi_2O_6 + 2\,HCl \\ &2\,CaF_2 + MgSiO_3 + SiO_2 + 2\,H_2O \quad \rightleftarrows Ca_2MgSi_2O_7 + 4\,HF\,. \end{aligned}$$

The oxidation state of the atmosphere and the oxygen amount in it is determined by the pyrite–anhydrite buffer (Lewis and Kreimendahl 1980, Barsukov et al. 1980):

$$FeS_2 + 2\,CaCO_3 + \tfrac{7}{2}O_2 \rightleftarrows 2\,CaSO_4 + FeO + 2\,CO_2$$

which results in the equilibrium oxygen mixing ratio of 3×10^{-25}. This ratio has been evaluated by a chemical indicator on board the V 13, V 14 (Florensky et al. 1983) made from $Na_4V_2O_7$ which is white and becomes dark blue or black in the presence of CO due to the formation of V_2O_4 and V_2O_3 by reactions

$$\begin{aligned} &Na_4V_2O_7 + CO_2 + CO \rightarrow V_2O_4 + 2\,Na_2CO_3 \\ &Na_4V_2O_7 + 2\,CO \qquad \rightarrow V_2O_3 + 2\,Na_2CO_3\,. \end{aligned}$$

This indicator was reproduced on the Venusian surface images and its albedo is consistent with $f_{CO} \gtrsim 10$ ppm in accord with the gas chromato-

Table 35. Composition (% mass) of rocks on Venus measured by V13 and V14 X-ray fluorescent spectroscopy (Surkov et al. 1983b)

Species	V13	V14
MgO	11.4±6.2	8.1±3.3
Al_2O_3	15.8±3.0	17.9±2.6
SiO_2	45.1±3.0	48.7±3.6
SO_3	1.62±1.0	0.88±0.77
Cl	0.3	0.4
K_2O	4.0±0.63	0.2±0.07
CaO	7.1±0.96	10.3±1.2
TiO_2	1.59±0.45	1.25±0.41
MnO	0.2±0.1	0.16±0.08
FeO	9.3±2.2	8.8±1.8
Total	96	96

graphic measurements. This gives $f_{O_2} < 10^{-23}$ in equilibrium between CO and O_2, thus confirming the above estimate.

The CO and H_2O mixing ratios are determined by the reactions (Lewis 1970)

$$3\,FeMgSiO_4 + CO_2 \rightleftarrows 3\,MgSiO_3 + Fe_3O_4 + CO$$

$$Ca_2Mg_5Si_8O_{22}(OH)_2 \rightleftarrows 3\,MgSiO_3 + 2\,CaMgSi_2O_6 + SiO_2 + H_2O$$

though the values obtained (5×10^{-5} and 5×10^{-4}, respectively) exceed the measured ones.

Composition of Rocks. The assumption of similarity of rocks on Earth and Venus may serve as an initial approximation only. Some important experimental results have been achieved in this field by γ-radiation spectroscopy on board the V8, V9, V10 (Surkov 1983) which yielded amounts of natural radioactive elements (*U, Th, K*). These elements are concentrated in the surface rocks on the Earth and are much less abundant in the interior as a result of heating of the interior and differentiation of terrestrial materials. Their amounts on Venus are close to those for terrestrial magnetic rocks, thus, indicating the evolution of Venus to be similar to that of the Earth.

The composition of surface rocks on Venus was studied by X-ray fluorescent spectroscopy on board the V13, V14 (Surkov et al. 1983b). This technique was considered in Chap. 3.9. The measured composition is given in Table 35.

Other important information concerning the rocks and interior of Venus has been obtained from the V9, V10, V13, V14 surface images, PV and V15, V16 radar studies, and measurements of components of gravitational potential from orbital parameters. These data show that most of the surface is covered by rocks similar to terrestrial basalts.

Sulfur Cycles (von Zahn et al. 1983, Prinn 1984a, b). Prior to the PV and V 11 and V 12 missions carbonylsulfide COS was considered as the main sulfur-bearing component in the atmosphere of Venus. It is evident now that sulfur dioxide dominates in it. Amounts of COS and H_2S are still controversial: those measured by the PV instruments are $\lesssim 3\%$ of SO_2, and the V 13, V 14 ones are comparable with SO_2.

The problem is resolved by including two atmospheric and one geologic cycle of sulfur. According to radar measurements there is some evidence of pyrite FeS_2 being present at some locations. The following reactions result in the atmospheric sulfur formation

$$FeS_2 + 2H_2O \rightarrow FeO + 2H_2S + 1/2\,O_2$$
$$FeS_2 + 2CO_2 \rightarrow FeO + 2COS + 1/2\,O_2 .$$

Atmospheric photochemistry tends to transform the reduced components to oxidized ones by means of free oxygen formed by CO_2 photolysis. This results in the net photochemical reactions

$$H_2S + 3CO_2 \rightarrow H_2O + SO_2 + 3CO$$
$$COS + 2CO_2 \rightarrow SO_2 + 3CO .$$

There are many paths for these reactions to proceed, e.g.:

$$\begin{array}{ll}
2(CO_2 + h\nu & \rightarrow CO + O) \\
O + O + M & \rightarrow O_2 + M \\
COS + h\nu & \rightarrow CO + S \\
S + O_2 & \rightarrow SO + O \\
SO + O + M & \rightarrow SO_2 + M \\
\hline
COS + 2CO_2 & \rightarrow SO_2 + 3CO .
\end{array}$$

For H_2S:

$$\begin{array}{ll}
3(CO_2 + h\nu & \rightarrow CO + O) \\
O + O + M & \rightarrow O_2 + M \\
H_2S + h\nu & \rightarrow H + HS \\
HS + O & \rightarrow SO + H \\
H + O_2 + M & \rightarrow HO_2 + M \\
H + HO_2 & \rightarrow H_2O + O \\
SO + O + M & \rightarrow SO_2 + M \\
\hline
H_2S + 3CO_2 & \rightarrow H_2O + SO_2 + 3CO .
\end{array}$$

These reactions as well as thermochemical reactions in the atmosphere near the surface with free sulfur formed, which partly restore H_2S and COS, represent the so-called slow atmospheric cycle of sulfur. Fast atmospheric sulfur cycle transforms SO_2 to sulfuric acid with subsequent restoration of SO_2 in the lower atmosphere:

$$\begin{aligned} &CO_2 + h\nu &&\rightarrow CO + O \\ &SO_2 + O + M &&\rightarrow SO_3 + M \\ &H_2O + SO_3 &&\rightarrow H_2SO_4 \\ &H_2SO_4 &&\rightarrow H_2O + SO_3 \\ &SO_3 + CO &&\rightarrow SO_2 + CO_2 \,. \end{aligned}$$

A sink of the atmospheric sulfur occurs via formation of anhydrite

$$SO_2 + CaCO_3 \rightarrow CaSO_4 + CO\,.$$

Conversion of anhydrite to pyrite completes the geologic cycle

$$2\,CaSO_4 + FeO + 7\,CO \rightarrow FeS_2 + 2\,CaCO_3 + 5\,CO_2\,.$$

Thus, the formation of atmospheric sulfur proceeds at locations enriched by pyrite and having high radar reflectivity (e.g., Theia Mons) due to high electric conductivity of pyrite. Its sink proceeds at locations enriched by calcium, e.g., the V13, V14 landing sites. Percentages of COS and H_2S depend on ratios of the geologic cycle rate to that of the slow atmospheric cycle and on the COS/SO_2 and H_2S/SO_2 ratios formed by the slow atmospheric cycle. Due to the very low rate to the geologic cycle the slow atmospheric cycle should account for these ratios in the atmosphere. It is easy to understand that fluctuations of volcanic activity should correlate with the atmospheric sulfur mixing ratio and the contribution of H_2S and COS in it.

Thermochemical Equilibrium of the Atmospheric Mixture. The approximation of the chemical equilibrium is valid for its time, $\tau_c = (\Sigma_i k_i n_i)^{-1}$ being less than mixing time, $\tau_m = H^2/K^{-1}$. Here k_i is the rate coefficient of the i reaction removing the component under consideration and n_i is the number density of another partner in this reaction. Most reactions of stable components typical of the lower atmosphere (e.g., the reactions of CO and O_2, H_2 and O_2, H_2 und Cl_2, etc.) need activation energy of the order of 50 kcal. Therefore, $k_i \simeq 10^{-10} \exp(-25000/T)\ cm^3\,s^{-1}$ may be considered as a typical value. Figure 174 shows chemical and mixing times in the troposphere for

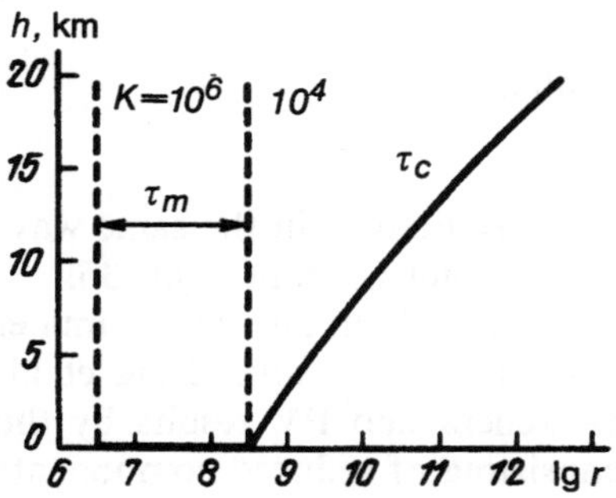

Fig. 174. Comparison of the chemical and mixing times, τ_c and τ_m in the lower atmosphere of Venus (Krasnopolsky and Parshev 1979b)

$n_i \simeq 10^{-5}$ [CO_2] and $K = 10^4 - 10^6$ cm^{2}s^{-1}. The mixing time is less than the chemical one throughout the lower atmosphere and almost equal to it near the surface. The planetary surface may catalyze those chemical reactions which establish the thermochemical equilibrium. This effect is not fully comprehended, and a qualitative estimate of it can be made by reducing the chemical time by two orders of magnitude. Then the thermochemical equilibrium is reached near the surface independently of the eddy mixing assumed, the latter prevailing throughout the lower atmosphere. Therefore, atmospheric constituent mixing ratios are determined by thermochemical equilibrium near the surface and are constant up to the cloud layer (Krasnopolsky and Parshev 1979b).

Let us consider first O, C, S compounds: CO_2, CO, O_2, SO_3, SO_2, S_2, COS, and CS_2. Others may be shown as existing in negligible quantities. Obviously one can adopt mixing ratios of three components out of the above mentioned containing three kinds of atoms to determine the system. All experiments give consistent information on f_{SO_2}, CO_2 being another evident candicate. The following system of equilibrium reactions can be used:

$$2CO + SO_2 = 2CO_2 + \tfrac{1}{2}S_2 \; ; \; \lg K_p = 9.14 \text{ at } 740\,\text{K}$$
$$3CO + SO_2 = 2CO_2 + COS; \quad 11.77$$
$$CO + \tfrac{1}{2}O_2 = CO_2 \; ; \quad 15.5$$
$$SO_2 + \tfrac{1}{2}O_2 = SO_3 \; ; \quad 2.1$$
$$CS_2 + O_2 = CO_2 + S_2 \; ; \quad 26.9 .$$

We analyzed this problem after preliminary results of the V11, V12 and PV missions became known. The main contradiction was the detection of CO and the absence of O_2 in the Venera gas chromatograms and vice versa in the PV ones. To solve it we derived the sulfur mixing ratio, $f_{S_2} \simeq 10^{-7}$, from the V11, V12 optical spectrometric data (see Chap. 3.5). Calculated mixing ratios (Table 36) confirmed the Venera data. It became known that the contradiction was caused by a small misleading error in a computer program of the PV gas chromatograph analysis and the corrected program shows f_{CO} close to that given in Table 36.

Hydrogen compounds can be calculated by including the water vapor reactions:

$$CO + H_2O = CO_2 + H_2 \; ; \; \lg K_p = 0.8 \text{ at } 740\,\text{K}$$
$$COS + H_2O = CO_2 + H_2S; \quad 2.02 .$$

Nitrogen, chlorine, and fluor compounds were derived in the same way. Components with mixing ratios less than 10^{-10} are not given in Table 36.

These calculations are very simple and were carried out also by Oyama et al. (1979), Barsukov et al. (1980), Khodakovsky (1982), and Craig et al. (1983). Khodakovsky tried to reconcile the Venera and PV results by the assumption of upward atmospheric flows consisting of reduced components

Table 36. Composition of the lower atmosphere of Venus under conditions of thermochemical equilibrium (Krasnopolsky and Parshev 1979)

Gas	Mixing ratio	Gas	Mixing ratio
CO_2	0.96	CO	1.5×10^{-5}
N_2	3.4×10^{-2}	S_2	1×10^{-7}
H_2O	3×10^{-5}	COS	2×10^{-5}
SO_2	1.5×10^{-4}	H_2S	5×10^{-8}
HCl	1×10^{-6}	H_2	3×10^{-9}
HF	1×10^{-8}	other	$<10^{-10}$

at some locations, while at other locations downward flows enriched by oxidizing species dominate. This idea, which was repeatedly published in a number of papers, employs large mixing ratios of H_2O and O_2, $\sim 10^{-3}$ and $\sim 10^{-4}$, respectively, in the downward flow replaced by $f_{H_2O} \sim 10^{-4}$ and by $f_{CO} \simeq 10^{-5}$ in the upward flow without any indication of how and from what these quantities may be obtained. However, approximate evaluations ($\phi \simeq nK/H$) show that this assumption suggests unknown chemical sources in the clouds and on the surface which are ~300 times as powerful as the photolytic formation of atomic oxygen. This is only one of the difficulties presented by this idea.

Confirming our calculations Craig et al. (1983) show that $f_{SO_2} \simeq 180$ ppm and $f_{O_2} \simeq 16$ ppm measured by the PV gas chromatograph result in $f_{SO_3} \simeq 1000$ ppm. It should lead to the formation of an extremely dense and low lying sulfuric acid cloud layer as well as to penetration of SO_3 into the upper cloud layer with a subsequent formation of large concentrations of SO_2. It should be noted that both possibilities present difficulties.

The Law of Constancy of the Elementary Composition. In the absence of condensation and sedimentation the one-dimensional description of a flux in the lower atmosphere where $K \gg D$ is given by

$$\phi_i = -K\left(\frac{dn_i}{dz} + \frac{n_i}{H}\right) = -Kn_a\frac{df_i}{dz}$$

where n_i, n_a are the species and atmospheric number densities, respectively, and H is the number density scale height.

If this description is applied to an element i which cannot originate or disappear in chemical reactions, then it follows from the continuity equation that $\phi_i =$ const. in the case of a time independent solution $[(\partial n_i)/\partial t = 0]$. In the case of a flux of the element on or from the surface being much less than $(n_i H)/K$, where n_i is the element number density at the upper boundary of the region under consideration, ϕ_i may be neglected, $(df_i)/dz = 0$ and $f_i =$ const. Similarly, if a horizontal flux of the element i is much less than $(n_i H)/K$, then $f_i =$ const. for a two- and three-dimensional solution as well. The law of constancy of the elementary composition is violated in the presence of molecular diffusion and condensation since the expression for the flux takes a form different from $\phi_i = -Kn_a(df_i)/dz$.

CO, COS, SO₂, Sₙ. Our assumption of a chemical composition of the atmosphere independent of height is a mere approximation which cannot be applied to all the components. Let us now consider the substances which have been experimentally found to feature a dependence of the mixing ratio on the altitude.

The decrease of f_{CO} with height from 30 ppm at $h > 30$ km to 15 ppm below 25 km, discovered in both gas chromatographic experiments, is accompanied by a decrease in f_{O_2} from 40 to 16 ppm at 52 and 42 km according to the PV gas chromatographic data. However, this effect cannot be attributed to the reaction

$$CO + O_2 \rightarrow CO + O\,, \qquad k = 4.2 \times 10^{-12} e^{-24000/T}\,\mathrm{cm^3\,s^{-1}}$$

which results in a very long lifetime of CO and O_2 far in excess of the mixing time. If all CO and O_2 formed by CO_2 photolysis got into the lower atmosphere, then their mixing ratios would be ~1 ppm due to this reaction only.

A comparison of the measurements and the results of the calculation of thermochemical equilibrium (Table 36) readily shows the approximate constancy of the sum $f_{CO} + f_{COS} = 3.5 \times 10^{-5}$. Then the altitude dependence of f_{CO} can be explained by the gradual conversion of CO to COS. Two problems arise here: the value of $f_{COS} = (5-10)$ ppm at 22 km which exceeds the upper limit of 2 ppm, and a source of sulfur to give 20 ppm of COS. Free sulfur is far from being abundant enough to account for 20 ppm, and if, however, sulfur is delivered by the consecutive reduction of SO_2 by CO, then it is necessary to spend three, and not one molecule of CO for the formation of each molecule of COS.

Our qualitative analysis of the problem has yielded the following solution (Krasnopolsky and Parshev 1983). Due to the rapid photolysis of S_3 by visible light, the thermochemical equilibrium of the primary forms of sulfur S_2, S_3, and S near the surface is appreciably disturbed and transformed to the photochemical equilibrium described by the following reactions

$$S_2 + S_2 \rightleftarrows S_3 + S$$

$$S_3 + M \rightleftarrows S + S_2 + M$$

$$S_3 + h\nu \rightarrow S_2 + S\,.$$

Then with $[S_3] = 6 \times 10^9\,\mathrm{cm^3}$ (see Chap. 3.5) at 12 km the equilibrium quantities of S_2 and S turn out to be 10 and 100 times as high, respectively, as in the case of thermochemical equilibrium, and $f_{S_2} = 10^{-7}$ in accord with our original estimate (Table 36). The change in the relative distribution of S, S_2, and S_3 affects the concentrations of COS by the processes

$$S + COS \rightleftarrows CO + S_2$$
$$S_3 + CO \rightleftarrows COS + S_2$$
$$S + CO + M \rightleftarrows COS + M\,.$$

The second of these processes is most important, and $[COS]/[CO] = K[S_3]/[S_2]$. Allowing for the photolysis of S_3 reduces this ratio ten times, and gives $f_{COS} \simeq 2$ ppm at the surface. Thus, the diminution of f_{CO} may be attributed to its transformation to CO_2 via reactions with SO_2 and then with SO. The variation of f_{CO} from 30 ppm at 40 km to 15 ppm at the surface corresponds to the diminution of f_{SO_2} by 7 ppm or by ~5% and to a growth of f_{S_2} by 3 ppm and of f_{COS} up to 2 ppm. And, in fact, an approximate calculation of the photochemical equilibrium at 12 km where the maximum of S_3 is located (Chap. 3.5) also indicates a rapid increase of f_{S_2} from the surface with $f_{S_2} \simeq 4$ ppm at 12 km.

H_2O, O_2 and H_2. It is difficult to explain the variations of the f_{H_2O} from 2×10^{-4} at 50 km to 2×10^{-5} at the surface. To do this it is necessary to add in some way considerable quantities of hydrogen. The most obvious possibility

$$CO + H_2O \rightarrow CO_2 + H_2$$

does not apply because of the following: $f_{H_2} \simeq 10^{-4}$ near the surface contradicts the measurements, the equilibrium of the reaction is shifted to the left and, at thermochemical equilibrium, $f_{H_2} = 6 f_{CO} f_{H_2O} \simeq 3 \times 10^{-9}$ near the surface, the variation of f_{CO} by 15 ppm is appreciably less than the change in the f_{H_2O} by 180 ppm. The vertical profile of f_{H_2O} (Table 20) results in a constant number density of $\sim 10^{16}$ cm^{-3} which presumes a source of $\sim 10^{14}$ cm^{-2} s^{-1} near the clouds and a corresponding sink on the surface. Only vulcanism may supply this source equal to $\sim 10^{10}$ g s^{-1} on the global scale. However, this hypothesis implies severe constraints on the atmospheric and lithospheric dynamics.

Explaining the gas chromatographic measurements of O_2 presents some difficulties too. It is interesting that $f_{O_2} = 44$ ppm at 52 km and 16 ppm at 42 km correspond to a constant number density of $[O_2] = 8 \times 10^{14}$ cm^{-3}. This may occur in the presence of a source and a sink below the bottom of the particular region, h_0; then the solution of the continuity equation at $h - h_0 > 2H$ is equal to $n = \phi H/K$. Substituting $[O_2] = 8 \times 10^{14}$ cm^{-3} and $K = 2 \times 10^4$ cm^{-3} yields $\phi = 5 \times 10^{12}$ cm^{-2} s^{-1}, i.e., equal to the production of O_2 by CO_2 photolysis. However, the bulk of CO_2 photolysis occurs above 65 km, and the measured oxygen amounts result in the absence of reactions of CO_2 formation down to 35 km, hence, $f_{CO} \simeq 600$ ppm and $f_{O_2} \simeq 300$ ppm should be at 65 km, which disagrees with the spectroscopic measurements ($f_{CO} = 50$ ppm and $f_{O_2} < 0.3$ ppm) that pertain to this altitude.

The V13, V14 gas chromatographic measurements confirm the value of ~10 ppm for molecular hydrogen in the lower atmosphere which was deduced from the interpretation of ion mass 2 measurements by the PV orbiter ion mass spectrometer to be H_2^+ (Kumar et al. 1981). As was mentioned above, the equilibrium H_2 mixing ratio is less by three orders of magnitude than this value and the interpretation of ion mass 2 being D^+ ions (McElroy et al. 1982a) seems more convincing.

Thermochemical Equilibrium Condensates in the Cloud Layer. The first attempt at this analysis was made by Lewis (1970). Recently Volkov et al. (1982) calculated mixing ratios of ~150 species for thermochemical equilibrium with the atmospheric constituents shown in Table 36. They assumed an elemental composition of the Venusian crust similar to that on Earth. Only seven components, As_4O_6, Sb_4O_6, $SbCl_3$, S_n, Se_2, Te_2, Hg, appear to have mixing ratios exceeding 0.1 ppm. The upper limit of mercury measured by V 12, V 14 is three orders of magnitude less than the calculated value ($f_{Hg} < 10^{-8}$ and 10^{-5}, respectively). The calculated Se_2, Te_2, and Sb_4O_6 should produce condensation layers at ~20 km which have not been observed. Therefore, only three candidates can be considered: S_n, As_4O_6, and $SbCl_3$. $SbCl_3$ forms SbOCl at 50 km via the reaction

$$SbCl_3 + H_2O \rightarrow SbOCl + 2HCl .$$

Despite the fact that this approach is a useful tool to study the chemical composition of the clouds, the results obtained cannot be accepted as unambiguous. Actually four components out of the calculated seven can be present in the Venusian atmosphere with abundances far less than calculated. On the other hand, thermochemical equilibrium amounts of sulfuric acid are negligible, though its presence in the clouds is beyond doubt. Lines of Sb 3.65 keV and As 10.5 keV were not observed by the V 12 and V 14 spectrometers, and their upper limits should be close to that of Hg.

4.8 Neutral Atmospheric Photochemistry Above 50 km. Main Problems, Previous Results, Main Chemical Reactions

We now begin to consider processes which determine the chemical composition of the atmosphere and cloud layer. Here, we follow methods, initial data, results, and conclusions of the papers by Krasnopolsky and Parshev (1980a, b, 1981). After that we will consider the paper by Winick and Stewart (1980) and an updating of the photochemical theory of the Venusian atmospheric made by Yung and DeMore (1982).

To construct a photochemical model of the Venusian atmospheric composition is a very challenging and important task. Based only on the number densities of several major components (CO_2, H_2O, HCl, SO_2) on the lower boundary, such a model enables calculations of fundamental data on the composition of the atmosphere that should agree with the whole range of experimental findings. The solution of the problem should, inter alia, give answers to the following questions:

1. Why are number densities of CO_2 photolytic products so low?
2. How is the Venusian cloud layer formed, what processes underly the formation of sulfuric acid which is absent in gas release from the lithosphere?
3. Why are number densities of sulfur-bearing gases which can be regarded as initial materials and intermediate links in H_2SO_4 formation very low in the top clouds?
4. Why does the mixing ratio of water in the top clouds exceed by two orders of magnitude that in the subcloud atmosphere?
5. Do chlorine and sulfur reach the upper atmosphere?

All the questions are somehow related to the photochemistry of CO_2, H_2O, HCl, and SO_2. Although other gases are present in the Venusian atmosphere as well, they are less important due to either their lower chemical activity (nitrogen, noble gases) or their low mixing ratios (HF). Hence, for the sake of simplicity we will not consider nitrogen photochemistry in the CO_2 atmosphere. Since basic processes for fluorine and chlorine are almost the same, while their number densities differ by two orders of magnitude, the former can be neglected (see also Parisot and Moreels 1984).

The aeronomy of CO_2, H_2O, and HCl on Venus down to 62 km was given by Sze and McElroy (1975). Sulfuric compounds were taken into account there only as a boundary condition for O_2; they were treated by Prinn (1975, 1978) and Wofsy and Sze (1975) in greater detail. In these studies a semi-analytical approach to the problem was used. The analysis was made only for a few reactions, e.g., excluding those of chlorine compounds, etc. According to Lewis (1970), COS was assumed to be a basic sulfur-bearing gas.

The above papers were based mainly on the data about the atmospheric composition obtained by ground-based spectroscopy in the visible and infrared regions (see Chap. 3.3) and the OAO-2 orbiter data in the range of 200 – 350 nm.

We are now at a qualitatively new level of knowledge about the Venusian atmospheric composition. So it is vital that an aeronomical model of the atmospheric composition be constructed, which could serve as a theoretical foundation for the data obtained. The present day level of our knowledge makes it possible to consider together sulfur and chlorine compounds, which were treated previously only individually. Indeed, the neglect of one of the above groups of compounds can introduce an essential error. We then consider a number of new processes and components.

Basic Chemical Processes Involving Hydrogen. Nearly all solar radiation absorption at $\lambda < 202$ nm results in CO_2 photolysis. The mean daytime photolytic rate of CO_2 per column is equal to the flux of solar quanta, i.e., $1.1 \times 10^{13}\,cm^{-2}\,s^{-1}$. In a pure CO_2 atmosphere almost all oxygen atoms formed enter the reaction:

$$O + O + CO_2 \rightarrow O_2 + CO_2$$

which should eventually lead to the accumulation of considerable amounts of CO and O_2. In the real atmospheres of Mars and Venus, catalytic reactions in which odd hydrogen H^*/H, OH, HO_2, H_2O_2 is involved tend to suppress that process (see Chap. 2.7).

The presence of water vapor with a mixing ratio, f_{H_2O}, of $\sim 10^{-4}$ in the Martian atmosphere ensures the amount of odd hydrogen required to reduce f_{CO} and f_{O_2} down to about 10^{-3}. In the top clouds on Venus, $f_{H_2O} \simeq 10^{-5}$, $f_{CO} = 4 \times 10^{-5}$ and $f_{O_2} < 3 \times 10^{-7}$. Thus, there is a considerable lack of H^*, while on the other hand, such low mixing ratios of O_2 and CO are difficult to explain even for a high enough hydrogen amount. The difficulties can be partially overcome by taking into account the HCl photolytic process which supplies a major part of the hydrogen atoms along with chlorine atoms into the atmosphere.

Reactions Involving Chlorine. Reactions of chlorine atoms with oxygen (O, O_2, and O_3) produce the oxides ClO and ClO_2 which are involved in numerous interactions with each other and with other components. As odd hydrogen, odd chlorine (Cl, ClO, and ClO_2) catalyzes O_2 formation via reactions, the final phases of which are:

$$O + ClO \rightarrow O_2 + Cl$$
$$O_3 + ClO \rightarrow O_2 + ClO_2$$

and O_2 destruction via

$$H + O_2 + CO_2 \rightarrow HO_2 + CO_2$$
$$HO_2 + Cl \rightarrow ClO + OH\,.$$

Odd hydrogen and chlorine do not vanish in all the above processes. Their sink occurs by the reactions:

$$H + HO_2 \rightarrow H_2 + O_2$$
$$H + HO_2 \rightarrow H_2O + O$$
$$HO_2 + OH \rightarrow H_2O + O_2$$
$$HO_2 + Cl \rightarrow HCl + O_2\,.$$

The rate of these processes, however, is not high because of low number densities, hence, each atom of H and Cl in odd form enters a large number of CO_2 and O_2 formation reactions.

The entire system of reactions among C, O, H, and Cl compounds considered by Sze and McElroy (1975) covers 47 processes. In their paper, however, a large part of the reactions that determine the balance of the most important components of the atmosphere have maximum rates at the lower boundary (62 km) where the validity of the photochemical scheme is doubtful because sulfur compounds are not taken into account and consequently, the boundary conditions are chosen arbitrarily. For instance, a value of the HO_2 number density equal to $10^9\ cm^{-3}$ at 62 km determines the balance of CO_2 and O_2. However, this value is obtained under the assumption of photochemical equilibrium at the lower boundary. As will be seen from the results of our calculations, the presence of sulfur and the effect of transport decrease the HO_2 number density by two orders of magnitude, and the HO_2 maximum value is $10^7\,cm^{-3}$. This implies that the hydrogen cycle is much less important in O_2 destruction.

Chemical reactions in the Venusian atmosphere and their rate coefficients are given in Table 37.

As known from the data on the Martian atmosphere, O_2 formation in processes 12, 18 (see Table 37) can be prevented if the eddy mixing coefficient is large enough, $K \sim 10^8\,cm^2\,s^{-1}$, thus providing a rapid supply of atomic oxygen to the region of high concentrations of odd hydrogen ($\sim$75 km for Venus) where the reaction $CO+OH \rightarrow CO_2+H$ is effective. However, observations (see Chap. 3.13) do not show such high values of K in the mesosphere. Therefore, much atomic oxygen on both planets is spent to form O_2, and to fit the measurements emphasis should be placed on the processes of O_2 destruction. If on Mars such processes are, above all, $2HO_2 \rightarrow H_2O_2+O_2$ with following H_2O_2 photolysis, on Venus they are reactions with chlorine compounds which enhance CO binding with O_2 molecules.

In our opinion, the following reactions with chlorine compounds should be added to the list Sze and McElroy (1975) suggested. The reactions

$$Cl_2+H \rightarrow HCl+Cl$$
$$Cl+H_2O_2 \rightarrow HCl+HO_2$$
$$H+Cl+CO_2 \rightarrow HCl+CO_2$$

ensure a sink of odd chlorine and hydrogen. In fact, photolytic rates for Cl_2 and H_2O_2 are much higher than for HCl, so that Cl_2 and H_2O_2 can be regarded as active components, despite the fact that the numbers of H and Cl atoms they include are even. As will be seen below, the first of the above reactions is the major destruction process for odd chlorine and hydrogen.

The reaction $CO_2^+ + HCl \rightarrow HCO_2^+ + Cl$ is similar to reaction 53. It is also a source of H and Cl.

Previously, only chlorine compounds with oxygen were considered. Of similar importance and interest are also compounds with carbon monoxide: COCl and phosgene $COCl_2$. COCl accelerates reactions of CO with O and O_2, reducing the amount of free oxygen in the atmosphere:

$$CO+Cl+CO_2 \rightarrow COCl+CO_2$$
$$COCl+O \rightarrow CO_2+Cl$$
$$COCl+O_2 \rightarrow CO_2+ClO\ .$$

Phosgene is formed in the $COCl+Cl_2$ reaction; it is less active than COCl, but also enhances CO_2 formation in the reaction $COCl_2+O \rightarrow CO_2+Cl_2$. Unfortunately, reliable information on rate coefficients of the reactions with COCl and $COCl_2$ have not been available for a long time.

Table 37. Chemical reactions in the atmosphere of Venus above 50 km and their rate coefficients used by Krasnopolsky and Parshev 1980a, b, 1981(KP) and Yung and DeMore 1982 (YD)[a]

No.	Reactions	Rate coefficients KP	Rate coefficients YD
1.	$CO_2+h\nu \rightarrow CO+O\ (1675<\lambda<2075\ Å)$ $CO+O(^1D)\ (\lambda<1675\ Å)$	1.6×10^{-6}	1.5×10^{-12}
2.	$H_2O+h\nu \rightarrow OH+H$	6×10^{-6}	2.2×10^{-11}
3.	$HCl+h\nu \rightarrow H+Cl$	2×10^{-6}	10^{-7}
4.	$SO_2+h\nu \rightarrow SO+O$	1.9×10^{-4}	1.5×10^{-4}
5.	$O_2+h\nu \rightarrow O+O$	4×10^{-6}	7.4×10^{-10}
6.	$O_3+h\nu \rightarrow O_2(^1\Delta g)+O(^1D)$	8×10^{-3}	8.6×10^{-3}
7.	$H_2O_2+h\nu \rightarrow OH+OH$	1.8×10^{-4}	10^{-4}
8.	$Cl_2+h\nu \rightarrow Cl+Cl$	2.5×10^{-3}	2.4×10^{-3}
9.	$SO_2+h\nu \rightarrow {}^3SO_2\ (3200<\lambda<3900\ Å)$	see text	–
10.	$SO_2+h\nu \rightarrow {}^1SO_2\ (2180<\lambda<3200\ Å)$	see text	–
11.	$CO+O+CO_2 \rightarrow CO_2+CO_2$	2×10^{-37}	$6.5\times10^{-33}e^{-2180/T}$
12.	$O+O+CO_2 \rightarrow O_2+CO_2$	$7\times10^{-31}/T$	$8.6\times10^{-28}T^{-2}$
13.	$O+O_2+CO_2 \rightarrow O_3+CO_2$	$3\times10^{-34}\exp(520/T)$	1.35×10^{-33}
14.	$H+O_2+CO_2 \rightarrow HO_2+CO_2$	$2\times10^{-31}(273/T)^{1.3}$	$3.6\times10^{-29}T^{-1}$
15.	$O+HO_2 \rightarrow OH+O_2$	2.5×10^{-11}	3.1×10^{-11}
16.	$CO+OH \rightarrow CO_2+H$	$2.2\times10^{-13}\exp(-100/T)$	1.4×10^{-13}
17.	$H+O_3 \rightarrow OH+O_2$	$1.3\times10^{-10}\exp(-500/T)$	$1.4\times10^{-10}e^{-480/T}$
18.	$O+OH \rightarrow O_2+H$	2×10^{-11}	3.8×10^{-11}
19.	$HO_2+HO_2 \rightarrow H_2O_2+O_2$	$3\times10^{-11}\exp(-500/T)$	2.3×10^{-12}
20.	$H+HO_2 \rightarrow H_2+O_2$	$2\times10^{-11}\exp(-350/T)$	1.4×10^{-11}
21.	$H+HO_2 \rightarrow OH+OH$	$4.2\times10^{-10}\exp(-950/T)$	3.2×10^{-11}
22.	$H+HO_2 \rightarrow H_2O+O$	$8\times10^{-11}\exp(-500/T)$	9.4×10^{-13}
23.	$HO_2+OH \rightarrow H_2O+O$	5×10^{-11}	8×10^{-11}
24.	$H+H+CO_2 \rightarrow H_2+CO_2$	2.6×10^{-32}	$6.6\times10^{-31}T^{-0.65}$
25.	$Cl+Cl+CO_2 \rightarrow Cl_2+CO_2$	2.6×10^{-32}	$1.2\times10^{-33}e^{900/T}$
26.	$H_2+Cl \rightarrow HCl+H$	$8\times10^{-11}\exp(-2480/T)$	$4.7\times10^{-11}e^{-2340/T}$
27.	$H+Cl_2 \rightarrow HCl+Cl$	$3\times10^{-10}\exp(-1000/T)$	$1.5\times10^{-10}e^{-593/T}$
28.	$HCl+H \rightarrow H_2+Cl$	$10^{-11}\exp(-1600/T)$	$2.4\times10^{-11}e^{-1740/T}$
29.	$H+Cl+CO_2 \rightarrow HCl+CO_2$	2.6×10^{-32}	10^{-32}
30.	$O_3+Cl \rightarrow ClO+O_2$	$2.2\times10^{-11}\exp(-170/T)$	$2.7\times10^{-11}e^{-257/T}$
31.	$Cl+O_2+CO_2 \rightarrow ClO_2+CO_2$	1.7×10^{-33}	$3.3\times10^{-30}T^{-1.3}$
32.	$ClO_2+Cl \rightarrow Cl_2+O_2$	1.5×10^{-10}	–
33.	$ClO_2+Cl \rightarrow ClO+ClO$	1.5×10^{-12}	–
34.	$O+ClO \rightarrow O_2+Cl$	$3.4\times10^{-11}\exp(75/T)$	$7.5\times10^{-11}e^{-120/T}$
35.	$O_3+ClO \rightarrow ClO_2+O_2$	2×10^{-15}	–
36.	$ClO+ClO \rightarrow Cl_2+O_2$	2.3×10^{-14}	–
37.	$ClO+ClO \rightarrow ClO_2+Cl$	$1.2\times10^{-12}\exp(-1260/T)$	–
38.	$ClO+ClO+CO_2 \rightarrow O_2+Cl+CO_2$	10^{-31}	–
39.	$O+Cl_2 \rightarrow ClO+Cl$	$10^{-20}\exp(-1560/T)$	$4.2\times10^{-12}e^{-1370/T}$
40.	$HCl+OH \rightarrow H_2O+Cl$	$2\times10^{-13}\exp(-310/T)$	$3\times10^{-12}e^{-425/T}$
41.	$O+HCl \rightarrow OH+Cl$	$1.8\times10^{-12}\exp(-2260/T)$	$10^{-11}e^{-3370/T}$
42.	$ClO_2+CO_2 \rightarrow Cl+O_2+CO_2$	$4\times10^{-12}\exp(-2500/T)$	$2.7\times10^{-9}e^{-2650/T}$

[a] Photolytic rates are given in s^{-1} for global averaged conditions at 200 km by KP and at 70 km by YD. Rate coefficients of the second- and the third-order reactions are in $cm^3\ s^{-1}$ and $cm^6\ s^{-1}$, respectively

Table 37 (continued)

No.	Reactions	Rate coefficients KP	Rate coefficients YD
43.	$CO+ClO \rightarrow CO_2+Cl$	1.7×10^{-15}	$10^{-12}e^{-3700/T}$
44.	$CO+ClO_2 \rightarrow CO_2+ClO$	10^{-15}	–
45.	$Cl+HO_2 \rightarrow HCl+O_2$	10^{-13}	$1.8\times10^{-11}e^{180/T}$
46.	$Cl+HO_2 \rightarrow OH+ClO$	10^{-13}	$2.2\times10^{-10}e^{-1000/T}$
47.	$OH+H_2 \rightarrow H_2O+H$	$7\times10^{-12}\exp(-2000/T)$	$1.8\times10^{-11}e^{-2330/T}$
48.	$O+H_2 \rightarrow OH+H$	$5\times10^{-11}\exp(-5100/T)$	$1.6\times10^{-11}e^{-4570/T}$
49.	$OH+O_3 \rightarrow 2\cdot O_2+H$	$1.3\times10^{-12}\exp(-950/T)$	$1.9\times10^{-12}e^{-1000/T}$
50.	$HO_2+O_3 \rightarrow 2\cdot O_2+OH$	$1.3\times10^{-12}\exp(-820/T)$	$1.4\times10^{-14}e^{-600/T}$
51.	$O+O_3 \rightarrow O_2+O_2$	$1.3\times10^{-11}\exp(-2140/T)$	$1.5\times10^{-11}e^{-2218/T}$
52.	$Cl+H_2O_2 \rightarrow HCl+HO_2$	8×10^{-13}	$1.1\times10^{-11}e^{-980/T}$
53.	$CO_2^+ +H_2 \rightarrow HCO_2^+ +H$	10^{-9}	–
54.	$CO_2^+ +HCl \rightarrow HCO_2^+ +Cl$	10^{-9}	–
55.	$HCO_2^+ +\bar{e} \rightarrow CO_2+H$	10^{-6}	–
56.	$CO+Cl+CO_2 \rightarrow COCl+CO_2$	3×10^{-33}	$1.3\times10^{-34}e^{1000/T}$
57.	$COCl+O_2 \rightarrow CO_2+ClO$	2×10^{-14}	–
58.	$COCl+O \rightarrow CO_2+Cl$	10^{-13}	3×10^{-11}
59.	$COCl+Cl_2 \rightarrow COCl_2+Cl$	$6\times10^{-13}\exp(-1400/T)$	$6\times10^{-13}e^{-1400/T}$
60.	$COCl+Cl \rightarrow CO+Cl_2$	$6\times10^{-10}\exp(-400/T)$	10^{-11}
61.	$COCl+CO_2 \rightarrow CO_2+CO+Cl$	$10^{-12}\exp(-3200/T)$	$6\times10^{-11}e^{-2250/T}$
62.	$COCl+H \rightarrow CO+HCl$	10^{-13}	10^{-11}
63.	$COCl_2+O \rightarrow CO_2+Cl_2$	10^{-14}	–
64.	$S+O_2 \rightarrow SO+O$	2.2×10^{-12}	2.2×10^{-12}
65.	$SO+OH \rightarrow SO_2+H$	1.2×10^{-10}	1.2×10^{-10}
66.	$SO_2+HO_2 \rightarrow SO_3+OH$	3×10^{-16}	10^{-18}
67.	$SO_2+OH+CO_2 \rightarrow HSO_3+CO_2$	1.6×10^{-31}	$4.2\times10^{-32}e^{1000/T}$
68.	$SO_2+O+CO_2 \rightarrow SO_3+CO_2$	10^{-32}	$8\times10^{-32}e^{-1000/T}$
69.	$SO+O \rightarrow SO_2$	$2.5\times10^{-16}\exp(-300/T)$	–
70.	$SO+O_2 \rightarrow SO_2+O$	$7.5\times10^{-13}\exp(-3250/T)$	$6\times10^{-13}e^{-3300/T}$
71.	$SO+O_3 \rightarrow SO_2+O_2$	$2.5\times10^{-12}\exp(-1050/T)$	$2.5\times10^{-12}e^{-1100/T}$
72.	$SO+O+CO_2 \rightarrow SO_2+CO_2$	5×10^{-31}	6×10^{-31}
73.	$SO+SO \rightarrow SO_2+S$	3×10^{-15}	8.3×10^{-15}
74.	$S+O+CO_2 \rightarrow SO+CO_2$	2.5×10^{-31}	–
75.	$S+S+CO_2 \rightarrow S_2+CO_2$	1.5×10^{-30}	–
76.	$S+HO_2 \rightarrow SO+OH$	2.5×10^{-11}	3.1×10^{-11}
77.	$SO_3+H_2O \rightarrow H_2SO_4$	10^{-12}	9×10^{-13}
78.	$OH+HSO_3 \rightarrow H_2SO_4$	10^{-13}	–
79.	S_n(aerosol) $+O \rightarrow S_{n-1}$(aerosol) $+SO$	10^{-16}	–
80.	$H_2O_2+SO_3 \rightarrow H_2SO_4+O$	10^{-12}	–
81.	$SO+O_2+CO_2 \rightarrow SO_3+CO_2$	$3\times10^{-32}\exp(-3250/T)$	–
82.	$SO_2+OH \rightarrow HSO_3$	5×10^{-13}	–
83.	$SO+O_2 \rightarrow SO_3$	$10^{-12}\exp(-3250/T)$	–
84.	$ClO+SO_2 \rightarrow SO_3+Cl$	10^{-15}	10^{-18}
85.	$ClO_2+SO_2 \rightarrow SO_3+ClO$	10^{-15}	–
86.	$HSO_3+Cl \rightarrow HCl+SO_3$	10^{-13}	–
87.	$S+OH \rightarrow SO+H$	2×10^{-11}	3.8×10^{-11}
88.	$O(^1D)+CO_2 \rightarrow O+CO_2$	1.8×10^{-10}	$6.8\times10^{-11}e^{117/T}$
89.	$O(^1D)+H_2O \rightarrow OH+OH$	3×10^{-10}	2.8×10^{-10}
90.	$O(^1D)+H_2 \rightarrow OH+H$	3×10^{-10}	2×10^{-10}

Table 37 (continued)

No.	Reactions	Rate coefficients KP	Rate coefficients YD
91.	$O(^1D)+HCl \rightarrow OH+Cl$	3×10^{-10}	1.4×10^{-10}
92.	$COCl+O \rightarrow CO+ClO$	–	3×10^{-12}
93.	$O(^1D)+CO \rightarrow O+CO$	8×10^{-11}	–
94.	$O_2(^1\Delta g)+CO_2 \rightarrow O_2+CO_2$	3×10^{-20}	3×10^{-20}
95.	$^1SO_2 \rightarrow SO_2+h\nu$ $^3SO_2+h\nu$	$\gamma=0.1$	–
96.	$^1SO_2+CO_2 \rightarrow SO_2+CO_2$ $^3SO_2+CO_2$	$\gamma=0.1$	–
97.	$^1SO_2 \rightarrow SO_2$ $^3SO_2$	$\gamma=0.1$	–
98.	$^3SO_2 \rightarrow SO_2+h\nu$	1.1×10^3	–
99.	$^3SO_2+CO_2 \rightarrow SO_2+CO_2$	5×10^{-14}	–
100.	$^3SO_2+SO_2 \rightarrow SO_3+SO$	$5\times10^{-11}\exp(-1370/T)$	–
101.	$^3SO_2+CO \rightarrow CO_2+SO$	$8\times10^{-12}\exp(-2000/T)$	–
102.	$^3SO_2+O_2 \rightarrow SO_3+O$	$5\times10^{-11}\exp(-1370/T)$	–
103.	$ClO+h\nu \rightarrow Cl+O$	–	5×10^{-3}
104.	$SO+h\nu \rightarrow S+O$	–	2.7×10^{-4}
105.	$H_2O_2+O \rightarrow OH+HO_2$	–	$2.8\times10^{-12}e^{-2125/T}$
106.	$H_2O_2+OH \rightarrow H_2O+HO_2$	–	$7.6\times10^{-12}e^{-670/T}$
107.	$Cl+OH \rightarrow HCl+O$	–	$10^{-11}e^{-2970/T}$
108.	$ClO+OH \rightarrow Cl+HO_2$	–	9.1×10^{-12}
109.	$S+O_3 \rightarrow SO+O_2$	–	1.2×10^{-11}
110.	$SO+HO_2 \rightarrow SO_2+OH$	–	2.3×10^{-11}
111.	$SO+ClO \rightarrow SO_2+Cl$	–	2.3×10^{-11}
112.	$SO_2+H_2O_2+\text{aerosol} \rightarrow H_2SO_4$	–	4.3×10^{-5}
113.	$SO_2+Cl+M \rightarrow ClSO_2+M$	–	4.6×10^{-33}
114.	$SO_3+SO \rightarrow SO_2+SO_2$	–	2×10^{-15}
115.	$NO+h\nu \rightarrow N+O$	–	1.7×10^{-6} $\exp(-4.2\times10^{-23}$ $N_{CO_2})$ $+1.2\times10^{-6}$ $\exp(-8.4\times10^{-22}$ $N_{CO_2})$
116.	$NO_2+h\nu \rightarrow NO+O$	–	10^{-2}
117.	$NO_3+h\nu \rightarrow NO_2+O$	–	2×10^{-2}
118.	$HNO+h\nu \rightarrow H+NO$	–	10^{-3}
119.	$HNO_2+h\nu \rightarrow OH+NO$	–	2×10^{-3}
120.	$HNO_3+h\nu \rightarrow OH+NO_2$	–	5.4×10^{-5}
121.	$N+O \rightarrow NO+h\nu$	–	2.5×10^{-17}
122.	$N+O+M \rightarrow NO+M$	–	$1.9\times10^{-31}T^{-1/2}$
123.	$N+O_2 \rightarrow NO+O$	–	$4.4\times10^{-12}e^{-3270/T}$
124.	$N+O_3 \rightarrow NO+O_2$	–	10^{-15}
125.	$N+OH \rightarrow NO+H$	–	5.3×10^{-11}
126.	$N+NO \rightarrow N_2+O$	–	2.1×10^{-11}
127.	$NO+O+M \rightarrow NO_2+M$	–	2.4×10^{-31}
128.	$NO+O_3 \rightarrow NO_2+O_2$	–	$2.3\times10^{-12}e^{-1450/T}$
129.	$NO+H+M \rightarrow HNO+M$	–	$1.5\times10^{-32}e^{300/T}$
130.	$NO+OH+M \rightarrow HNO_2+M$	–	1.3×10^{-30}

Table 37 (continued)

No.	Reactions	Rate coefficients	
		KP	YD
131.	$NO+HO_2 \to NO_2+OH$	–	$3.5\times10^{-12}e^{250/T}$
132.	$NO+ClO \to NO_2+Cl$	–	$6.5\times10^{-12}e^{280/T}$
133.	$NO_2+O \to NO+O_2$	–	9.3×10^{-12}
134.	$NO_2+O+M \to NO_3+M$	–	1.8×10^{-31}
135.	$NO_2+OH+M \to HNO_3+M$	–	5.2×10^{-30}
136.	$NO_2+SO \to NO+SO_2$	–	1.4×10^{-11}
137.	$HNO+O \to NO+OH$	–	10^{-13}
138.	$HNO+H \to H_2+NO$	–	10^{-13}
139.	$HNO+Cl \to HCl+NO$	–	10^{-13}
140.	$HNO+HNO \to N_2O+H_2O$	–	4×10^{-15}
141.	$HNO_2+OH \to H_2O+NO_2$	–	6.6×10^{-12}
142.	$HNO_3+OH \to H_2O+NO_3$	–	1.5×10^{-14}
143.	$COCl_2+h\nu \to COCl+Cl$	–	5×10^{-5}
144.	$HOCl+h\nu \to OH+Cl$	–	4×10^{-4}
145.	$ClO+HO_2 \to HOCl+O_2$	–	$4.6\times10^{-13}e^{710/T}$
146.	$COCl+O_2+CO_2 \to ClCO_3+CO_2$	–	$\dfrac{5.7\times10^{-15}e^{500/T}}{10^{17}+0.05\,M}$
147.	$ClCO_3+O \to CO_2+Cl+O_2$	–	10^{-11}
148.	$ClCO_3+Cl \to CO_2+Cl+ClO$	–	10^{-11}
149.	$ClCO_3+H \to CO_2+Cl+OH$	–	10^{-11}
150.	$COCl+COCl \to COCl_2+CO$	–	10^{-11}
151.	$S+CO_2 \to SO+CO_2$	–	10^{-20}

Laboratory measurements of some reaction rates performed in past years resulted in some rate coefficients which differ significantly from those accepted in our papers. An especially great difference is for reactions *R*40, 41, 46. It is known also that the reaction of COCl with O_2 (*R*57), which is the most important in our list of photochemical processes, results in formation of $ClCO_3$ with subsequent destruction by chlorine atoms

$$COCl+O_2+M \to ClO_3+M$$
$$ClCO_3+Cl \to CO_2+ClO+Cl$$
$$\text{net } COCl+O_2 \to CO_2+ClO\,.$$

Hence, the net result of this interaction coincides with our assumption. The main consequences of changes of rate coefficients will be discussed further.

Reactions Involving Sulfur Compounds. Prinn (1975) analyzed nine reactions involving sulfur compounds, the source of which he believed to be COS. Now sulfur dioxide is well established as a main sulfur-bearing species in the Venusian troposphere at least above 25 km. Sulfur dioxide transforms into sulfuric acid and solid sulfur in the following way:

$$SO_2+h\nu \to SO+O$$
$$SO_2+O+CO_2 \to SO_3+CO_2$$

$$SO_3 + H_2O \rightarrow H_2SO_4$$
$$SO + SO \rightarrow SO_2 + S\ .$$

The last reaction also yields S_2O and S_2O_2, thus reducing formation of a elemental sulfur. We do not consider these components. As a result of SO_2 photolysis a large amount of freee oxygen is formed, exceeding its production and transport from the above-cloud atmosphere where CO_2 photodissociation occurs. Actually CO_2 and SO_2 photolytic rates equal fluxes of quanta in the ranges of $\lambda < 202$ nm and 202–218 nm, being 1.1×10^{13} and 3.6×10^{13} $cm^{-2}s^{-1}$, respectively, under mean day side conditions on Venus. Atomic oxygen transforms SO_2 into sulfuric anhydride SO_3 which actively absorbs water, thus producing sulfuric acid. Both a part of the oxygen atoms that have time to form molecules O_2 and O_2 molecules forming above the cloud layer and moving down to that layer via eddy mixing, participate in SO_2 oxidization through HO_2 or the reaction:

$$O_2 + SO \rightarrow SO_2 + O\ .$$

The amount of active hydrogen in the cloud layer also decreases because of H_2SO_4 formation with the OH involved. Free sulfur is formed in the last of the processes mentioned.

Only basic reactions of the cycle under discussion are given above. On the whole, there are many more reactions and they are summarized in Table 37. We have not taken into account reactions with chlorine-sulfur compounds, whose rate coefficients for the most part are unknown. Only two of these reactions are included, for no new components are formed in the case:

$$SO_2 + ClO \rightarrow SO_3 + Cl$$
$$HSO_3 + Cl \rightarrow SO_3 + HCl\ .$$

Excited Atoms and Molecules and Their Reactions. Energy released in many reactions considered is sufficient to excite molecules and atoms formed in those reactions. Out of the whole variety of possible excited states, only four were taken for calculation: $O(^1D)$, $O_2(a^1\Delta g)$, $SO_2(a^3B_1)$, $SO_2(A^1B_1)$. For the sake of brevity, a superscript will denote multiplicity, for instance, $^1SO_2$.

Ozone photolysis is the basic source of $O(^1D)$. $O(^1D)$ atoms are subject to quenching by CO_2; besides they decompose H_2O and H_2, forming H and OH. To these well-known reactions we have added the following:

$$O(^1D) + HCl \rightarrow OH + Cl$$

which is similar to the reaction with H_2.

The state $O_2(^1\Delta g)$ accounts for a very intense airglow of the 1.27 μm band (0, 0) which is the strongest emission in the airglow of the Earth (Evans et al. 1968), of Venus (Connes et al. 1979), and Mars (Noxon et al. 1976). Ozone photolysis is mainly responsible for this band excitation. There are, however, many reactions among those given in Table 37 in which $O(a^1\Delta g)$ formation is possible. This problem will be discussed in greater detail below.

Solar radiation absorption in the near UV region leads to the excitation of two metastable states of SO_2. The $^1SO_2$ state is formed at $\lambda < 320$ nm and then it passes – either with or without radiation – to the corresponding ground state or to $^3SO_2$. Similar transitions may occur in the case of quenching as well, and the $^3SO_2$ yield in both processes is the same and equal to 0.1 (Strickler et al. 1974, Rao et al. 1969a, b). $^3SO_2$ molecules originate from the absorption of 320 – 390 nm radiation and a considerable part of them passes to the ground state due to radiation and collisions with CO_2. The other part of $^3SO_2$ reacts with CO, SO_2, and O_2 either giving up or adding an oxygen atom (Rao et al. 1969a, b, Jackson and Calvert 1971)

$$^3SO_2 + CO \rightarrow SO + CO_2$$
$$^3SO_2 + SO_2 \rightarrow SO + SO_3$$
$$^3SO_2 + O_2 \rightarrow SO_3 + O\ .$$

Despite the fact that this part is small its contribution is not negligible due to the great flux of solar radiation in the near UV region.

4.9 Radiative Transfer and Aerosol Transport in the Cloud Layer

Photolysis and Photoexcitation Rates. The data on the solar radiation obtained by Makarova and Kharitonov (1972) for $\lambda > 320$ nm, Broadfoot (1972) for 210 – 320 nm, and Widing et al. (1970) and Higgins (1976) for $\lambda < 210$ nm were used to calculate photolytic and photoexcitation rates. For global averaging, the solar zenith angle was taken to be equal to 60° and the calculated values were halved. Solar radiation in a range of 145 – 210 nm, which is the most important in our study, was overestimated by Widing et al. (1970) by a factor of 2 when compared to the data by Mount et al. (1980). Therefore, it is better to consider the results as a mean day side rather than global average. Absorption cross-sections were measured by Shemansky (1972) and Nakata et al. (1965) for CO_2, by Thompson et al. (1963) and Hudson (1971) for H_2O, O_2 and SO_2 ($\lambda < 340$ nm), by Griggs (1968) for O_3, by Romand and Vodar (1948) for HCl, by Rohatgi and De More (1978) for H_2O_2, by Seery and Britton (1964) for Cl_2, and by Vikesland and Strickler (1974) for SO_2 (340 – 390 nm). Contrary to CO_2 absorption, which becomes noticeable only at $\lambda < 205$ nm, i.e., at wavelengths shorter than a dissociation threshold of 225 nm, SO_2 absorption begins at 390 nm though the dissociation threshold is 219 nm. Therefore, it is necessary to investigate the SO_2 dissociation quantum yield versus wavelength dependence. From the data obtained by Okabe (1971) it may be assumed equal to unity for $\lambda < 217$ nm and zero for $\lambda > 217$ nm.

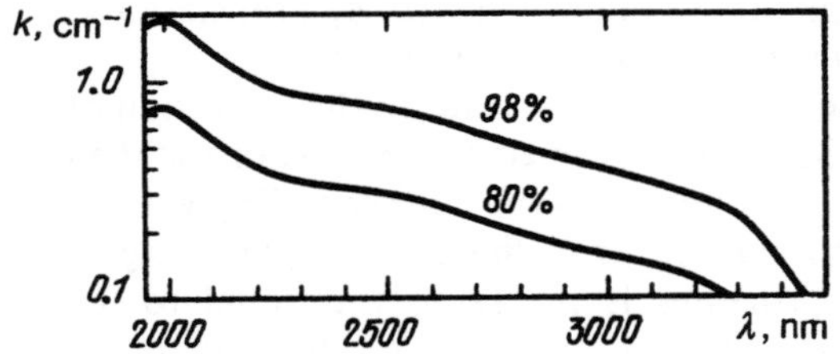

Fig. 175. Absorption coefficient of sulfuric acid solution with a concentration of 98% and 80%

In our case the calculation of photolytic rates is highly complicated due to the following circumstances. First, contrary to most calculations of the Martian and Venusian aeronomy, the absorption of gases other than CO_2 should be taken into account. Really the atmosphere cannot be taken as optically thin for $\lambda > 202$ nm in the considered range of heights down to 50 km. This is most evident for SO_2 absorption for which as it will be shown the oxygen production is about twice that for CO_2 absorption. The atmospheric optical thickness at 50 km for $\lambda \simeq 220$ nm turns out to be of the order of a thousand. Second, unlike CO_2, we do not know the height distribution of SO_2 and other components and, therefore, the photolytic rates cannot serve as the initial data for the computation and they are continuously refined during computation as the component altitude profiles, initially given, approaches the final result. The third difficulty is associated with the necessity of accounting for light scattering in the cloud layer. The detailed description of this process is possible only by the solution of the radiative transfer equation. However, an approximate description of this process can be used. In our case only the effect of scattering in the cloud layer on the photolytic process is of interest.

The cloud layer consisting of drops of about 80% solution of sulfuric acid with a radius, r, of ~ 1 μm will be considered. It is known that for the optically thick atmosphere the scattering multiplicity is $q = (1-\omega)^{-1/2}$. On the other hand, $1-\omega = \frac{2}{3}kr$ (Chap. 3.2) or $q = (\frac{2}{3}kr)^{-1/2}$ where k is the material absorption coefficient, ω is the single scattering albedo. To obtain the dependence $k(\lambda)$ for sulfuric acid, we made measurements for 98 and 80% solution. Figure 175 shows the results obtained. It is of interest that the absorption of the 80% solution is 2.5 times as small as for the 98% solution. Figure 175 indicates that the scattering multiplicity in the cloud layer consisting of only sulfuric acid drops should be very high, of the order of 100 in the region of $\lambda \sim 200$ nm. The real value is lower and can be estimated from a known planet albedo by means of an asymptotic formula (Chap. 3.2).

For the estimation, the region of 340 – 390 nm where the planet albedo is $A \simeq 0.5$ is considered. Calculations with a scattering optical thickness of clouds of about 30 and a parameter g of the Henyey-Greenstein phase function (Chap. 3.2) of 0.7, give for q a value of 10.

The scattering multiplicity approaches unity for $\omega < 0.7$, i.e., for the values of optical thickness of absorbing gas of more than 10 at 50 km. For adopted SO_2 densities at 50 km (see below), this condition corresponds to $\sigma_{SO_2} > 10^{-20}$ cm^2 or $\lambda \leqslant 320$ nm. In the region of 340 – 390 nm, $\sigma_{SO_2} \simeq 10^{-22}$ cm^2 and $q = 10$ which is consistent with the accepted boundary conditions. Therefore, we shall assume that SO_2 absorption in this region of the spectrum which leads to the excitation of $^3SO_2$ has a cross-section ten times as high as the real cross-section and the processes of multiple scattering may be neglected. For $\lambda < 340$ nm, this approximation is valid for real absorption cross-sections as in this case the scattering multiplicity is close to unity.

For other components, real cross-sections will be taken as they absorb in a rather short-wave region of the spectrum or they are mainly in the overcloud atmosphere.

Aerosol Components. The photochemical formation of sulfuric acid and elemental sulfur is followed by their condensation to aerosol particles.

It is assumed that there is a phase equilibrium between the H_2SO_4 solution and H_2O, H_2SO_4 vapors in the atmosphere. In this case, almost all sulfuric acid molecules should be in the liquid phase and, consequently, H_2SO_4 vapor can be neglected. In fact, no absorption bands of *gaseous* H_2SO_4 have ever been observed in the Venus atmosphere. Then the concentration of the H_2SO_4 solution drops is determined by the temperature and density of water vapor, and the gas and aerosol phases contribute to the total amount of water in the unit volume according to phase equilibrium curves.

The relation $\lg p(\text{atm}) = -5166/T + 8.783$ describes the pressure of sulfur saturated vapor (Mills 1974). Sulfur amount in the gas phase turns out to be comparable with that in the solid phase only near the lower boundary of 50 km. Therefore, for simplicity we assume that all sulfur passes into the solid phase having achieved an S_2 state.

For both kinds of aerosol a gradual sedimentation of particles occurs with the velocity:

$$V = \frac{2}{9} \varrho g \frac{r^2}{\eta} \left(1 + \frac{l}{r}\right) (1.257 + 0.4\, e^{1.1 r/l}) \, .$$

This is the Stokes-Davis formula where ϱ is the particle matter density, g is the gravitational acceleration, η is the atmospheric viscosity, and l is the length of the molecule free path. Atmospheric mixing counteracts the particle sedimentation.

The complex structure of the cloud layer is associated with various types of physical and chemical interactions of aerosol particles among each other and with the atmospheric gas components. Of all these reactions, we will consider only the oxidation reaction of solid sulfur particles by atomic oxygen. For the S_2 free molecule, this reaction rate coefficient is about 3×10^{-12} cm^3 s^{-1} (Baulch et al. 1976), while for the particles with the radius $r \simeq 1$ μm it should be about 10^{-16} cm^3 s^{-1} per one molecule.

To determine the velocity of aerosol sedimentation it is assumed that the radii of sulfuric acid and sulfur particles are the same and $r = 1$ μm for

$h > 65$ km, $r = 4$ μm for $h < 60$ km and change linearly at 60 – 65 km. Such a dependence $r(h)$ describes roughly the results of particle size spectrometry of the Pioneer Venus and supposes that sulfuric acid, sulfur, and products of their interaction with other components belong to the second and third modes of size distribution. We shall consider this problem when discussing the calculation results.

4.10 Boundary Conditions

To obtain the boundary conditions the lower atmospheric chemical composition should be known. For the calculation of the photolytic products of CO_2, H_2O, HCl, and SO_2 the number densities of these components at 50 km must be known. The choice of the position of the lower boundary at 50 km is conditioned by the requirement to give a more detailed description of all photochemical processes in the atmosphere including the processes of the cloud layer formation, but excluding the thermochemical reactions which can play an important role in the subcloud atmosphere. We considered thermochemistry of the Venusian lower troposphere in Chap. 4.7.

Boundary conditions for CO_2 at 50 km present no difficulties. For the upper cloud layer, the mixing ratio f_{HCl} was measured to be 6×10^{-7} (Connes et al. 1967). As our calculation shows, f_{HCl} varies slightly with height. Therefore, we assumed this value at 50 km. As far as SO_2 content is concerned, the results of gas chromatography on Venera 12 (Gelman et al. 1979) and Pioneer Venus (Oyama et al. 1979) agree well and in the lower atmosphere $f_{SO_2} \simeq 1.3 \times 10^{-4}$. As will be shown below, the model considered is very sensitive to the ratio $[H_2O]/[SO_2]$ at the lower boundary (50 km). Therefore, we will assume that $f_{SO_2} = 1.3 \times 10^{-4}$ at this level and will change H_2O densities to fit experimental data, in particular f_{H_2O}, f_{SO_2}, and sulfuric acid concentration in the upper cloud layer and amounts of exospheric hydrogen.

Having obtained the conditions at the lower boundary for the main components we should now determine these conditions for photolytic products. Among them there are stable long-lived molecules, such as CO, O_2, H_2, etc. Direct measurements of f_{CO} by the V12 and PV gas chromatographs agree well with ground-based spectroscopy and result in $f_{CO} = 4 \times 10^{-5}$. For other components there are no experimental data or they are doubtful. To overcome this difficulty, we have decided not to use the number densities of these components from Table 36 and to work out a unified method for the prescribing of boundary conditions for daughter components, both stable and chemically active atoms and radicals of O, OH, Cl, S, HO_2, COCl, etc.

Boundary conditions are usually given in the form of either number density, flux, or photochemical equilibrium of a component (widespread in the case of prescribing of number densities). None of these forms can be used for all the considered compounds because for active compounds the photochemical equilibrium is preferable (however, at 50 km their production is so small that the downward transport is more essential), for stable components diffusive equilibrium is preferred, whereby many intermediate cases exist.

We have solved the problem assuming that for each of the daughter components the conditions at 50 km were kept constant down to the surface where their concentrations were taken to be equal to zero. Then a solution of the continuity equation at 50 km with constant coefficients p, L, K, H is

$$n(z_0) = \frac{P}{L} - \frac{2(\phi_0 H/K + P/L)}{1+\sqrt{1+4H^2L/K}}.$$

It is easy to see that $4H^2L/K = 4\tau_m/\tau_c$ where τ_m and τ_c are the mixing and photochemical equilibrium times. The solution will be considered in two extreme cases: (1) the predominance of transport $\tau_m/\tau_c \ll 1$ when the formation of particles inside the region greatly exceeds their influx, i.e., $pH/\phi_0 \ll 1$; (2) the predominance of photochemistry, $\tau_m/\tau_c \gg 1$. In the first case $n_0 = -\frac{\phi_0 H}{k}$, in the second case $n_0 = \frac{P}{L} - \frac{\phi_0}{\sqrt{KL}} \simeq \frac{P}{L}$, i.e., it gives the conventional condition of photochemical equilibrium. So the assumptions made and the solution obtained for the lower boundary incorporate cases of photochemical and diffusion equilibrium along with all intermediate cases.

It is absolutely clear that the obtained form of boundary conditions can be applied only to components which originate in the atmosphere above the boundary, i.e., for the products of photochemical reactions.

At an upper boundary of 200 km the conditions of diffusion equilibrium were taken for all components except hydrogen – the equality of fluxes zero.

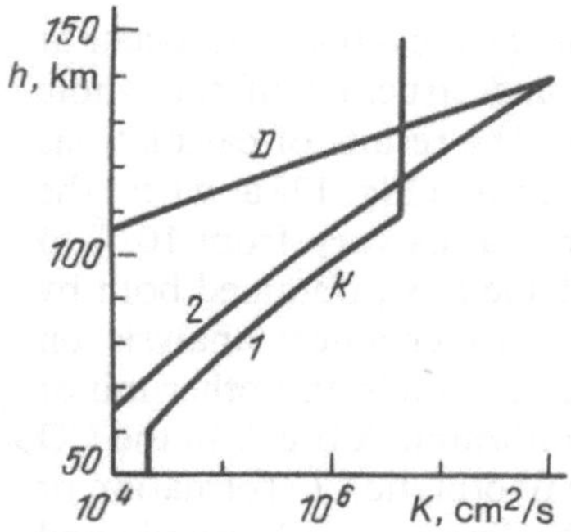

Fig. 176

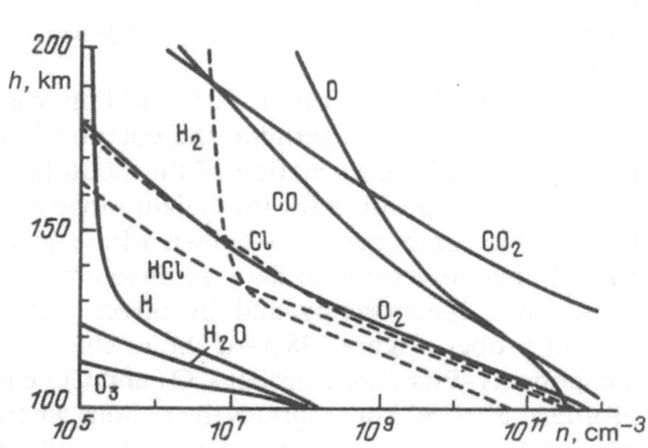

Fig. 177

Fig. 176. Altitude profile of the eddy and molecular diffusion coefficients adopted in the calculations (*1*) and recommended by von Zahn et al. 1979 (*2*)

Fig. 177. Chemical composition of the upper atmosphere. For the *dashes*, the number densities are a factor of 10^3 less than shown (Krasnopolsky and Parshev 1980b, 1981)

For hydrogen atoms on Venus, nonthermal escape should be taken into consideration which prevails over conventional thermal escape (Chap. 4.4). The nonthermal escape rate was assumed to be $10^7\,cm^{-2}\,s^{-1}$ according to Kumar et al. (1978). This value was recently confirmed by McElroy et al. (1982a).

Eddy Diffusion. The profile of K was prescribed in the half-logarithmic scale as the designated (Fig. 176). Its lower part at 50–60 km corresponds to minimum values $K = 2 \times 10^4\,cm^2\,s^{-1}$. Above 60 km we have taken $K \sim n^{-1/2}$ according to von Zahn et al. (1979). The upper part of the K profile is taken to be constant and equal to $4 \times 10^6\,cm^2\,s^{-1}$. This $K(Z)$ dependence corresponds approximately to that discussed in Chap. 3.13, being somewhat stronger for $Z < 120$ km and weaker for $Z > 120$ km up to the turbopause. The turbopause is at 128 km in our case, which is lower than recommended in Chap. 3.13. The reasons for the choice of the $K(Z)$ dependence will be given during the discussion of the model calculation results.

4.11 Atmospheric Composition at 50 to 200 km (Results of Calculations)

Upper Atmosphere. Vertical profiles of number densities of some upper atmospheric constituents are shown in Fig. 177. Those of CO and O agree well with the preliminary data of the PV measurements which were available at the time of preparation of the paper under consideration. However, there is some difference with the latest results: at 150 km the calculated [O] and [CO] are by factors of 1.5–2 and 5–6, respectively, less than measured ones. This discrepancy can be avoided by the assumption of $K = 10^{13} M^{-1/2}\,cm^2\,s^{-1}$ (see Chap. 4.5).

CO_2 and CO Equilibrium. The most important chemical reactions occur at 50 to 100 km. They determine the composition and structure of the whole atmosphere and the formation of the cloud layer. The results of calculations of a model composition for this height range are given in Fig. 178a, where the most stable components are shown whose mixing ratios vary from 10^{-6} to 10^{-4}. Some observational data on them are available now, obtained both by cloud layer spectroscopy and by direct methods of chemical analysis on descent probes. Figure 178b–d shows chlorine compounds and other minor components. First, we consider CO_2 and CO equilibrium. A break in the CO profile at 80 to 90 km separates the regions where photolytic CO formation or transport from the lower atmosphere prevails. Binding CO with atomic and molecular oxygen occurs mainly at 55 to 90 km through catalytic cycles described in Chap. 4.8.

Figure 179 depicts a height dependence of CO_2 formation rates in those cycles. The mean daytime CO_2 photolytic rate per column constitutes $1.1 \times 10^{13}\,cm^{-2}\,s^{-1}$. Most important is the cycle involving carbonyl chloride

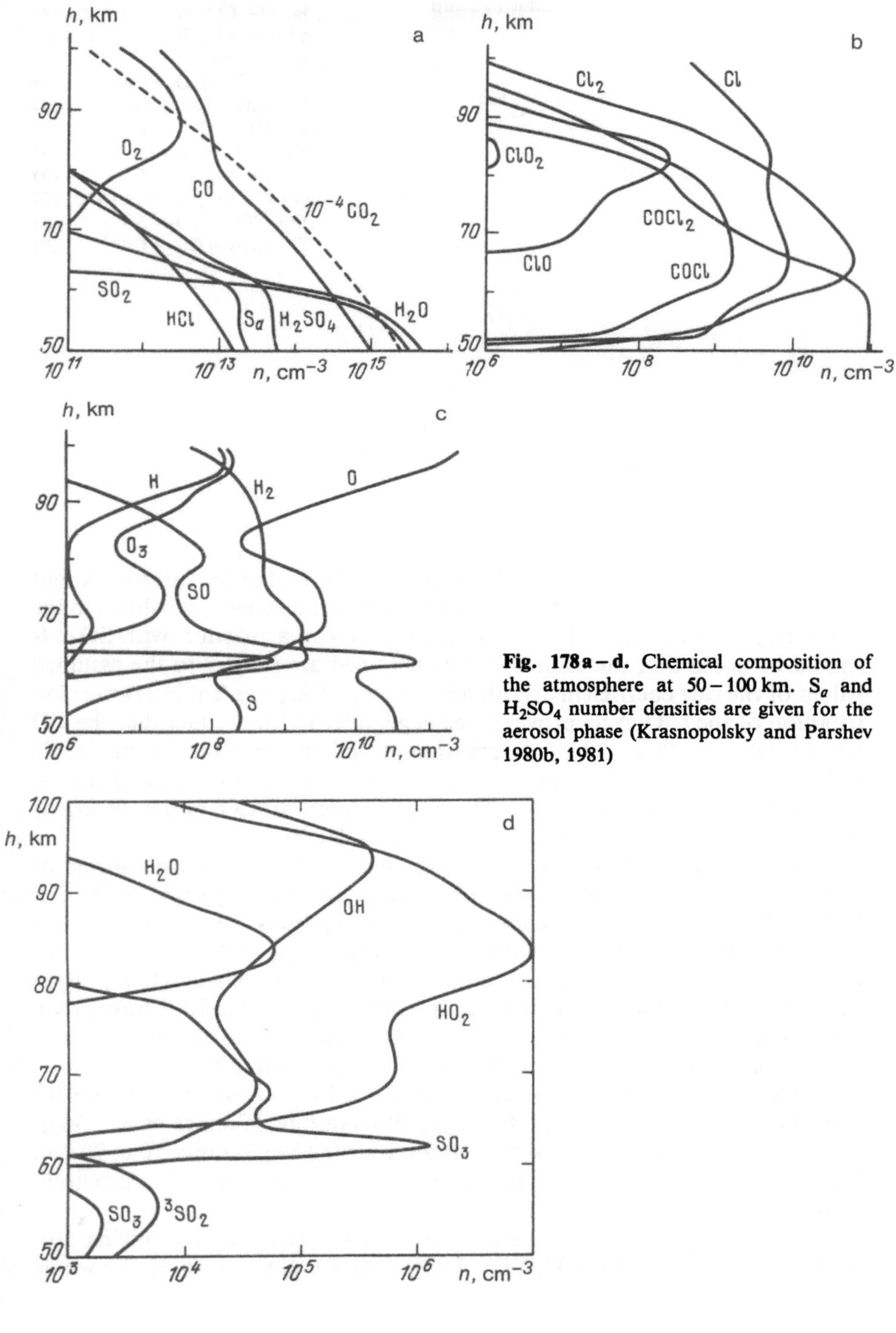

Fig. 178a–d. Chemical composition of the atmosphere at 50–100 km. S_a and H_2SO_4 number densities are given for the aerosol phase (Krasnopolsky and Parshev 1980b, 1981)

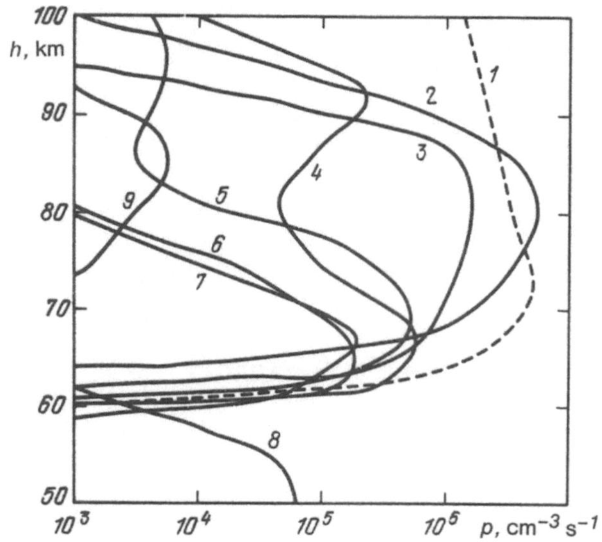

Fig. 179. CO_2 photolytic rate (*1*) and rates of CO_2 formation by the reactions:
$COCl + O_2 \rightarrow CO_2 + ClO$ (*2*)
$CO + ClO \rightarrow CO_2 + Cl$ (*3*)
$CO + OH \rightarrow CO_2 + H$ (*4*)
$COCl + O \rightarrow CO_2 + Cl$ (*5*)
$COCl_2 + O \rightarrow CO_2 + Cl_2$ (*6*)
$O + CO + CO_2 \rightarrow CO_2 + CO_2$ (*7*)
$CO + {}^3SO_2 \rightarrow CO_2 + SO$ (*8*)
$CO + ClO_2 \rightarrow CO_2 + ClO$ (*9*)

COCl and O_2 which provides 60% of all the CO_2 molecules formed. About 2% of the carbon monoxide formed in the upper atmosphere goes down to the lower troposphere where it transforms into CO_2 in a reaction with SO_3. It should be noted that the estimates thus obtained are subject to the assumed values of the rate coefficients. With a new value of k_{43} (see Table 37) the contribution of the ClO cycle proves to be negligible. From the fact that the CO mixing ratio at 50 to 70 km is constant it follows that the spectroscopic estimate $f_{CO} = 4 \times 10^{-5}$ (Connes et al. 1968) for the top cloud should also be valid near its lower boundary. The total excess of CO above 80 km is 0.8×10^{19} cm^{-2}. It should be accumulated in ~20 days.

The calculated CO profile agrees well with microwave measurements of CO ($J = O \rightarrow 1$) 2.6 mm absorption line (Chap. 3.4) which give $f_{CO} \simeq 10^{-3}$ at 100 km. However, we do not know a process which leads to very low values of f_{CO} at 70 – 80 km (Fig. 93). Diurnal variations of f_{CO} at 100 km need two-dimensional models similar to that of Dickinson and Ridley (1977) to be explained. The nighttime excess of f_{CO} predicted by that model is larger than the measured one (Fig. 93). The daytime maximum of f_{CO} at 80 km may result from the ceasing of the atmospheric superrotation at this altitude.

Hydrogen Chloride. We assumed $f_{HCl} = 6 \times 10^{-7}$ at the lower boundary according to the results of spectroscopic observations (Connes et al. 1968), since the HCl mixing ratio is fairly constant over the height range 50 to 90 km. Figure 180 demonstrates basic processes of HCl formation and destruction. Their net vertical rate is 4.5×10^{11} cm^2 s^{-1}.

With f_{HCl} being constant, the conclusion might be made that mixing processes are more rapid than HCl-involving chemical reactions over the whole

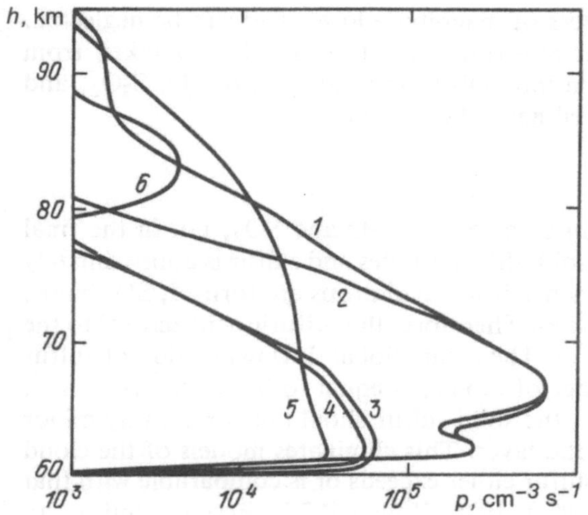

Fig. 180. Sinks and sources of HCl:

$H + Cl_2 \rightarrow HCl \rightarrow HCl + Cl$ (*1*)
$O + HCl \rightarrow OH + Cl$ (*2*)
$HCl + H \rightarrow H_2 + Cl$ (*3*)
$H_2 + Cl \rightarrow HCl + H$ (*4*)
$HCl + h\nu \rightarrow H + Cl$ (*5*)
$Cl + HO_2 \rightarrow HCl + O_2$ (*6*)

height range 50 to 100 km. This is not the case, however, because at 65 km the photochemical equilibrium time is three times as small as that of mixing. Despite this, f_{HCl} deviation from the constant value is not calculated, which is attributed to a very similar shape of HCl total destruction and formation profiles. The large value of equilibrium time (about a terrestrial month at 65 km and even more at higher altitudes) indicates that the only reason for local HCl variations, if they do take place, could be changing conditions in the lower troposphere.

Sulfur and Sulfur Dioxide. The source of sulfuric compounds in the atmospheric region discussed is sulfur dioxide SO_2. Its mixing ratio at 50 km was taken as $f_{SO_2} = 1.3 \times 10^{-4}$ according to the data of gas chromatography in the lower troposphere.

SO_2 destruction occurs as a result of photolysis at $\lambda < 217$ nm. With the CO_2 atmospheric opacity at altitudes of 60 to 70 km and $z_0 = 60°$ taken into account, the mean global solar radiation flux in this spectral region is equal to 1.8×10^{13} quanta $cm^{-2} s^{-1}$, i.e., three times as high as that of CO_2. For low values of the eddy mixing coefficient, *K*, photolysis occurs in a very thin layer with $\sigma[SO_2] H_{SO_2} s z_0 \simeq 1$; with $\sigma \simeq 3 \times 10^{-18}$ cm^2 (Thompson et al. 1963) and $H_{SO_2} = 1.5$ km we get $[SO_2] = 10^{12}$ cm^{-3}. The reverse:

$$SO + O + CO_2 \rightarrow SO_2 + CO_2$$

reduces the "quantum yield" of SO_3:

$$SO_2 + O + CO_2 \rightarrow SO_3 + CO_2 .$$

The smaller *K*, the narrower the layer where photolysis occurs and the lower the SO_3 yield. In our case the mean global total SO_3 production is 1.1×10^{13} $cm^{-2} s^{-1}$ for $K = 2 \times 10^4$ $cm^2 s^{-1}$. The influx of free oxygen accompanying

CO_2 dissociation is by two orders of magnitude lower and can be neglected. Therefore, oxygen required for SO_2 transformation into SO_3 is taken from other SO_2 molecules which turn into sulfur and partly into S_2O, S_2O_2, and other components not considered here. As a result:

$$3\,SO_2 \rightarrow 2\,SO_3 + S\,.$$

Part of the S atoms is again oxidized into SO and SO_2, but in the final analysis the production of aerosol H_2SO_4 particles and sulfur is approximately determined by this relation. When sulfuric acid drops are formed, SO_3 binds, on the average, two H_2O molecules. Therefore, the sulfuric acid aerosol to the sulfur aerosol mass ratio is ~7.5. The mean global downward flux of sulfur particles, in terms of the number of atoms, is equal to $0.5 \times 10^{13}\ cm^{-2}\,s^{-1}$, i.e., half the H_2SO_4 flux. Thus, the solid sulfur should be a relatively minor component of the Venusian cloud layer. This eliminates models of the cloud layer where the abundance of sulfur either exceeds or is comparable with that of sulfuric acid (Hapke and Nelson 1975, Young 1977). Besides, sulfur can react with chlorine compounds forming volatile components, thus decreasing the amount of aerosol. This class of reactions is not considered in our calculations.

Figure 178a shows the calculated profile of the aerosol sulfur density S_a. The latter is almost constant and equals $10^{-9}\ g\,cm^{-3}$ at midcloud altitudes of 51 to 58 km. The density is equal to about 15% of the whole amount of the third mode aerosol with $r \simeq 4\ \mu m$. The amount of sulfur the model yields corresponds to the density of its saturated vapor (Mills 1974) at 51 km where $T = 330$ K. Therefore, all sulfur particles here will transform into gas. That height exactly coincides with the lower boundary of the middle cloud layer. Since no changes in the fine-dispersion fraction of clouds (mode 1) occur on that boundary, sulfur if it forms should belong in the third mode of particles in the middle cloud layer in accord with our initial assumption about the size of sulfur particles.

Figure 178a also shows height dependence of $[SO_2]$. Above 55 km the SO_2 number density sharply decreases, which generally agrees with the interpretation of spectroscopic observations (see Chaps. 3.8 and 3.10) giving the level of $[SO_2] = 10^{12}\ cm^{-3}$ at 63 km. However, the mean SO_2 scale height is smaller than the measured one. The approximate profile of SO_2 can be obtained in the following way. SO_2 is destroyed in a very thin layer, and below that layer the SO_2 flux can be assumed to be constant and equal to $\phi_0 = 1.1 \times 10^{13}\ cm^{-2}\,s^{-1}$:

$$\phi_0 = -K\left[\frac{dn}{dz} + \frac{n}{H}\right] = -K[CO_2]\frac{df_{SO_2}}{dz}$$

hence, $f_{SO_2}(z) = f^0_{SO_2} - \dfrac{\phi_0 H}{K[CO_2]_0}(e^{z/H} - 1)$. The height at which $\dfrac{f_{SO_2}}{f^0_{SO_2}} \rightarrow 0$ is equal to (if measured from the lower boundary)

$$z_{max} = H \ln\left(\frac{Kn_0}{\phi_0 H} + 1\right).$$

It should correspond to the height of a layer where photolysis occurs, that is, to the level $n = 10^{12}\,cm^{-3}$, the value is much less than $n_0 = 3.5 \times 10^{15}\,cm^{-3}$ and corresponds to the condition $f_{SO_2}/f^0_{SO_2} \to 0$. The scale height, H, of the atmosphere in the region of 50 to 60 km is 8 km and for $K = 2 \times 10^4\,cm^2\,s^{-1}$ we get $z_{max} = 68$ and 63 km for the mean global and daytime values of ϕ_0, respectively. The latter corresponds to our calculations and coincides with their result.

The additional source of aerosol from sulfuric acid and elemental sulfur is $^3SO_2$ excitation which mainly occurs at altitudes of the middle cloud layer (51 to 58 km) and is mostly due to 330 to 400 nm radiation. The flux of quanta in this spectral region is very intense and constitutes $7 \times 10^{15}\,cm^{-2}\,s^{-1}$, with global averaging taken into account. In our calculational scheme about one-half of that flux is absorbed with $^3SO_2$ excitation. In fact, this part is smaller due to light reflection and absorption by other, not accounted for, absorbers (Chaps. 3.8 and 3.10). Almost all the molecules $^3SO_2$ are quenched in collisions with CO_2; the radiative deactivation rate is $5 \times 10^{12}\,cm^{-2}\,s^{-1}$ and the net rate of the process

$$^3SO_2 + SO_2 \to SO_3 + SO$$

is $4 \times 10^{12}\,cm^{-2}\,s^{-1}$, that is, about 20% of the total production of sulfuric acid and sulfur. This value is somewhat overrated, and this source is not significant.

Owing to drastic decrease of the SO_2 content at 55 to 62 km, sulfur dioxide cannot be spectroscopically observed at phase angles more than 90° (Barker 1979). With increasing phase angle the upper regions of the cloud layer increase their contribution to the reflected radiation. At the same time the SO_2 abundance becomes negligibly small towards these upper regions.

Water and Sulfuric Acid. As has already been mentioned, we have varied the mixing ratio of water vapor at the lower boundary to choose the solution which best fits the experimental data. It proves that fairly small variations of H_2O/SO_2 at 50 km cause considerable variations of the mixing ratios of H_2O, SO_2, H_2SO_4, H_2, H, and other products of H_2O photolysis in the overcloud atmosphere. This gives a possible explanation of long-term variations of SO_2 and upper haze (see Chap. 3.8). According to the ground-based spectroscopic observations $f_{H_2O} = 10^{-6}$ to 10^{-5} at 62 to 64 km (Chap. 3.3). To get the calculated values in this range we should adopt $f_{H_2O} = 1.9 \times 10^{-4}$ at 50 km. The obtained value agrees well with the results of the spectroscopic measurement of f_{H_2O} on board the Venera probes (Chap. 3.5), it is consistent with the data of the first gas chromatographic analysis of Pioneer Venus, which was made for 54 km and gave $f_{H_2O} < 6 \times 10^{-4}$.

So, the mixing ratio of water vapor drastically decreases from 2×10^{-4} to about 10^{-6} at heights from 50 to 65 km in the Venusian atmosphere. In the

Earth troposphere f_{H_2O} decreases even more strongly (it is about 10^{-2} near the surface and about 10^{-5} in the tropopause at 10 – 12 km). In the case of the Earth's troposphere it is caused by a significant decrease of temperature and condensation of water vapor. No condensation could occur on Venus in the height range discussed. The f_{H_2O} decrease results from photochemical production of sulfuric acid. Drops of sulfuric acid are in phase equilibrium with water vapor, that is, the concentration of H_2SO_4 solution and water vapor pressure above the solution are unambiguously determined from specified H_2O and H_2SO_4 amounts.

According to our calculations, the aerosol density decreases above 60 km with a scale height of about 4 km, being nearly constant below this altitude. Direct measurements on board the Venera and Pioneer Venus probes give a similar dependence for the upper and middle cloud layer. From the calculations it follows that the unit optical depth is achieved at 72 km; observations give $\tau \simeq 1$ at 68 km. The net optical thickness of the upper and middle cloud layer is 20 which agrees well with the observational data.

Figure 181 gives the calculated concentration of sulfuric acid solution in the aerosol phase. On the average it is close to 85% which is in agreement with the observations of Pollack et al. (1978) and it corresponds to the monohydrate $H_2SO_4 \cdot H_2O$. As has been mentioned, in this case sulfuric acid particles should be solid in the top clouds which is inconsistent with the polarimetric observations of Hansen and Hovenir (1974). Overcooling or solidification without crystallization possibly occurs here.

Thus, the results of calculations and observations show satisfactory agreement if it is assumed that large particles in the middle cloud layer are either sulfuric acid or its products. Strong arguments in favor of this assumption were discussed in Chap. 3.10.

The mixing time in the cloud layer is about one-half of a terrestrial year. The settling time for particles with a radius of 4 μm is about a terrestrial month. It is much longer than the time of zonal circulation (4 days), hence, diurnal variations of the middle cloud layer must be low. In the top clouds the SO_2 lifetime relative to photolysis drastically decreases and becomes 2 h at 62 km in the daytime; therefore, the number densities of SO_2 and H_2O may grow at the heights of the upper cloud layer in the nighttime. This contradicts the interpretation of the PV 45 μm band measurements (Schofield et al. 1982, see Chap. 3.4). However, their interpretation was not confirmed by the spectroscopic observations of Aumann et al. (1982) and Ertel et al. (1984).

Molecular Oxygen. At 90 to 98 km the amount of O_2 in the atmosphere (Fig. 178 a) is half that of CO. This implies that above 90 km O_2 is a result of CO_2 photolysis and almost all oxygen atoms formed above 90 km have the possibility to form O_2 molecules. Below 90 km the reaction with COCl causes a rapid destruction of O_2. This is the only important reaction of O_2 destruction in our scheme of processes. For $h < 90$ km much atomic oxygen combines with CO rather than form O_2 molecules, owing to this the total rate of the process which COCl is less than the total rate of CO_2 photolysis (it is about

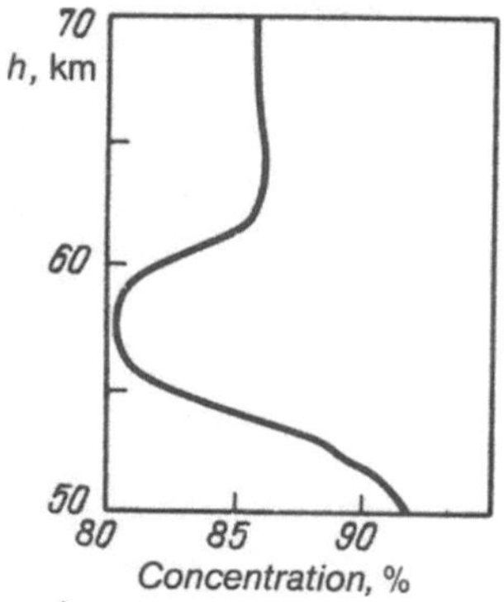

▲
Fig. 181. Concentration of sulfuric acid solution in the cloud drops vs height (Krasnopolsky and Parshev 1980)

Fig. 182 ►

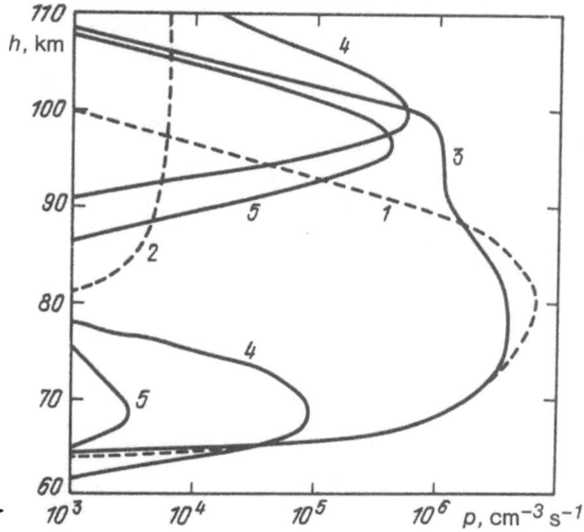

Fig. 182. Sinks and sources of O_2 (Krasnopolsky and Parshev 1980):
$COCl + O_2 \rightarrow CO_2 + ClO$ (*1*)
$O_2 + h\nu \rightarrow O + O$ (*2*)
$O + ClO \rightarrow O_2 + Cl$ (*3*)
$O + O + CO_2 \rightarrow O_2 + CO_2$ (*4*)
$O + OH \rightarrow O_2 + H$ (*5*)

60%). A major contribution to O_2 formation (Fig. 182) is made by the chlorine cycle (*R*34), 90%; the three body reaction contributes 6% and the hydrogen cycle, 4%. Molecular oxygen forms a layer 13 km thick at a height of 88 km, with the total amount of O_2 being 3.7×10^{18} cm^{-2}. The value is somewhat larger than the spectroscopic limit 1.5×10^{18} cm^{-2}, but agrees with its previous value, 5×10^{18} cm^{-2}, given by Traub and Carleton (1974).

Figure 182 indicates that O_2 destruction does not occur in the O_2 layer. Hence, the O_2 amount is inversely proportional to the mixing rate, and to be in accord with the new spectroscopic limit of O_2, K should be increased by a factor of 3 at 80–90 km.

In the model of Sze and McElroy (1975) processes of O_2 formation are the same as in our scheme; O_2 destruction, however, mainly occurs via the formation and photolysis of H_2O_2. The presence of sulfur compounds considerably diminishes the role of that cycle, and O_2 is destroyed while interacting with COCl. If this reaction is eliminated, appreciable amounts of O_2 will appear down to heights of about 62 km where O_2 can decompose in the following reaction:

$$S + O_2 \rightarrow SO + O\,.$$

The amount of O_2 in this case very much exceeds the spectroscopic limit, the CO flux to the subcloud atmosphere will become more intense, and the amount of aerosol sulfur will be strongly reduced.

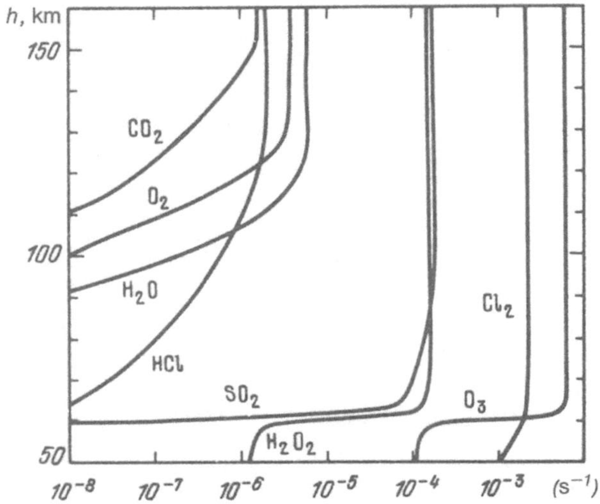

Fig. 183. Photolytic coefficients of different components in the Venusian atmosphere (Krasnopolsky and Parshev 1980b)

Calculations show that the overcloud atmosphere cannot be a source of molecular oxygen in the cloud layer detected by the gas chromatographic experiments ($f_{O_2} \sim 4 \times 10^{-5}$). If such O_2 concentrations are assumed to exist at the lower boundary, the height profile of O_2 will be similar to SO_2 and H_2O profiles. At the level of the SO_2 photolytic layer S atoms will rapidly react with O_2 and sharply decrease its abundance. The outcome of that sequence of reactions is:

$$2SO_2 + O_2 \rightarrow 2SO_3 .$$

Since $f_{O_2} < \frac{1}{2} f_{SO_2}$ at 50 km, aerosol sulfur will be produced photochemically, though in smaller amounts. The spectroscopic limit of O_2 abundance may not be exceeded with an appropriate choice of K. Sedimentation of sulfuric acid particles with an excess of oxygen as compared with SO_2 will compensate the upward flux of O_2. But if free oxygen exists in the atmosphere as a layer at 40 to 55 km, it should exist perhaps in a weakly bound form and with the same mixing ratio in the lower atmosphere where it should react with CO and vanish irreversibly. Hence, we have not found any explanations to the O_2 measurements in the cloud layer and have not used them.

Odd Hydrogen and Chlorine. Hydrogen and chlorine compounds could be divided into two groups. The first group involves H_2O, H_2, and HCl which are stable and chemically nonactive molecules. The other group includes chemically active components capable of rapid mutual transformations: H, OH, HO_2, Cl, ClO, ClO_2, COCl, and $COCl_2$. We assume that H_2O_2 and Cl_2 with a high rate of photolysis (Fig. 183) also belong to that group.

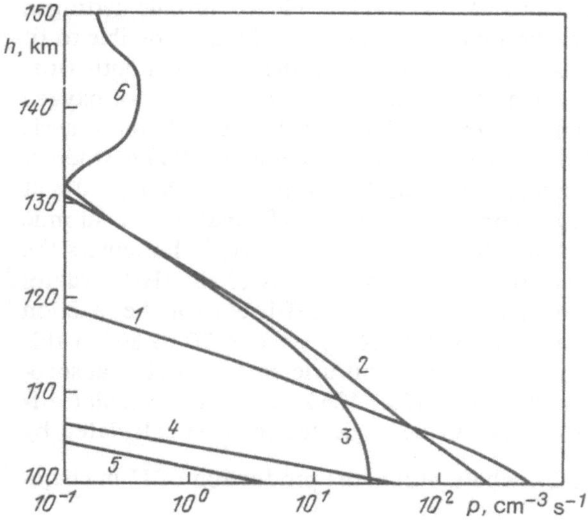

Fig. 184. Sinks and sources of atomic hydrogen in the upper atmosphere (Krasnopolsky and Parshev 1980b):

$O + HCl \rightarrow OH + Cl$ (*1*)
$HCl + h\nu \rightarrow H + Cl$ (*2*)
$H_2O + h\nu \rightarrow H + OH$ (*3*)
$H + Cl_2 \rightarrow HCl + Cl$ (*4*)
$H + HO_2 \rightarrow H_2O + O$ (*5*)
$HCO_2^+ + e \rightarrow CO_2 + H$ (*6*)

Odd hydrogen and chlorine (see Chap. 1.7) act as catalysts in the processes of CO and O_2 formation. We have already discussed their role, the calculated number densities are given in Fig. 178.

It is of interest that number densities of H* are much less than of Cl*, i.e., $[H^*] \ll [Cl^*]$, although equal amounts of H* and Cl* are formed and destroyed. This is attributed to the reaction:

$$OH + HCl \rightarrow H_2O + Cl$$

where H* is destroyed and Cl* is formed. Hence, the excess of Cl* which is twice the column rate of the process is transported in the form of $COCl_2$ into the lower troposphere where the reversal occurs at a high temperature:

$$COCl_2 + H_2O \rightarrow CO_2 + 2HCl\,.$$

The $COCl_2$ flux is $2.4 \times 10^9\,cm^{-2}\,s^{-1}$.

Below 95 km the photochemical equilibrium of H* sets up when net rates of formation and destruction reactions are equal. Above 100 km (Fig. 184) only H* formation processes occur and the sink of H atoms thus formed is provided either by transport into the lower layers or by nonthermal escape. The flux of nonthermal escape is taken as equal to $10^7\,cm^{-2}\,s^{-1}$ according to the calculations of Kumar et al. (1978) and McElroy et al. (1982a). In the present model, processes of nonthermal escape of atomic hydrogen forming as a result of molecular hydrogen dissociation in reactions with O^+ ion are negligibly small due to the low number density of molecular hydrogen at the turbopause level. The mixing ratio of water vapor decreases not only in the cloud layer remaining unchanged above it as Kumar et al. (1978) assumed based on the model of Sze and McElroy (1975), but continues to decrease above the cloud layer also down to 10^{-7} at 100 km. Hence, atomic hydrogen production by water photolysis becomes less intense, making it possible to fit the atomic hydrogen concentration at 200 km for the flux of nonthermal escape ($10^7\ cm^{-2}\ s^{-1}$) which is mainly caused by collisions with hot oxygen atoms formed by O_2^+ recombination and by charge exchange with ionospheric protons. The downward hydrogen flux reaches its maximum at 96 km where it is $4.7 \times 10^8\ cm^{-2}\ s^{-1}$. The boundary between the fluxes on which $\phi_H = 0$ is at 115 km. For $h \gtrsim 115$ km H_2O contributes to the formation of atomic hydrogen about twice as much as HCl. The amount of H_2O that enters the upper atmosphere is rather sensitive to small variations of the $[H_2O]$ values assumed for 50 km and to the choice of the rate coefficient for the reaction between sulfur particles and atomic oxygen (*R*83) since at 70 to 80 km this reaction eventually leads to the formation of sulfuric acid with H_2O absorption. Atomic hydrogen concentrations at 100 – 125 km, where molecular diffusion and H destruction processes can be neglected, can be calculated by solving the equation $\frac{d\phi}{dz} = p(z)$, where $p(z)$ is the production of H in chemical reactions (Fig. 184); $\phi = -K[CO_2]\frac{df}{dz}$; the boundary condition is $\phi_H(\infty) = 10^7\ cm^2\ s^{-1}$. It is more difficult to formulate the condition at the lower boundary; the assumption $f = 0$ at 80 km and $p(z) = 0$ at 80 to 100 km is made. In fact, H production over this interval is sufficiently high, but it is compensated by H destruction. The assumption $f = 0$ at 80 km only slightly affects H concentration at 100 to 125 km. Under those assumptions the equation is easily integrated:

$$f(z) = \int_{80\,\mathrm{km}}^{z} \frac{dx}{K(x)n(x)} \left[\int_x^\infty p(y)\,dy - 10^7\right] = -\int_{80\,\mathrm{km}}^{z} \frac{\phi(x)\,dx}{K(x)n(x)}; \quad n = [CO_2]\,.$$

Hence, $[H] = 2\phi H/K$ at 100 km. The flux ϕ at 100 km, equal to the net production of H per column above that level (Fig. 184), is $2.3 \times 10^8\ cm^{-2}\ s^{-1}$. Then $[H] = 1.3 \times 10^8\ cm^{-3}$ at 100 km agrees excellently with the model. The calculated number densities of atomic hydrogen fit the measurements (Chaps. 3.11 and 4.3). The amount of Cl_2 in the top cloud layer is by more than an order of magnitude lower than the spectroscopic limit (Table 20).

4.12 O_2 1.27 μm and $O(^1D)$ 630 nm Airglow. Photolytic Rates

O_2 1.27 μm. High-resolution ground-based observations (Connes et al. 1979) made it possible to measure the airglow of the Venusian atmosphere in the 1.27 μm band (0, 0) $O_2\,(a^1\Delta_g \rightarrow X^3\Sigma_g^+)$. The airglow intensity turned out to be very strong and almost equal on the day and night sides of the planet, i.e., 1.5

and 1.2 *MR*, respectively (Chap. 3.11). Prior to that the observations of this emission on Mars had been reported, having an intensity of 3 *MR* at low latitudes (Chap. 1.3). In the terrestrial atmosphere the intensity of airglow in this band is 20 *MR* (Evans et al. 1968). Thus, the 1.27 μm airglow intensity exceeds that of all other emissions on all three terrestrial planets. The day airglow of 1.27 μm on the Earth and Mars is due to ozone photolysis whose $O_2(^1\Delta_g)$ yield is close to unity. Almost similar intensities of Venusian airglow on the day and night side have led Connes et al. (1979) to the conclusion that the reaction ($O_3 + Cl \rightarrow ClO + O_2$) is the source of $O_2(^1\Delta_g)$.

To convincingly explain the processes of $O_2(^1\Delta_g)$ excitation on Venus, it is necessary to perform aeronomical calculations of diurnal variations of the components, using the available experimental data as a basis. The calculations present some difficulties because of the complicated character of the atmospheric dynamics (four-day circulation at heights lower than 80 km and winds from the terminator to the antisolar point in the upper atmosphere). Nevertheless some semiquantitative estimates can be made on the basis of this model.

Energy conditions and the rule of spin conservation allows $O_2(^1\Delta_g)$ to form in the following reactions:

$$O_3 + h\nu \rightarrow O_2 + O \quad (1.2 \times 10^{12}\,cm^{-2}\,s^{-1}) \tag{1}$$
$$O_3 + Cl \rightarrow O_2 + ClO \quad (1.3 \times 10^{12}) \tag{2}$$
$$O + ClO \rightarrow O_2 + Cl \quad (2.2 \times 10^{12}) \tag{3}$$
$$O + O + CO_2 \rightarrow O_2 + CO_2 \quad (3.5 \times 10^{11}) \tag{4}$$
$$O + HO_2 \rightarrow O_2 + OH \quad (3.2 \times 10^{11}) \tag{5}$$
$$Cl + ClO_2 \rightarrow O_2 + Cl_2 \quad (3 \times 10^{11}) \tag{6}$$
$$O_3 + H \rightarrow O_2 + OH \quad (1.4 \times 10^{11})\,. \tag{7}$$

Given in the brackets are column rates of those processes above 85 km. Below it, $O_2(^1\Delta_g)$ quenching on CO_2 becomes rather significant. Figure 185 gives height profiles of those processes. There are other reactions producing $O_2(^1\Delta_g)$, but their contribution should be small.

Diurnal variations of the O_2 Herzberg second band system airglow show that at 100 km the nighttime concentrations of atomic oxygen should be almost equal to daytime ones (Chap. 3.11). In the upper atmosphere gas masses enriched by light components are transferred from the day to the night side and then go down, compensating the absence of both dissociation and some chemical processes in the night side atmosphere. Hence, the rates of reactions (4–7) can be almost equal on the day and night sides. The simplest explanation of the low diurnal variations of the 1.27 μm airglow is that the absence of ozone photolysis on the night side is replaced by the enhanced rate of reaction (2) followed by (3), so that $\gamma_2 + \gamma_3 \simeq \gamma_1 = 1$ where γ is the $O_2(^1\Delta g)$ yield. We now try to fit the calculations and measurements by choosing a quenching rate coefficient:

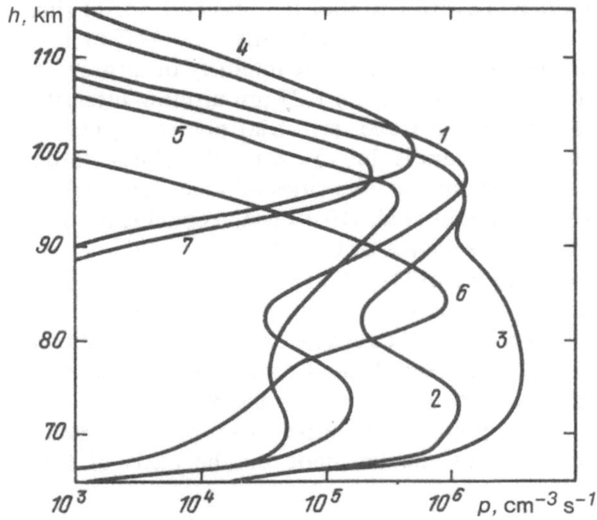

Fig. 185. Reaction rates of possible $O_2(^1\Delta_g)$ formation; *numbers of curves correspond* to the text (Krasnopolsky and Parshev 1980)

$$O_2(^1\Delta_g) + CO_2 \rightarrow O_2 + CO_2$$

for which laboratory measurements give values in the interval 1.5×10^{-20} to 4×10^{-18} $cm^3\,s^{-1}$ (see Noxon et al. 1976). With only the first three processes taken into account, an agreement is reached at $k \simeq 5 \times 10^{-20}$ $cm^3\,s^{-1}$. The calculated airglow profile is shown in Fig. 186.

Correcting for the overestimated solar radiation flux in our calculations (Chap. 4.8) results in a quenching rate coefficient of 3×10^{-20} $cm^3\,s^{-1}$ in accord with the measurements of the Martian airglow (Chap. 1.3).

Let us compare the results obtained with the data of NO nightglow interpretation (Chap. 3.11). This interpretation gives the downward flux of atomic oxygen on the night side to be $\phi = 10^{12}$ $cm^{-2}\,s^{-1}$. With the three body recombination (reaction 4) assumed as a source of 1.27 μm airglow it needs a quantum yield $\gamma \simeq 2$ which is unacceptable. However, $\gamma \simeq 2$ can be realized by a two-step process involving reactions (2) and (3) as suggested in our calculations. Though $\gamma_2 + \gamma_3 \simeq 1$ is recommended, a twofold difference can be attributed to the limited accuracy of both our and Stewart's et al. (1980) models as well as of both O_2 1.27 μm and NO airglow measurements.

$O(^1D - {}^3P)$ 630 nm Airglow. While the $O(^1D)$ production maximum occurs at 100 km, which corresponds to the main peak of the ozone layer, the high quenching rate of this state results in a significant shift of the $O(^1D)$ concentration maximum to the upper atmosphere. Most effective in $O(^1D)$ excitation here are the CO_2 photodissociation at $\lambda < 168$ nm and the dissociative recombination of O_2^+ ions which are formed in the process of CO_2^+ and atomic oxygen charge exchange. $O(^1D)$ yield in the case of O_2^+ recombination is close to unity; this is why each CO_2 ionization event produces an atom

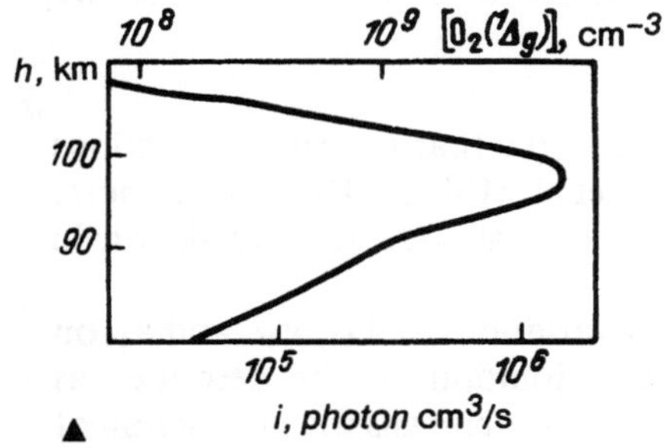

▲ Fig. 186

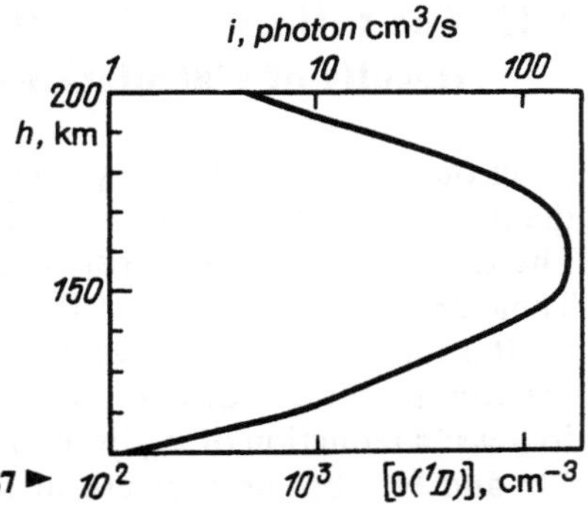

Fig. 187 ►

Fig. 186. Altitude profile of $O_2(^1\Delta_g)$ number density (Krasnopolsky and Parshev 1980)

Fig. 187. Density and volume emission rate of atomic oxygen dayglow at 630 nm (Krasnopolsky and Parshev 1980)

$O(^1D)$. Thus, the final result of dissociation and ionization is the same. Figure 183 shows height dependence of the sum of the CO_2 ionization and dissociation rates. It is not difficult to calculate $O(^1D)$ concentrations taking into account quenching on CO_2 and CO as well as radiative deactivation ($A = 10^{-2}$ s^{-1}). Figure 187 shows calculational results. Atoms of $O(^1D)$ form a layer ~30 km thick with a maximum at 160 km. The zenith intensity of the $O(^1D-^3P)$ 6300 Å airglow is 700 *R* at $z_0 = 60°$. The airglow intensity is 25 *kR* at the planet limb with the sighting height 140 km. The airglow can be detected by a satellite with a pericenter height of about 150 km, as by Pioneer Venus. Observations from the satellite with a pericenter height of about 1500 km will be very complicated because of the small angle (~40′) between the observed layer and the visible planet limb, the brightness of the latter being 2×10^3 *MR* nm^{-1}. Hence, the 630 nm day airglow was not detected on Venera 9 and Venera 10 (Krasnopolsky et al. 1976).

Height Dependence of Photolytic Rates. These data are depicted in Fig. 183. A group of components (CO_2, HCl, H_2O, O_2) can be singled out which are decomposed only by a far UV radiation. Their dissociation rate is of the order of 10^{-6} to 10^{-5} s^{-1} in the upper atmosphere and the values are very small at heights lower than 100 km. Dissociation rates decrease with height due to CO_2 absorption. The other group of species (SO_2, H_2O_2, Cl_2, O_3) is subject to photolysis by the middle and near UV radiation and has rates of the order of 10^{-4} to 10^{-2} s^{-1}, decreasing sharply at 60 km because of the strong SO_2 absorption. In the cloud layer the dissociation of those components is possible mainly due to radiation in the visible spectral region.

4.13 The Influence of Some Reaction Rate Coefficients on the Results of Calculations

After our calculations were published we discovered that for certain reactions the rate-coefficients recommended by Baulch et al. (1980) differed from ours. The discrepancies were most severe for reactions 40, 41, 45, and shown in Table 38 (Krasnopolsky and Parshev 1983).

It is interesting to analyze the influence of variations of k_{40}, k_{41}, and k_{45} on the results of the calculations. Basically the situation can be described as follows: a reduction of k_{41} by an order of magnitude curtails the production of Cl* and H*, but an increase of k_{40} by ~ 10 times preserves [Cl*], which is determined by the balance between the generation of Cl* in reaction 40 and its transport to the lower atmosphere. For the same reason, low concentrations of H* do not affect [Cl*] even when k_{45} increases by a factor of 400. The curtailed production of H* in reaction 41 is modified by the photolysis of HCl. Thus, [H*] is reduced by more than an order of magnitude, [Cl*] by less than a factor of 2, and consequently, the overall number densities of the various forms of oxygen at 60 to 80 km increase by less than $\sqrt{2}$. Above 80 km, [O_2] as basically specified by transport and by reaction 57, is virtually unchanged. Since the main quantity of O_2 is located here, its total amount remains almost the same.

In a more detailed quantitative analysis of the problem, total production and loss rates can be assumed equal, except for the doubled rate P of reaction 40 in the vertical column. This process occurs in a layer at 63 to 70 km, therefore, above 70 km f_{Cl^*} is constant, while below 63 km $[Cl^*] = PH/K$ is also constant. Within the context of this dependence, an equilibrium between the different forms of Cl* exists; this also determines the number densities of H*, O, O_2, and O_3 at 60 to 80 km. Only the main processes were utilized and ratios were calculated between the concentrations of the various components under the new and the old values of the rate coefficients. It turned out that at heights of 60 to 80 km, the overall quantity of Cl* had been reduced by a factor of 1.75, that of O by 1.14, O_2 had been increased by a factor 1.3. Variations of [H*] are more complex, as shown in Fig. 188.

Table 38. Rate coefficients of reactions 40, 41, and 45 used by Krasnopolsky and Parshev (1980a, b, 1981) and recommended by Baulch et al. (1980), $k(KP)$ and $k(B)$, and their values at 250 K

No.	Reaction	$k(KP)$	$k(B)$	$k(KP)$ at 250 K	$k(B)$ at 250 K
40	$OH+HCl \rightarrow H_2O+Cl$	$2\times10^{-13}e^{-310/T}$	$3\times10^{-12}e^{-425/T}$	5.8×10^{-14}	5.5×10^{-13}
41	$O+HCl \rightarrow OH+Cl$	$1.8\times10^{-12}e^{-2260/T}$	$10^{-11}e^{-3370/T}$	2.1×10^{-16}	1.4×10^{-17}
45	$HO_2+Cl \rightarrow HCl+O_2$	10^{-13}	$1.8\times10^{-11}e^{180/T}$	10^{-13}	3.7×10^{-11}

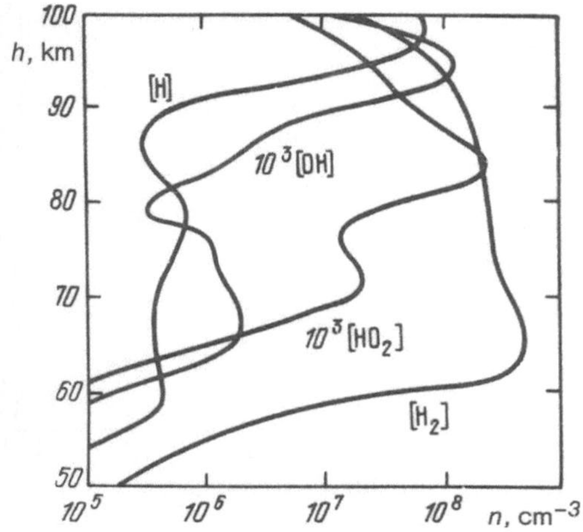

Fig. 188. H_2 and odd hydrogen number densities for k_{40}, k_{41}, k_{45} recommended by Baulch et al. (1980) (Krasnopolsky and Parshev 1983)

So the change of number densities of chlorine- and oxygen-containing components is not significant. Odd hydrogen undergoes great variations, but its role in the atmospheric chemistry of Venus is small in our model. However, the efficiency of the COCl cycle was overestimated in our calculation due to some confusion in the data. Therefore, the results described in Chap. 4.11 should be considered as the extreme case of very rapid removal of O_2 and formation of CO_2. Some parts of the results and conclusions are not affected strongly by this cycle, e.g., formation of sulfuric acid, altitude profiles of H_2O, SO_2, CO, etc.

4.14 Photochemistry of the Venusian Mesosphere as Considered by Winick and Stewart (1980)

This paper represents calculations of the atmospheric composition at 58 to 96 km (Fig. 189). A list of atmospheric constituents was similar to that used in our calculations, except for COCl and $COCl_2$, which were not considered by the authors. Some consequences of excitation of $^1SO_2$ and $^3SO_2$ were discussed, but sulfuric acid formation by these species appeared, to be small, and therefore, these excited molecules were not introduced in the calculations. The list of reactions included about half the number of reactions considered by us due to the absence of those species and reactions which seemed to be insignificant in the Venusian atmosphere. On the contrary, photolyses of SO, ClO, and ClO_2 were considered, and more up-to-date reaction rate coefficients were used. As in our calculations, Winick and Stewart employed a simplified version of radiative transfer in the cloud layer. Fluxes of atomic oxygen and other components with short lifetimes were adopted to be zero at both

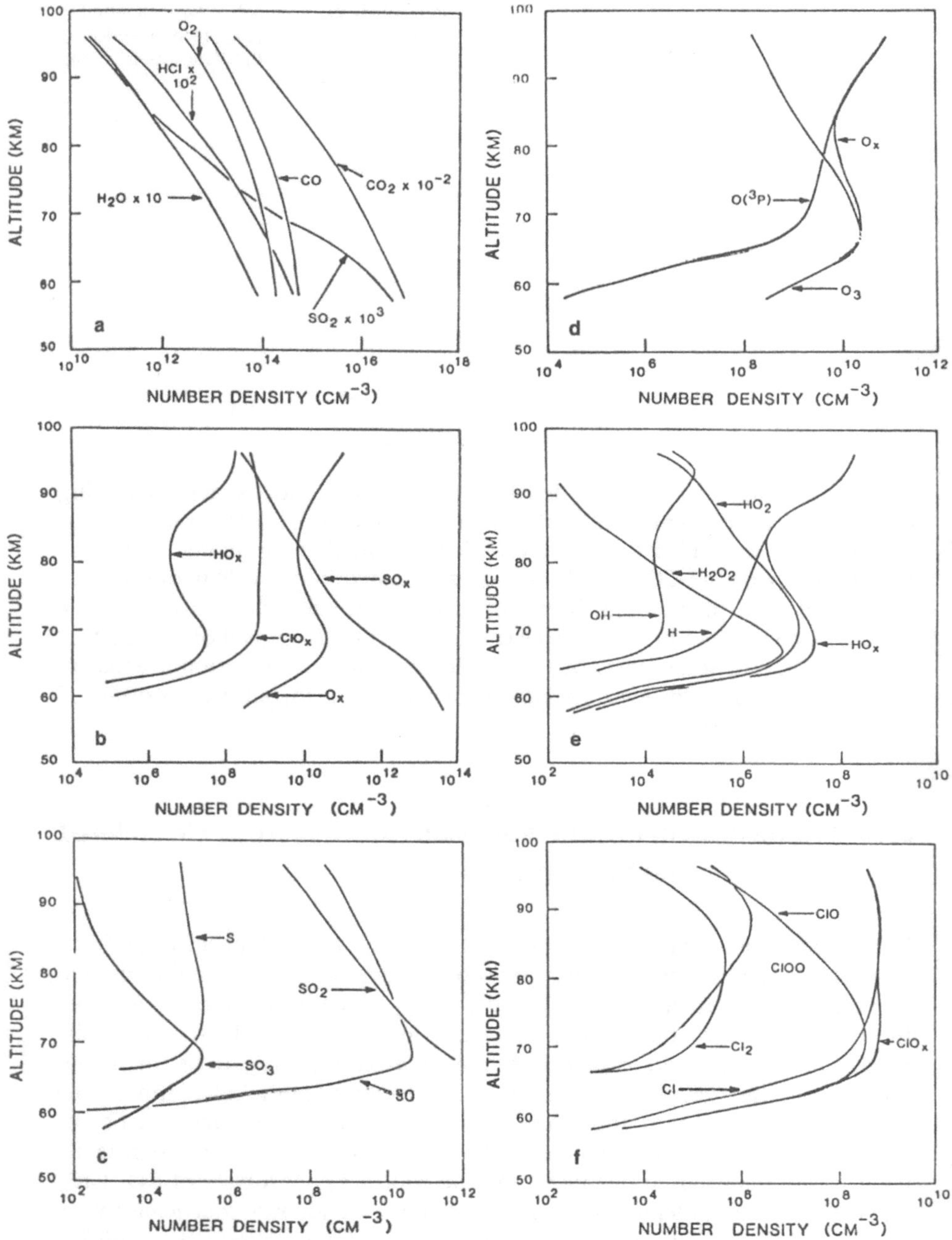

Fig. 189a–f. Chemical composition of the Venusian atmosphere at 58–96 km calculated by Winick and Stewart (1980): **a** parent and long-lived species; **b** group species; **c** SO_x; **d** O_x; **e** HO_x; **f** ClO_x

boundaries. At the upper boundary fluxes of CO and O_2 were assumed to equal the CO_2 photolysis above 96 km and a half of it multiplied by a factor which accounted for a part of the atomic oxygen converted to O_2 above 96 km. Fluxes of other stable components were assumed to be zero at 96 km. At the lower boundary $f_{HCl} = 4 \times 10^{-7}$, $f_{H_2} = 2 \times 10^{-7}$, f_{SO_2} was a fitting parameter. Unfortunately, conditions for CO and O_2 cannot be clearly understood from the text. The water vapor mixing ratio was assumed to be 10^{-6} throughout 58 to 96 km.

The new value of the $CO + ClO \rightarrow CO_2 + Cl$ reaction (Table 37) rate coefficient precludes formation of CO_2 by the ClO cycle. With the COCl cycle disgarded the most important path of the CO_2 formation from its photolysic products is the hydrogen cycle with a small contribution of the three body reaction. Below 70 km sulfur compound number densities are large enough to provide a rapid removal of odd hydrogen. Therefore, the CO_2 formation does not balance its photolysis, and the excess of CO and O_2 forms downward fluxes of these species towards the lower atmosphere. The CO mixing ratio equals 64 ppm at 58 km and 300 ppm at the clouds tops, i.e., exceeds the spectroscopic estimate by a factor of 6. Atomic oxygen formed by CO_2 photolysis is a source of molecular oxygen via the ClO cycle (*R*34, Table 37). The O_2 removal occurs in the reactions involving S and SO (*R*64, 70, Table 37) with the formation of SO_3 and, in the final analysis, of sulfuric acid drops from SO_2. The sedimentation flux of sulfuric acid is 1.8×10^{11} cm^{-2} s^{-1} at 58 km, i.e., much less than the CO_2 photolytic rate (5×10^{12} cm^{-2} s^{-1}). Their difference is equal to the total rate of the reaction of CO and OH (*R*16, Table 37) and the O_2 downward flux. Perhaps this value defines the lower boundary conditions of CO and O_2. The molecular oxygen mixing ratio is 100 ppm at 70 km which is far in excess of the spectroscopic upper limit, 0.3 ppm. At 58 km, $f_{O_2} \simeq 30$ ppm which agrees with the gas chromatographic data. Large O_2 amounts result in large ozone number densities ($\sim 2 \times 10^{10}$ cm^{-3}) which should produce substantial absorption at 260 nm, contrary to the UV spectroscopic measurements. These shortcomings were recognized by Winick and Stewart, but they wanted to test how far one can go with only known reactions.

Good agreement with the observations was achieved for a sulfur dioxide number density profile with f_{SO_2} adopted to be 4 ppm at 58 km. This value is consistent with the experimental data. Sulfuric acid amounts equal those of the measured second mode particles at 58 to 62 km and differ slightly above 62 km. SO number densities have a maximum at 67 km where [SO] = 4×10^{10} cm^{-3} or $f_{SO} \simeq 2 \times 10^{-8}$, i.e., exceed the spectroscopic limit (Table 37) by an order of magnitude. In the model, $K = 2 \times 10^{13} M^{-1/2}$ cm^{-2} s^{-1} was adopted.

4.15 Analysis of Atmospheric Photochemistry on Venus by Yung and DeMore (1982)

Outline and Approach. The most comprehensive study of the mesospheric photochemistry on Venus at 58 to 110 km was made by Yung and DeMore

(1982) with different assumptions on the composition of the lower atmosphere. Chemical reactions and their coefficients used in this paper are listed in Table 37. The rate coefficients were chosen on the strength of up-to-date laboratory measurements. We employed the rate coefficients cited by Sze and McElroy (1975), Baulch et al. (1976), and Kondratiev (1971). Therefore, the differences in k in Table 37 reflect the progress made in this field over the last 5 years. The paper provides a more detailed and exact description of the interaction between COCl and O_2 which became possible thanks to recent laboratory measurements, including those done by DeMore et al. This reaction proves to proceed in a manner different from what has been described in earlier papers and was used by us,

$$\mathrm{COCl} + \mathrm{O_2} \rightarrow \mathrm{CO_2} + \mathrm{ClO} \quad \text{but} \tag{$R57$}$$

$$\mathrm{COCl} + \mathrm{O_2} + M \rightarrow \mathrm{ClCO_3} + M\,. \tag{$R146$}$$

COCl formation needs substantial amounts of chlorine atoms, therefore, it is followed by

$$\mathrm{ClCO_3} + \mathrm{Cl} \rightarrow \mathrm{CO_2} + \mathrm{Cl} + \mathrm{ClO}\,. \tag{$R148$}$$

The net reaction $R146$ and $R148$ is $R57$. The efficiency of molecular oxygen destroying by COCl is determined by the three body reaction

$$\mathrm{CO} + \mathrm{O_2} + \mathrm{Cl} \rightarrow \mathrm{CO_2} + \mathrm{ClO}$$

with a rate coefficient $k = k_{56}k_{57}/k_{61}$ or $k = k_{56}k_{146}M/k_{61}$ when using $R57$ or $R146$ and $R148$, respectively. Here k_{56} and k_{61} are the rate coefficients of the direct and inverse reactions

$$\mathrm{CO} + \mathrm{Cl} + M \rightleftarrows \mathrm{COCl} + M\,.$$

Our k was overestimated by two orders of magnitude due to confusion in k_{61} cited by Kondratiev (1971).

Other reactions with $ClCO_3$ considered by Yung and DeMore (1982),

$$\mathrm{ClCO_3} + \mathrm{O} \rightarrow \mathrm{CO_2} + \mathrm{Cl} + \mathrm{O_2} \tag{$R147$}$$

$$\mathrm{ClCO_3} + \mathrm{H} \rightarrow \mathrm{CO_2} + \mathrm{Cl} + \mathrm{OH} \tag{$R149$}$$

are less important. Net results of these reactions are $CO + O \rightarrow CO_2$ and $CO + O_2 + H \rightarrow CO_2 + OH$, respectively. The authors mention, but do not use the reactions

$$\begin{aligned} &\mathrm{ClCO_3} + \mathrm{NO} \rightarrow \mathrm{CO_2} + \mathrm{Cl} + \mathrm{NO_2} \\ &2\,\mathrm{ClCO_3} \quad\ \rightarrow 2\,\mathrm{CO_2} + 2\,\mathrm{Cl} + \mathrm{O_2}\,. \end{aligned}$$

The former can be important with $f_{NO} \gtrsim 3$ ppb which seems to be real (Chap. 4.6). In our opinion, the $ClCO_3$ photolysis, its collisional destruction and reaction with SO not mentioned in the paper merit attention as well.

Table 39. Main reactions of CO_2 and O_2 formation and O_2 removal (after Yung and DeMore 1982)

Formation of CO_2	Formation of O_2	Removal of O_2
$O+CO+CO_2 \rightarrow CO_2+CO_2$	$O+O+CO_2 \rightarrow O_2+CO_2$	$O_2+h\nu \rightarrow O+O$
$CO+OH \rightarrow CO_2+H$	$O+OH \rightarrow O_2+H$	$H_2O_2+h\nu \rightarrow OH+OH$
$COCl+O \rightarrow CO_2+Cl$	$O+ClO \rightarrow O_2+Cl$	$O_2+S \rightarrow SO+O$
$ClCO_3+Cl \rightarrow CO_2+Cl+ClO$	$O+NO_2 \rightarrow O_2+NO$	$O_2+SO \rightarrow SO_2+O$
		$HO_2+SO \rightarrow SO_2+OH$
		$HO_2+NO \rightarrow NO_2+OH$
		$COCl+O_2+M \rightarrow ClCO_3+M$

The excited states of SO_2 are not considered in the model. Yung and DeMore provide a detailed description of the nitrogen compound chemistry (33 reactions of N, NO, NO_2, NO_3, HNO, HNO_2, HNO_3, NOCl, N_2O) whose source is assumed to be lightning. With such a large number of components and reactions considered, a number of catalytic cycles in this system also increases (about 20 cycles of CO_2 formation from CO and O, CO and O_2, and O_2 destruction). These cycles are not considered here and only the main reaction of each cycle is given in Table 39. Table 39 shows that Yung and DeMore add three catalytic processes, two of them with NO, and the third is the reaction of HO_2 and SO, not considered by us. In their reaction list the rate coefficient of the reaction

$$S+CO_2 \rightarrow CO+SO\,, \quad k=10^{-20}\,cm^3\,s^{-1}$$

is open to question because the inverse reaction is exothermic ($\Delta E = 2.6$ kcal mode^{-1}) and similar to that of O_2 and CO, whose rate coefficient is $4.2\times10^{-12}e^{-2400/T}$ $cm^3\,s^{-1}$. Then $k=4.2\times10^{-12}e^{-25300/T}$ $cm^3\,s^{-1}$ and less than 10^{-45} $cm^3\,s^{-1}$ at $T \lesssim 300$ K.

For short-living components, zero fluxes are assumed at both boundaries. For long-living components, fluxes at the upper boundary are adopted to be zero except for $\phi_{CO_2} = -\phi_{CO} = -\phi_O = 10^{12}\,cm^{-2}\,s^{-1}$, $f_{HCl} = (4-8)\times10^{-7}$, $f_{NO}=0$ or 3×10^{-8}, f_{SO_2} is a fitting parameter. Conditions for atomic and molecular oxygen at the lower boundary are given by their velocities, $V=-\alpha K/H$ with $\alpha \leqslant 1$. Unfortunately, this very convenient and physically adequate form of boundary conditions is used very seldom.

Water vapor amounts are not treated here. Its altitude profile with constant $f_{H_2O} = 1$ ppm is assumed. Obviously this assumption, on the one hand, simplifies the calculations. However, on the other hand, some important aeronomic problems, such as the altitude behavior of f_{H_2O} and variations of concentration of sulfuric acid solution in the cloud layer remain unsolved. We discussed earlier (Chap. 4.11) a very high sensitivity of these values of the choice of boundary conditions and $K(h)$. The water vapor mixing ratio calculated by Krasnopolsky and Parshev (1980b, 1981) depends strongly on height.

The three body reactions of SO_2 with OH and Cl (*R*67, 113) form HSO_3 and $ClSO_2$ which do not participate in other chemical reactions considered, and it is suggested that both species are absorbed by sulfuric acid drops. These reactions are significant as H* and Cl* sinks are excluded from some variants of the calculations.

The main source of H* and Cl* is as usual the photolysis of HCl and the reaction of O and HCl. Mutual transformations of H* and Cl* occur in the reactions

$$OH + HCl \rightarrow H_2O + Cl$$
$$H + HCl \rightarrow H_2 + Cl$$
$$Cl + H_2 \rightarrow HCl + H \, .$$

If the first two reactions proceed at approximately equal rates, then $[H^*]/[Cl^*] \simeq 0.1\,[H_2]/[HCl]$, where 0.1 is the ratio of the rate coefficients. So the atmospheric photochemistry depends strongly on the H_2 amount in the lower atmosphere whose estimates vary from $f_{H_2} = 3 \times 10^{-9}$ for thermochemical equilibrium near the surface to 2×10^{-5} which was obtained from the V 13, 14 gas chromatography and the interpretation of ion mass 2 as H_2^+ in the ionosphere (see Chap. 4.4). Therefore, three models have been calculated.

Model A (Figs. 190, 191 and 192) assumes $f_{H_2} = 20$ ppm and $f_{NO} = 0$, reactions with COCl and NO_x are not involved, and the heterogeneous sink of H* and Cl* (*R*67, 113) is taken into account. Due to the large molecular hydrogen number densities, $[H^*] \gg [Cl^*]$ in the most important altitude region of 65 to 80 km. Atomic oxygen forms either CO_2 in the reaction *R*16 CO + OH or O_2 by three body collisions (*R*12) and *R*18 O + OH. The ratio of these reaction rates is determined by atmospheric mixing which is assumed to be $K = 3 \times 10^5\ cm^2\,s^{-1}$ above 64 km to retard the O_2 formation. Decrease of *K* to $4 \times 10^4\ cm^2\,s^{-1}$ at 58 km is necessary to put the SO_2 altitude profile in cor-

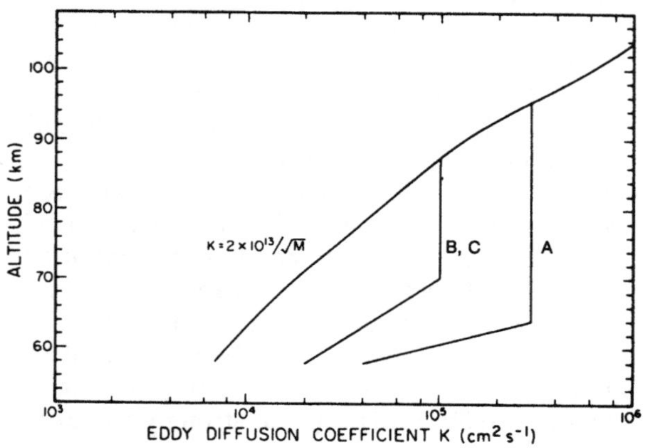

Fig. 190. Profiles of the eddy mixing coefficient used by Yung and DeMore (1982) in models *A, B, C*

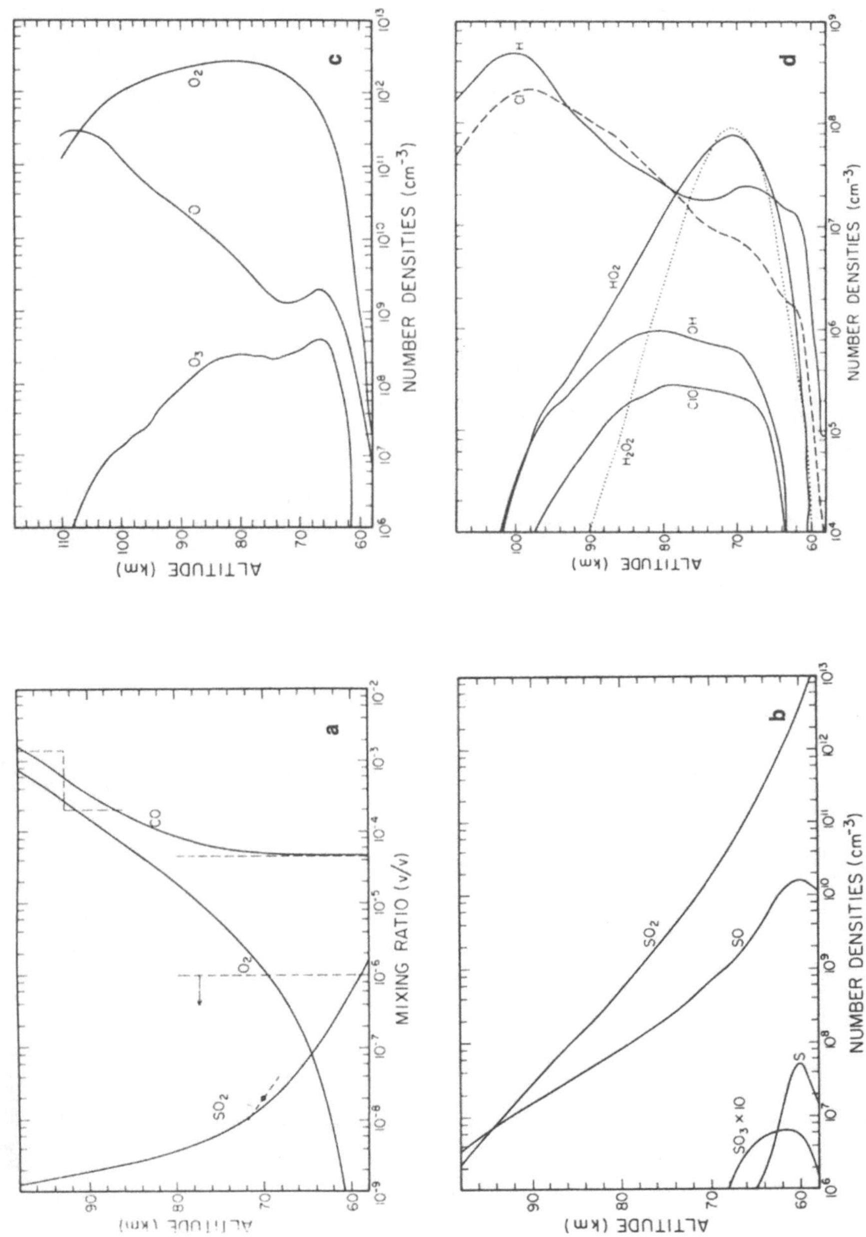

Fig. 191 a – d. Model *A* of the chemical composition of the Venusian atmosphere (Yung and DeMore 1982)

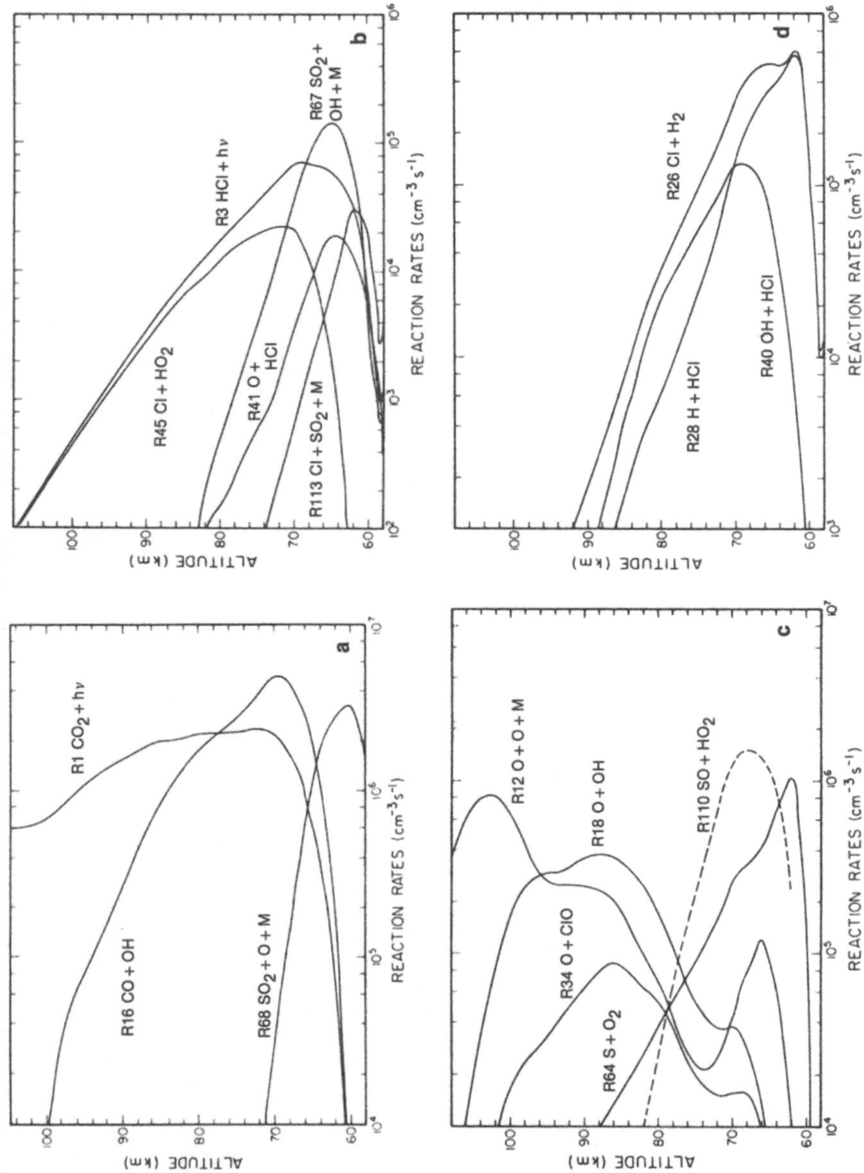
a
R1 CO2 + hν
R16 CO + OH
R68 SO2 + O + M
ALTITUDE (km)
REACTION RATES (cm-3 s-1)
b
R45 Cl + HO2
R3 HCl + hν
R41 O + HCl
R113 Cl + SO2 + M
R67 SO2 + OH + M
ALTITUDE (km)
REACTION RATES (cm-3 s-1)
c
R12 O + O + M
R18 O + OH
R34 O + ClO
R64 S + O2
R110 SO + HO2
ALTITUDE (km)
REACTION RATES (cm-3 s-1)
d
R26 Cl + H2
R28 H + HCl
R40 OH + HCl
ALTITUDE (km)
REACTION RATES (cm-3 s-1)

Fig. 192a–d

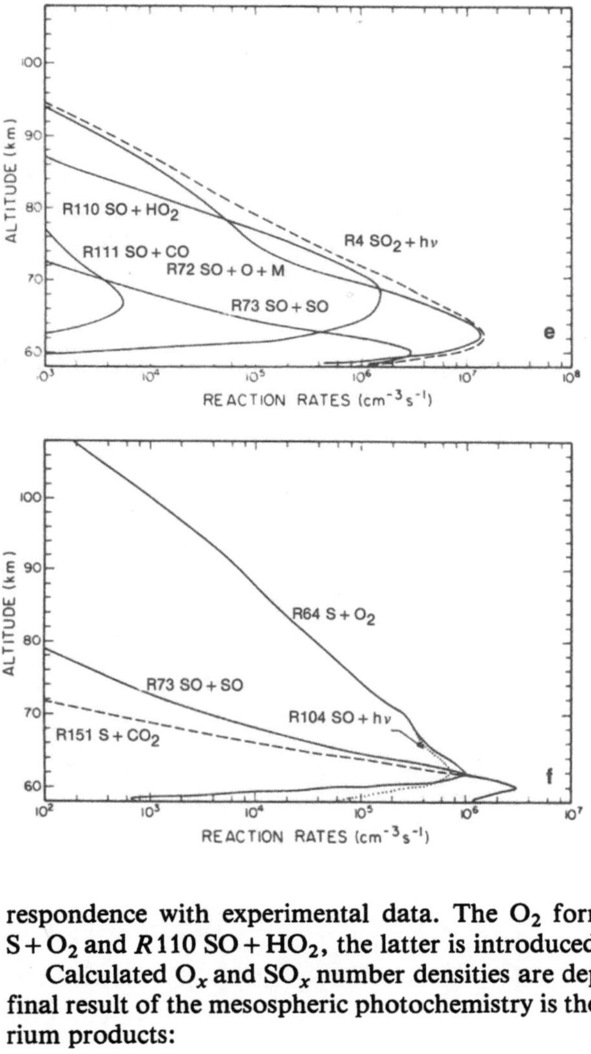

Fig. 192a–f. Model *A* by Yung and DeMore (1982). Sources and sinks of **a** oxygen; **b** HO_x and ClO_x; **c** O_2; **d** equilibriums between HO_x and ClO_x; **e** SO_2 and SO; **f** SO and S

respondence with experimental data. The O_2 formed is destroyed by *R*64 $S+O_2$ and *R*110 $SO+HO_2$, the latter is introduced for the first time.

Calculated O_x and SO_x number densities are depicted in Fig. 191 c, d. The final result of the mesospheric photochemistry is the formation of nonequilibrium products:

$$CO_2+SO_2+H_2O+h\nu \rightarrow CO+H_2SO_4 .$$

Upward fluxes of the component on the left are equal to downward fluxes of those on the right and constitute $2.2\times10^{12}\,cm^{-2}\,s^{-1}$ at 58 km. The cloud particle sedimentation flux referred to H_2SO_4 is $4\times10^{11}\,cm^{-2}\,s^{-1}$ at 58 km according to the measurements (see Chap. 3.10). Here the sedimentation velocity is $V+K/H$ where V is the Stokes velocity. Yung and DeMore give this flux as $10^{12}\,cm^{-2}\,s^{-1}$. Therefore, the production of sulfuric acid calculated for model *A* exceeds the measured one at 58 km by a factor of 5. The molecular oxygen column density obtained is $7.3\times10^{18}\,cm^{-2}$, i.e., five

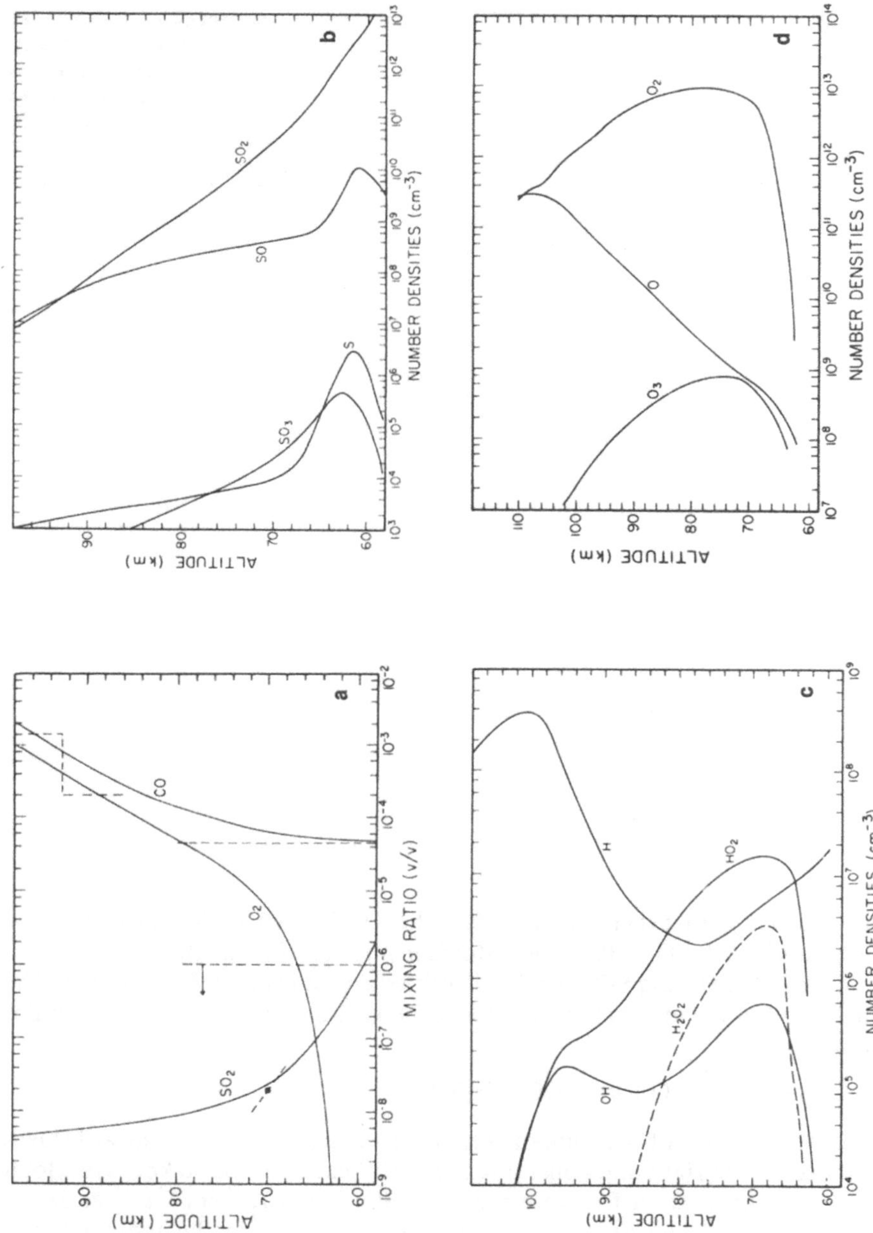
a
ALTITUDE (km)
MIXING RATIO (v/v)
SO2
O2
CO
b
ALTITUDE (km)
NUMBER DENSITIES (cm-3)
SO3
S
SO
SO2
c
ALTITUDE (km)
NUMBER DENSITIES (cm-3)
OH
H2O2
HO2
H
d
ALTITUDE (km)
NUMBER DENSITIES (cm-3)
O3
O
O2

Fig. 193a – d

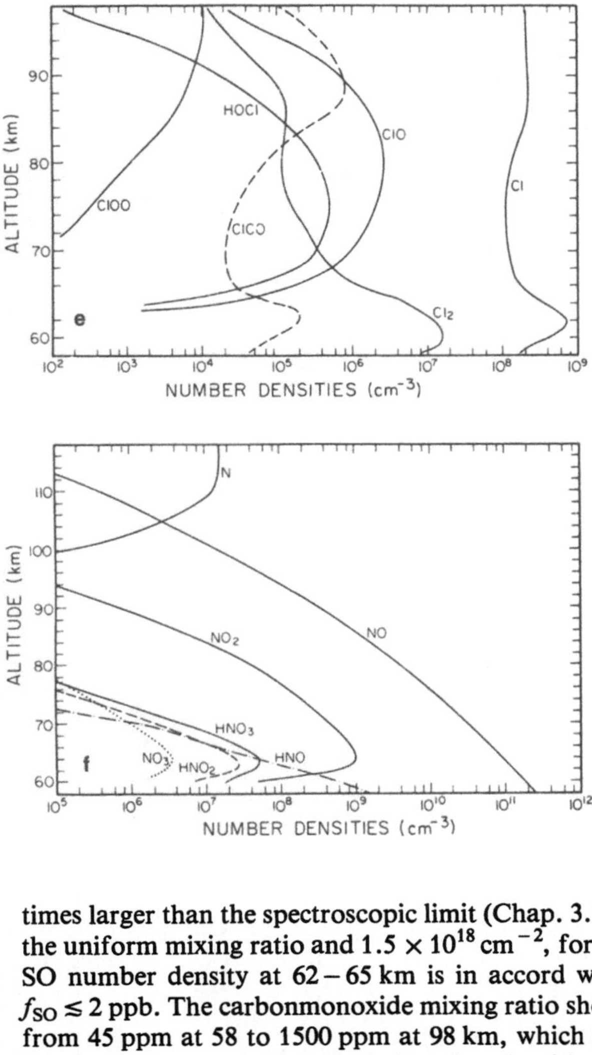

Fig. 193a–f. Model *B* of the chemical composition of the Venusian atmosphere (Yung and DeMore 1982)

times larger than the spectroscopic limit (Chap. 3.3) which equals 0.3 ppm for the uniform mixing ratio and $1.5 \times 10^{18}\,\mathrm{cm}^{-2}$, for the layer above 65 km. The SO number density at 62–65 km is in accord with the spectroscopic limit, $f_{SO} \lesssim 2$ ppb. The carbonmonoxide mixing ratio shows the increase with height from 45 ppm at 58 to 1500 ppm at 98 km, which agrees with the CO 2.6 mm line measurements (Chap. 3.5). Number densities and scale height of SO_2 at 70 km fit the measurements for appropriately chosen K and for $f_{SO_2} =$ 1.5 ppm at 58 km.

From the above it follows that the model *A* agrees much better with the experimental data than the model by Winick and Stewart which employ the same reactions and rate coefficient. It is attributed to the very large amounts of molecular hydrogen assumed in model *A* which exceed by two orders of magnitude those used by Winick and Stewart. Therefore, $[H^*] \gg [Cl^*]$ in model *A* contrary to the model of Winick and Stewart. With the COCl cycle being neglected in both models, odd chlorine accelerates O_2 formation by *R*34

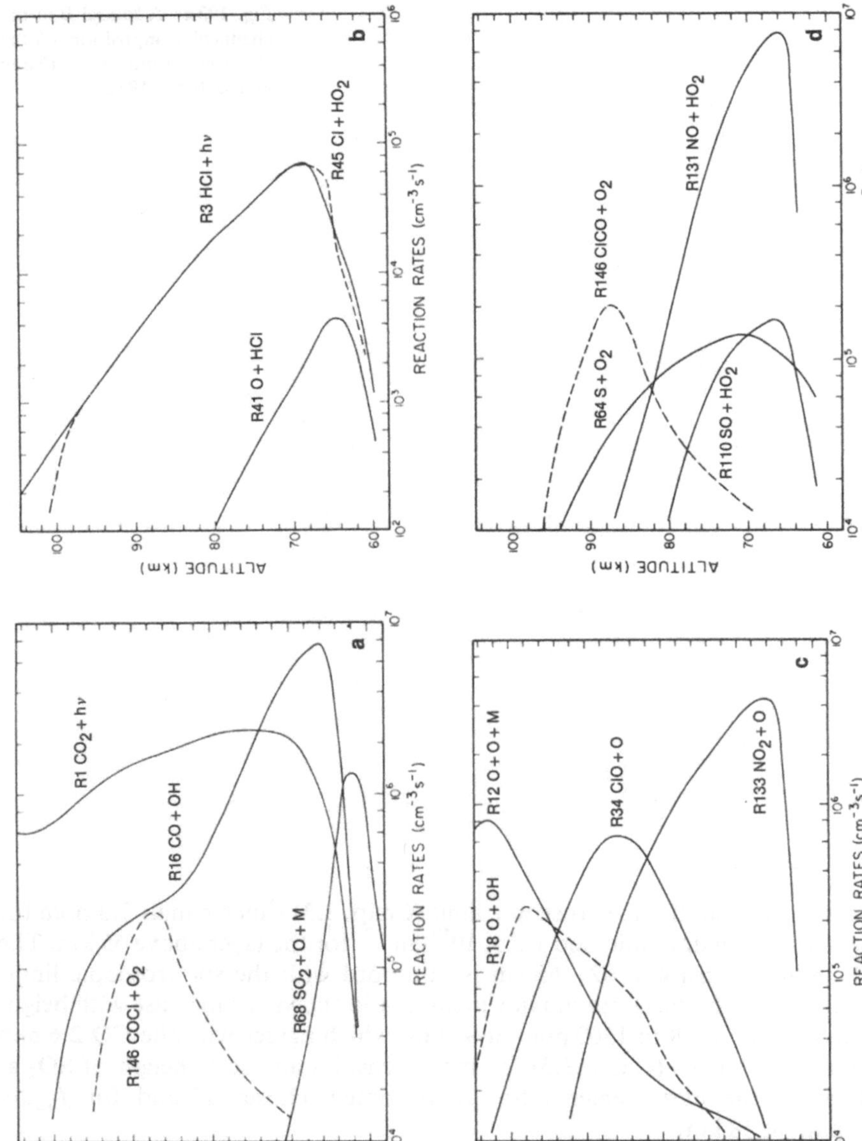

a
R1 CO2 + hν
R16 CO + OH
R146 COCl + O2
R68 SO2 + O + M
b
R3 HCl + hν
R45 Cl + HO2
R41 O + HCl
c
R12 O + O + M
R18 O + OH
R34 ClO + O
R133 NO2 + O
d
R131 NO + HO2
R146 ClCO + O2
R64 S + O2
R110 SO + HO2
ALTITUDE (km)
REACTION RATES (cm-3 s-1)

Fig. 194a–d

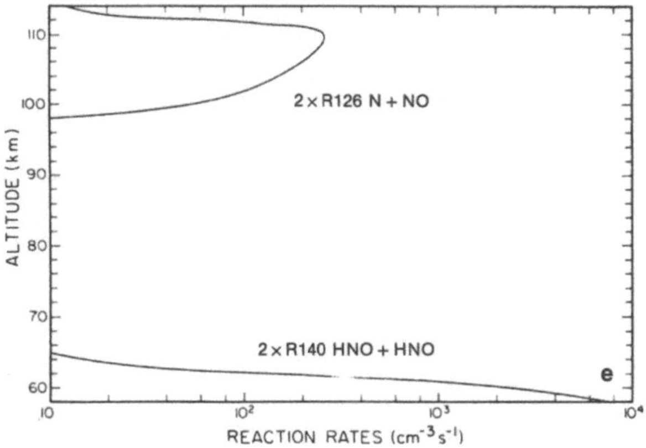

Fig. 194a–e. Model *B* by Yung and DeMore (1982): **a** sources and sinks of free oxygen; **b** HO_x and ClO_x; **c, d** O_2; **e** NO_x

O + ClO while odd hydrogen forms CO_2 rather than O_2. So large amounts of O_2 and CO are eliminated in model *A*.

Photochemical transformation of H_2 to H_2O via *R*40 OH + HCl occurs in the mesosphere. With *R*16 CO + OH being the main process of CO_2 formation, the water vapor production equals

$$\int k_{40}[\mathrm{HCl}]\,[\mathrm{OH}]\,dh = \alpha \int k_{16}[\mathrm{CO}]\,[\mathrm{OH}]\,dh = \alpha S\,.$$

Here, $\alpha = k_{40}\,[\mathrm{HCl}]/k_{16}\,[\mathrm{CO}] \simeq 10^{-2}$ and $S \simeq 0.6 \times 10^{13}\ \mathrm{cm^{-2}\,s^{-1}}$ is the total CO_2 photolytic rate. Therefore, the upward flux of H_2 from the lower atmosphere is equal to $\sim 10^{11}\ \mathrm{cm^{-2}\,s^{-1}}$. Variations of the f_{H_2} in the range of 10 to 50 ppm do not significantly affect model *A*.

Model B (Figs. 193 and 194) employs $f_{H_2} = 0.5$ ppm and $f_{NO} = 30$ ppb at 58 km and the full list of reactions except the heterogeneous sink of H* and Cl* via *R*67, 113. *K* is assumed to equal $10^5\ \mathrm{cm^2\,s^{-1}}$ down to 70 km with further decrease to $2 \times 10^4\ \mathrm{cm^2\,s^{-1}}$ at 58 km.

Of interest is the nitrogen chemistry of this model. $f_{NO} = 30$ ppb adopted in this model is the upper limit of the nitric oxide abundance in the cloud layer. This value coincides with that obtained in Chap. 4.6 for the energy release by lightning based on the V9, 10 optical measurements with the sink of odd nitrogen in the lower atmosphere disregarded.

Even an overestimated value of the atomic nitrogen flux at 110 km equal to $-10^{10}\ \mathrm{cm^{-2}\,s^{-1}}$ which is used in model *B* (the total downward flux of N at the night side is $-3 \times 10^9\ \mathrm{cm^{-2}\,s^{-1}}$, and ~10% of NO airglow occurs below 110 km, i.e., $\phi_N \simeq -3 \times 10^8\ \mathrm{cm^{-2}\,s^{-1}}$ at 110 km) cannot account for a substantial increase of odd nitrogen amount below 80 km.

Except for the reaction N + NO in the upper atmosphere, four sinks of odd nitrogen can be regarded as significant. The most important is predissociation of NO with the subsequent reaction of N and NO. Formation of HNO leads to

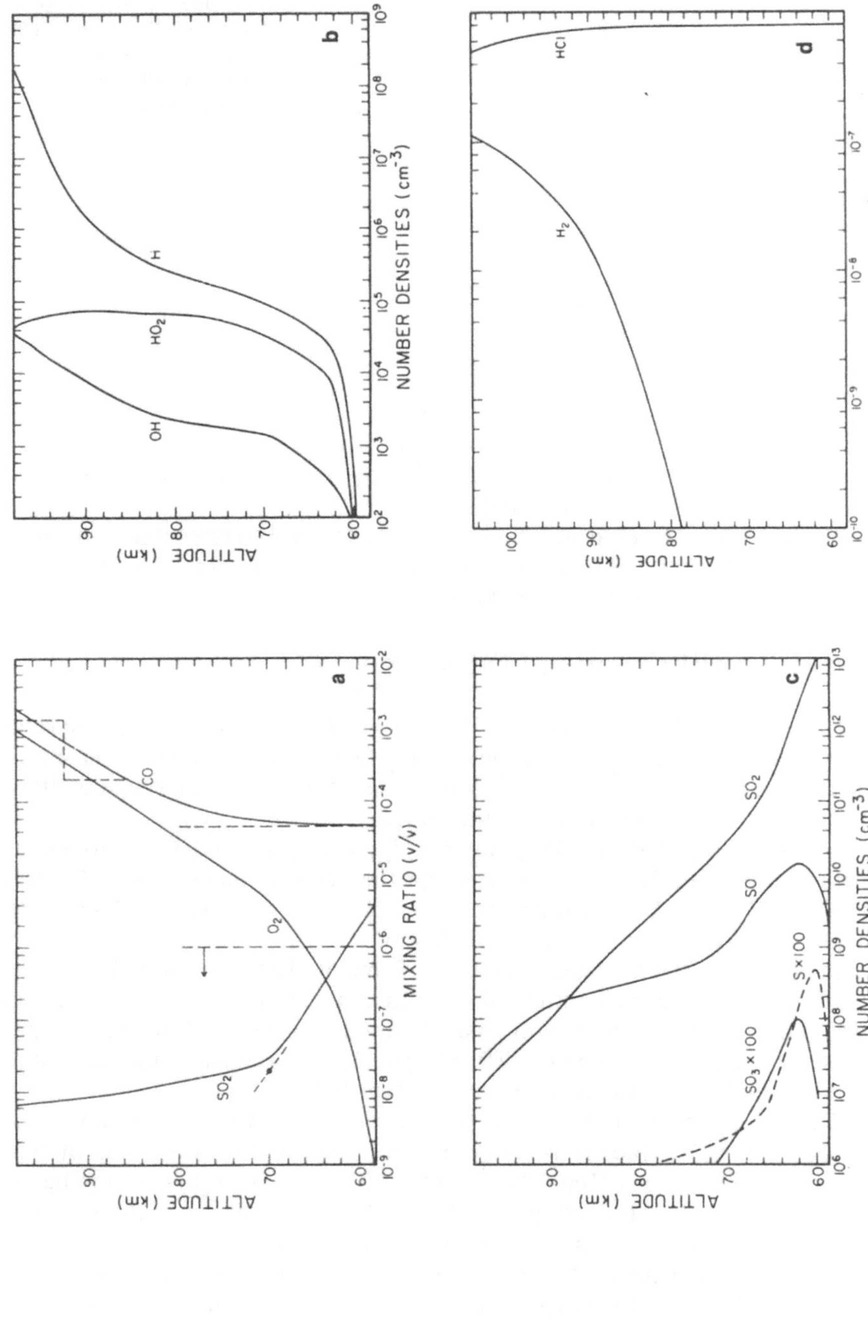

a
ALTITUDE (km)
MIXING RATIO (v/v)
CO
O2
SO2
b
ALTITUDE (km)
NUMBER DENSITIES (cm-3)
OH
HO2
H
c
ALTITUDE (km)
NUMBER DENSITIES (cm-3)
SO2
SO
SO3 × 100
S × 100
d
ALTITUDE (km)
MIXING RATIO (v/v)
HCl
H2

Fig. 195a – d

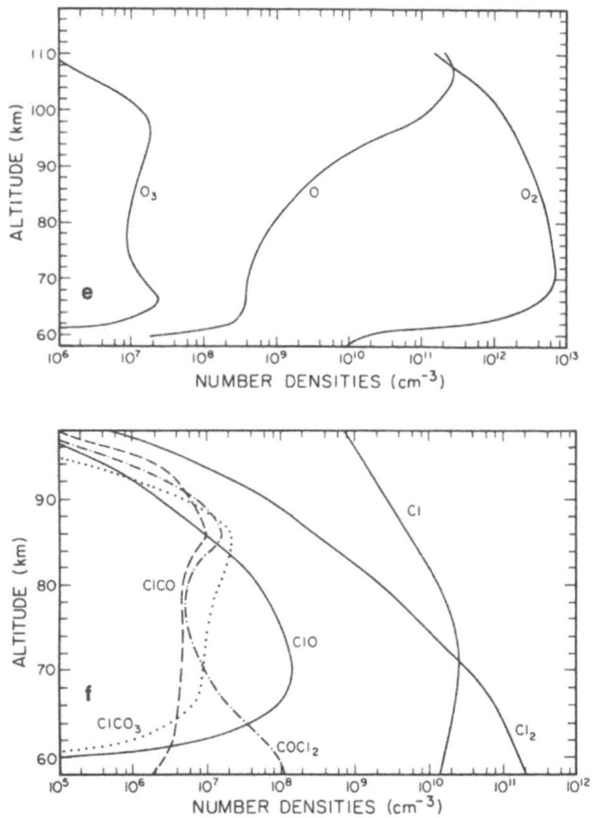

Fig. 195a–f. Model *C* of the chemical composition of the Venusian atmosphere (Yung and DeMore 1982)

$$HNO + HNO \rightarrow N_2O + H_2O$$
$$N_2O + h\nu \rightarrow N_2 + O(^1D)$$

and can remove NO_x in the cloud layer. The next possibility is the reaction $NO + NO \rightarrow N_2O + O$ with $k = 8 \times 10^{-10} e^{-2900/T} cm^3 s^{-1}$ which is extremely slow ($\tau \simeq 10^5$ yr) even near the surface. However, it may be catalyzed by the surface. Formation of nitrates on the surface can also be considered, but Prinn argued the instability of simple nitrates at high temperatures, thus eliminating this sink. So despite odd nitrogen being thermochemically unstable in the lower atmosphere, mechanisms of its effective removal are not known. Altitude profiles of various forms of odd nitrogen are shown in Fig. 193.

The H_2 mixing ratio adopted in model *B* is close to that in the model of Winick and Stewart. Therefore, $[H^*]/[Cl^*] \simeq 0.1$ here. With $[Cl^*] \simeq 10^8 cm^{-3}$ the COCl cycle is unimportant. Atomic oxygen produced by CO_2 photolysis forms O_2 via the three body reaction (~15%), *R*34 O+Cl (~15%), and the rest by *R*133 NO_2+O. The breaking of the O–O bond occurs through *R*131

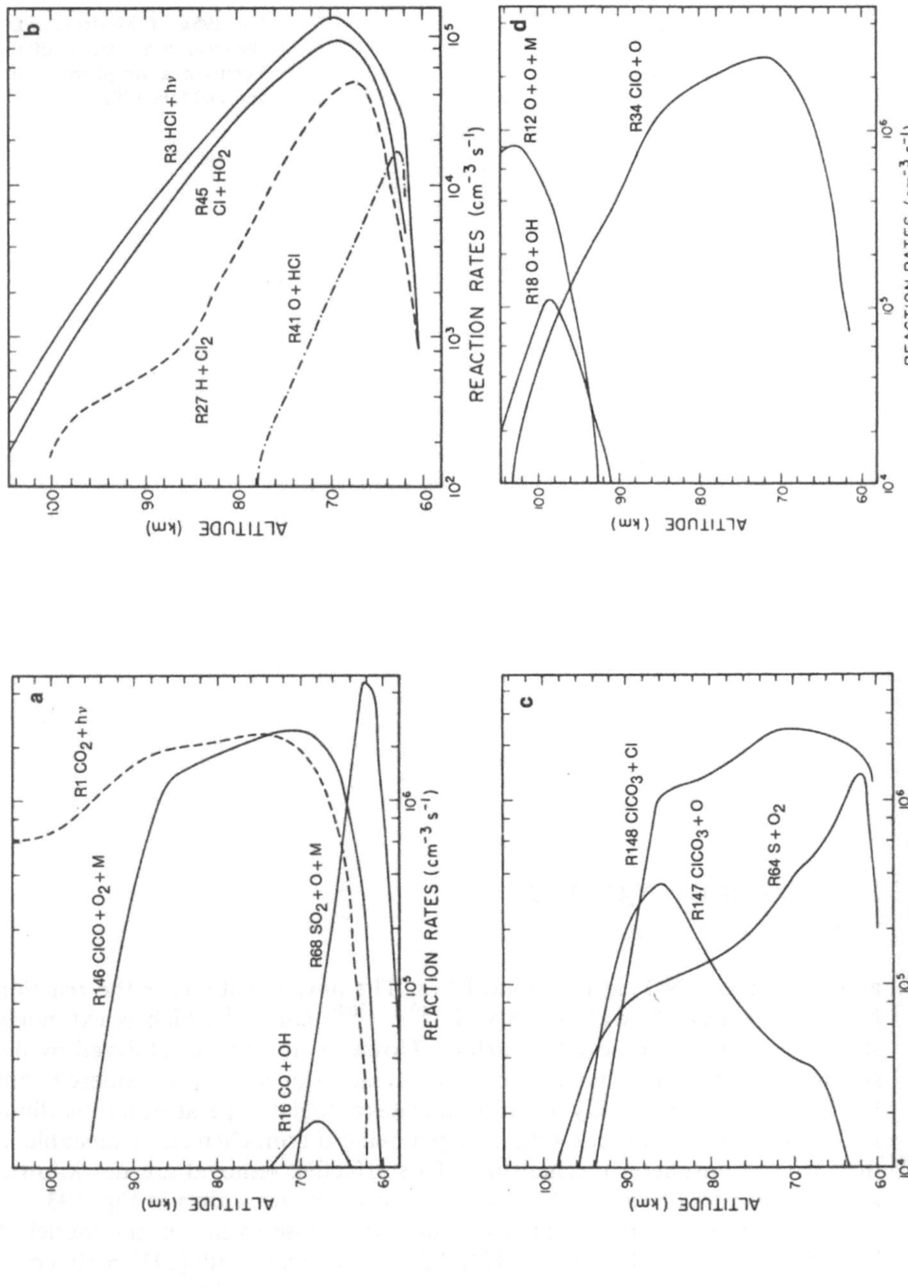
a
R1 CO2 + hν
R146 ClCO + O2 + M
R68 SO2 + O + M
R16 CO + OH
ALTITUDE (km)
REACTION RATES (cm^-3 s^-1)
b
R3 HCl + hν
R45 Cl + HO2
R27 H + Cl2
R41 O + HCl
c
R148 ClCO3 + Cl
R147 ClCO3 + O
R64 S + O2
d
R12 O + O + M
R18 O + OH
R34 ClO + O

Fig. 196a–d

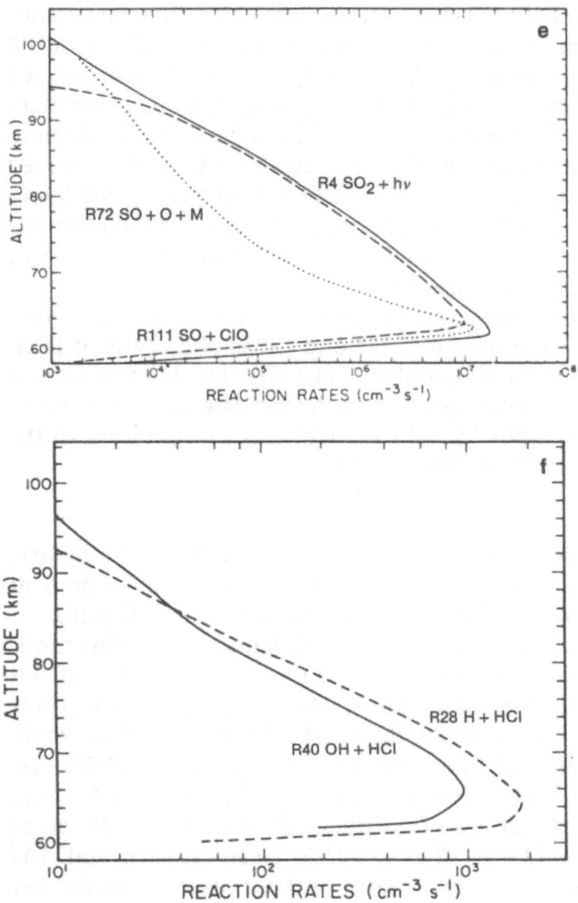

Fig. 196a–f. Model *C* by Yung and DeMore (1982): **a** sources and sinks of free oxygen; **b** HO_x and ClO_x; **c, d** formation of O_2; **e** equilibrium between SO_2 and SO; **f** production rates of H_2 and H_2O

$NO+HO_2$ with a subsequent formation of CO_2 by *R*16 CO+OH. The sulfuric acid precipitation flux is equal to $0.9\times10^{12}\,cm^{-2}s^{-1}$, while it should be $2.5\times10^{11}\,cm^{-2}s^{-1}$ with *K* assumed. As in model *A*, the upward H_2 flux is $10^{11}\,cm^{-2}s^{-1}$. The O_2 column density exceeds the spectroscopic limit by a factor of 15.

Model C (Figs. 195 and 196) suggests only photochemical production of molecular hydrogen as do our calculations. The $f_{HCl}=8\times10^{-7}$ in it and the nitrogen compounds are ignored. At the lower boundary, velocities of O and O_2 are assumed to be $0.65\,K/H$ rather than K/H. Odd hydrogen number densities are so small in this model that its role can be neglected. Free chlorine formed mainly as Cl_2 is far in excess of corresponding values in models *A* and *B* and has $f_{Cl_2}=25$ ppb at 58 km. However, this mixing ratio is too small (by a factor of 30) to produce substantial absorption in the near UV region.

*R*34 O + ClO is predominant in O_2 formations, while the three body association prevails above 95 km. O_2 destruction occurs almost totally in *R*148 $ClCO_2$ + Cl with some contribution of *R*64 S + O_2. The molecular oxygen column density is 1.8×10^{19} cm^{-2}, i.e., by an order of magnitude larger than the spectroscopic limit. CO_2 formation by the COCl cycles is in equilibrium with CO_2 photolysis at 65 to 90 km, while the excess of CO and O formed above 90 km produces the downward flow of CO and H_2SO_4 at the lower boundary. The equilibrium of CO at 65 to 90 km is not of photochemical nature since the mixing time is shorter than the photochemical time in this region (2×10^6 and 10^7 s at 80 km, respectively).

The molecular hydrogen mixing ratio becomes larger with height and equals 10^{-7} at 100 km. The main source of H* and Cl* is photolysis of HCl, while their sink occurs in *R*45 Cl + HO_2 and *R*27 H + Cl_2. The H_2O formation by *R*40 OH + HCl produces the hydrogen sink in the mesosphere with the subsequent absorption of water vapor by sulfuric acid and precipitation to the lower atmosphere. The net result of these processes is

$$2HCl + CO_2 + h\nu \rightarrow H_2O + CO + Cl_2 .$$

Free chlorine flux constitutes -3×10^9 cm^{-2} s^{-1} at the lower boundary. Sulfur dioxide number densities fit the observations for $f_{SO_2} = 4$ ppm at 58 km, and the SO mixing ratio is close to the spectroscopic limit (2 ppb).

It is evident that *R*148 $ClCO_3$ + Cl forms ClO which may react with atomic oxygen. This prevents the destruction of the O – O bond and the net of this cycle is CO + O → CO_2, as was discussed above. Therefore, to destroy O_2 the photolysis of ClO (*R*103) or its reaction with SO (*R*111) should follow *R*148.

Comparison of the Models. The amounts of CO, O_2, SO_2, and SO predicted by the models have been already discussed. The $O_2(^1\Delta_g)$ 1.27 μm airglow may serve as an additional test of adequacy for the models. It can be excited by processes *R*12 O + O + *M*, *R*15 O + HO_2, *R*30 O_3 + Cl, and *R*34 O + ClO. Yung and DeMore assume equal rates of these reactions at the day and night sides and their yield of $O_2(^1\Delta_g)$ to be two-thirds. The ozone photolysis enhances the oxygen dayglow. Then calculated day and night airglow intensities are 1.2 – 0.8, 1.5 – 0.8, and 0.9 – 0.76 *MR* for the models *A, B, C*, respectively, with a CO_2 quenching rate coefficient of $k_{CO_2} = 3 \times 10^{-20}$ cm^3 s^{-1}. All the models agree with the measurements; however, model *C* gives the best fit to the observation with $k_{CO_2} = 10^{-20}$ cm^3 s^{-1}: 1.5 and 1.3 *MR*. The rate coefficient of the three body association (*R*12) is larger than that in our calculations by a factor of 6 at 200 K. Therefore, the contribution of *R*12 appears to be more important and is responsible for about one-half of the airglow intensity. However, some laboratory measurements and terrestrial airglow observations indicate that the $O_2(^1\Delta_g)$ excitation occurs through energy transfer by collisions from other excited states. This problem is beyond the scope of our book.

It is significant that model *C* has the advantage over the other models in that it can explain diurnal variations of CO at 80 km with a midnight mini-

Table 40. Some reactions related to S_2O, $(SO)_2$, $H_2S_2O_2$, and $H_2S_2O_3$ and their rate coefficients (Yung and DeMore 1982)

Reaction	Rate coefficient
$S_2O+h\nu \rightarrow S+SO$	3.0×10^{-7}
$2SO+M \rightarrow (SO)_2+M$	4.4×10^{-31}
$S_2O+O \rightarrow 2SO$	1.5×10^{-12}
$S_2O+S \rightarrow S_2+SO$	1×10^{-13}
$S_2O+H_2O \rightarrow H_2S_2O_2$	1×10^{-20}
$(SO)_2+M \rightarrow 2SO+M$	9.0×10^{-23}
$(SO)_2+O \rightarrow S_2O+O_2$	3.3×10^{-14}
$(SO)_2+SO \rightarrow S_2O+SO_2$	3.3×10^{-14}
$(SO)_2+H_2O \rightarrow H_2S_2O_3$	1×10^{-20}

mum. Actually, in model *C*, CO is close to photochemical equilibrium at 80 km, while in model *A* and *B* the region of photochemical equilibrium is very narrow and shifted to 75 km.

The main reason for differences among the models is the assumed molecular hydrogen amounts, and the reliability of any model depends on the foundation of the adopted f_{H_2}. The thermochemical equilibrium constant of the reaction $H_2+CO_2 \rightleftarrows H_2O+CO$ depends strongly on temperature: $K=f_{H_2}/f_{H_2O}f_{CO}$ equals 6 at 750 K, 130 at 500 K, and 10^5 at 300 K. Therefore, f_{H_2} can be as large as ~10 ppm if thermochemical equilibrium occurs up to ~40 km. However, this possibility, which is very important for model *A*, should be rejected, because the f_{H_2} increase should inevitably lead to the f_{H_2O} decrease contrary to observations. Besides, chemical lifetimes exceed that of mixing. So some nonequilibrium chemistry should be taken into account to yield $f_{H_2} \simeq 10$ ppm, which can be initiated by photolysis of elemental sulfur by visible light. We note that $f_{H_2}=3\times10^{-9}$ (Chap. 4.7, Table 36) for thermochemical equilibrium near the surface. This value seems to be valid in the total lower atmosphere, therefore, $[H^*]/[Cl^*] \simeq 0.1\ [H_2]/[HCl] \simeq 3\times10^{-4}$ favors model *C*.

One may note that in all models, O_2 destruction cycles are not efficient enough, thus, preventing formation of elemental sulfur. The reaction of

$$S+CO_2 \rightarrow SO+CO$$

also suppresses its formation, though in our opinion it should be neglected. Yung and DeMore analyze qualitatively the reactions with the participation of S_2, S_2O, $(SO)_2$, $H_2S_2O_2$, and $H_2S_2O_3$ (Table 40). Approximate production rates of $H_2S_2O_2$, $H_2S_2O_3$, and S_2 of these reactions are 10^{10}, 3×10^9, and 10^9 $cm^{-2}s^{-1}$, respectively. They make a further formation of H_2S and H_2 possible:

$$H_2S_2O_3+H_2O \rightarrow H_2SO_4+H_2S$$

$H_2S + h\nu \rightarrow H + HS \quad (\lambda < 317\,\text{nm})$

$H + H_2S \rightarrow H_2 + HS$.

So this scheme may explain photochemical formation of reduced components, such as H_2S, H_2, and S_2. However, rates of the H_2 to H_2O transformation equal $\sim 10^{11}\,\text{cm}^{-2}\,\text{s}^{-1}$, i.e., exceed that of the $H_2S_2O_3$ formation by a factor of 30. A substantial development of this scheme is necessary to provide rather large H_2 amounts. One should bear in mind that, apart from the photochemical sink in the mesosphere, H_2 is also removed by thermochemical reactions in the lower atmosphere.

Model *C* can be valid only in the case of extremely low, perhaps less than 1 ppb, abundances of odd nitrogen. Odd nitrogen in model *C* would accelerate O_2 formation by the reaction *R*133 $NO_2 + O$, while O_2 removal by *R*131 $NO + HO_2$ would be negligible due to low number densities of odd hydrogen. Therefore, model *C* can be accepted only assuming that lightning does not occur in the Venusian atmosphere or a sink for NO_x formed by lightning occurs near the surface.

Photochemistry and Atmospheric Evolution. The water vapor mixing ratio equal to 10^{-4} corresponds to a column abundance of $6\,\text{g cm}^{-2}$, i.e., less by five orders of magnitude than that on the Earth. The reason for the change of f_{H_2O} from 20 ppm to ~ 200 ppm near the surface and at 50 km, respectively, is unknown therefore every value is this range can be assumed for the estimate of hydrogen amount on Venus taking into account the possible presence of H_2 and H_2S. Pollack and Yung (1980) argued that even carbonaceous meteorites captured by the planet are a source of water in an amount no less than 1% of the terrestrial value. Therefore, one can assume that at some stages of the atmospheric evolution the water vapor abundance on Venus exceeded the present one by three orders of magnitude. Then the thermochemical equilibrium molecular hydrogen mixing ratio could be ~ 10 ppm, which corresponds to model *A*. According to Kumar et al. (1981, see Chap. 4.4) the hydrogen escape rate should be $10^8\,\text{cm}^{-2}\,\text{s}^{-1}$ under these conditions. As Venus lost water vapor, the Venusian atmosphere passed an intermediate stage which can be described by model *B* with $f_{H_2} = 0.5$ ppm and water vapor abundance 50 times higher than the present one. With the further loss of water, the atmosphere came into a stage described by model *C*, the preferred model for the present atmosphere. Despite the very low H_2 abundance in the lower atmosphere, this model supplies by photochemical processes substantial amounts of H_2 and H in the upper atmosphere which form an escape flux of $10^7\,\text{cm}^{-2}\,\text{s}^{-1}$. Further loss of water vapor will result in the enrichment of the atmosphere with CO, Cl_2, and $COCl_2$. Problems of atmospheric evolution will be discussed in more detail in the next section.

4.16 Loss of Water from Venus and Its Atmospheric Evolution

Hydrogen and oxygen escape on Venus was considered in Chap. 4.4 as calculated by McElroy et al. (1982a, b). Their analysis implies the escape rates of H and O equal to 10^7 and $0.5 \times 10^7\ cm^{-2}\,s^{-1}$, respectively, while that of D is negligible. The escape rates being assumed proportional to water vapor amount, the present deuterium to hydrogen ratio, $D/H \simeq 10^{-2}$, results in an initial (i.e., which refers to the end of the acception of the planet) ratio, D/H, of 5×10^{-5} and an initial water vapor amount of $\sim 800\ g\ cm^{-2}$. Some aspects of this problem as well as the assumption made were considered in more detail in papers which will be discussed below.

A hydrogen dominated planetary thermosphere can be considered as a limiting case of very large water vapor amounts in the atmosphere. In the present atmosphere the mesopause is located at 110 km. Furthermore, a boundary exists at about 150 km between the isothermal atmosphere above which corresponds to optical thin conditions for solar UV radiation and the atmosphere with high temperature gradients below. It coincides with a boundary between the CO_2 and atomic oxygen dominated atmospheres. On the day side the exobase altitude is 190 km. Temperature gradients in the 110 – 150 km region are equal to a ratio of solar EUV heating rate to atmospheric thermal conductivity. Both values do not change markedly for the hydrogen dominated atmosphere, so dT/dz should be almost the same. However, the optical thin region should be located much higher due to the large scale height of hydrogen. Then its altitude equals $110 + (150 - 110)\ m_{CO_2}/m_H \simeq 1700$ km, its temperature is $170 + (290 - 170)\ m_{CO_2}/m_H \simeq 5000$ K. The position of the exobase is $1700 + (190 - 150)\ m_O/m_H \times 5000\ K/290\ K \simeq 12000$ km. Gravitational acceleration is much lower here than at $h \simeq 200$ km, therefore, the exobase should move to infinity. With the kinetic energy of hydrogen atoms greatly exceeding their gravitational energy, i.e., $(\gamma M m_H)/r \ll 3/2\,kT$, a hydrodynamic flow of the atmospheric gas to space forms. All this demonstrates that in this case the conventional approach which supposes a static atmosphere with minor deviations caused by the Jeans' escape is unacceptable.

Hydrodynamic flow was discussed in a number of papers. We shall follow that of Hunten (1982). Usually the Jeans' escape is given in the form

$$\phi_J = n_c v_J = n_c \frac{BU}{2\pi^{1/2}} (1 + \lambda_c) e^{-\lambda_c}$$

where $\lambda = (\gamma M m)/kTr$ is the ratio of gravitational energy to kT, $U = (2kT/m)^{1/2}$ is the mean molecular velocity, $n_{ct} H_{ct} \sigma = 1$ defines the critical level with $\sigma \simeq 3 \times 10^{-15}\ cm^2$ being a collisional cross-section and the subscripts c and ct referring to the escaping component and total atmospheric components at the critical level, respectively. B is the constant close to unity and indicates deviations from the Maxwell-Boltzmann distribution caused by escaping molecules.

It can be shown that in the case of hydrodynamic flow in the isothermal atmosphere the expanding gas reaches speed of sound at $\lambda = 2$. Thermal energy and gravitational energy of monoatomic gas are equal at $\lambda = \frac{3}{2}$, for diatomic gas at $\lambda = \frac{5}{2}$, therefore, the sonic level is between them. This demonstrates a similarity of the kinetic theory approach of Jeans and the hydrodynamic methods which were proposed by Walker (1983) as limiting cases of a common theory. He proved that if the sonic level is higher than the exobase, i.e., $\lambda_c > 2$, the Jeans' approach should be used; if lower, the hydrodynamic flow (blowoff) occurs. The latter equals $n\,V_s$ where V_s is the speed of sound and n is the number density at the sonic level $\lambda = 2$.

The blowoff of light gases should drag along heavier gases. In the very rapid blowoff they attain velocities close to those of light gases.

Some fractionation of elements and isotopes occurs caused by the hydrodynamic flow.

Isothermal expansion and blowoff require an infinite power input. An energetic limit of the thermospheric blowoff can be readily established by comparing the solar UV radiation absorbed and then released in heating and the gravitational energy to be overcome:

$$\phi \leqslant \frac{I \varepsilon r_1^2}{2 \gamma M m r_0} .$$

Here I is the solar EUV ($\lambda < 100$ nm) radiation, ε is the heating efficiency, r_1 is the radius of the level $\tau = 1$ in the EUV region, and r_0 is the planet radius. On Venus, this limit constitutes $2 \times 10^{12}\,\mathrm{cm^{-2}\,s^{-1}}$ for atomic hydrogen (Watson et al. 1981).

A diffusion limit on Jeans' and nonthermal escape fluxes can be readily obtained:

$$\phi_i \leqslant \frac{D n_i}{H} = \frac{2 \times 10^{19} f_i}{H}\,\mathrm{cm^{-2}\,s^{-1}}$$

where $D \simeq (2 \times 10^{19})/n_a\,\mathrm{cm^2\,s^{-1}}$ is the diffusion coefficient and f_i is the mixing ratio, H is the atmospheric-scale height in cm. All values refer to the homopause.

Thermospheric Models for Different Water Vapor Mixing Ratios at the Cold Trap. An extensive study of hydrodynamic flow in the earlier Venusian atmosphere was made by Kasting and Pollack (1983). This study gives very interesting results for the atmospheric evolution.

First of all they define atmospheric dynamics, chemistry, and thermal regime to perform self-consistent calculations of the atmospheric structure. After that the hydrogen escape flux becomes a subject of calculation. Then a relationship between the water vapor mixing ratios at the cold trap which is assumed to be the lower boundary of their model and in the lower atmosphere is analyzed. As a final step D/H variations due to preferential escape of H in some processes are discussed.

We shall not consider hydrodynamic equations and their transformation used in the paper. The vertical distribution of chemical constituents was described by coupled continuity and flux equations for a one-dimensional time-independent atmosphere with spherical symmetry. The eddy mixing coefficient was given by $K = 1.4 \times 10^{13} M^{-1/2}$ cm^2 s^{-1} in the range of 3×10^5 to 5×10^7 cm^2 s^{-1}. The molecular diffusion coefficients were specified for heavy and light (H, D, H_2, HD, He) constituents. Conventional chemistry of H_2O and CO_2 photolytic products defined the thermospheric composition. Rate coefficients of reactions with D compounds were taken as the same as for H components with a correction for reduced masses. The mass mixing ratios of CO_2, N_2, H_2O, and HDO and velocities of CO, O, O_2, H, H_2, D, and HD equal to K/H_a represented the lower boundary conditions. The H_2O mass mixing ratio was varied from 5×10^{-7} to 0.46, that of He was 10^{-10}, and $HDO/H_2O = 10^{-4}$. Production and loss rates along with the velocities of the heavy components were taken equal to zero at 600 km. The velocities of light components were assumed equal to a flow velocity at $\lambda = 0$.

The lower boundary conditions imply the conversion of H_2, O_2, and CO deposited into the lower atmosphere to H_2O and CO_2 with a neutral oxidation of the lower atmosphere. Watson et al. (1983) suggest that water vapor reacts with surface rocks in the following way

$$H_2O + 2FeO \rightarrow Fe_2O_3 + H_2$$

which may release copious amounts of molecular hydrogen with a subsequent increase of the hydrogen escape rate. Therefore, hydrogen escape rates calculated by the model of Kasting and Pollack should be considered as moderate.

Solar maximum fluxes (Mount et al. 1980) were used from 120 to 200 nm, while those for minimum solar activity multiplied by a factor of up to 16 were used below 120 nm. This factor reflects a presumably more intense solar UV radiation in the past than today. Calculations were performed for daytime mean conditions with a heating efficiency of 0.15, which is the lower limit estimated by Fox and Dalgarno (1981, see Chap. 4.1). A simplified version of radiative transfer with infrared cooling in the 15 μm CO_2 and 50 μm H_2O bands was developed for model calculations. The procedure adopted being applied to the present H_2O mass mixing ratio at the cold trap (5×10^{-7} or 1.2 ppm by volume at 90 km) results in the atmospheric temperature profile close to the measured one (Fig. 197). The calculated atmospheric composition (Fig. 198) is in reasonable agreement with observations as well.

Curves in Fig. 198 can be transformed to number densities with the temperature profile (Fig. 197) and the known atmospheric density at the cold trap. The temperature at the cold trap was taken equal to 170 K and independent of the water vapor mixing ratio. Then the atmospheric mass density at the cold trap is $\varrho_0 = \dfrac{0.95 \times 10^{-11}}{C_0(H_2O)}$ g cm^{-3} where $C_0(H_2O)$ is the water vapor

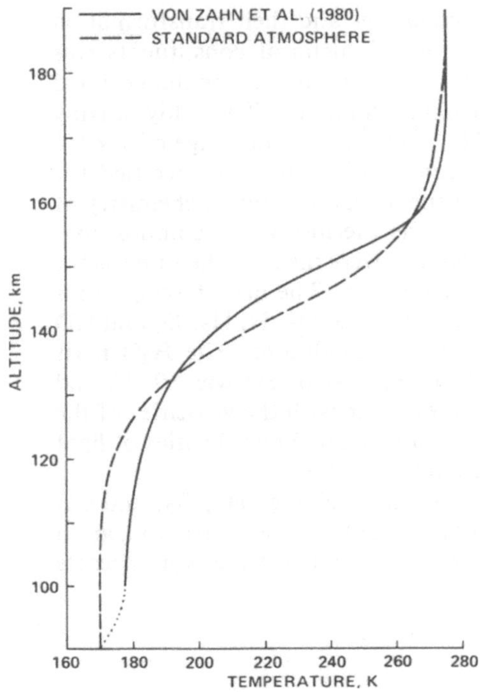

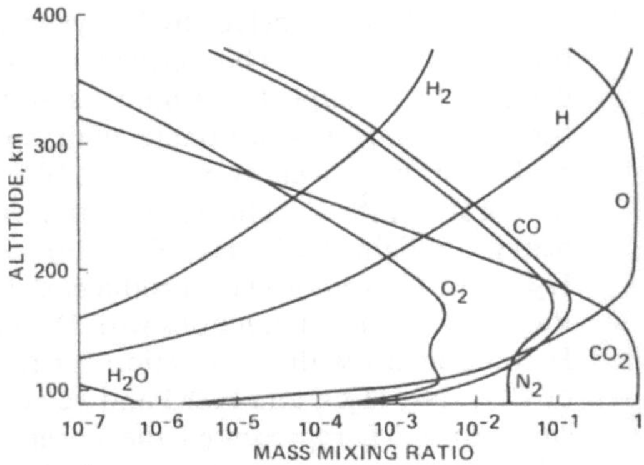

▲ **Fig. 198.** Mass mixing profiles for the present atmosphere (Kasting and Pollack 1983)

◄ **Fig. 197.** Calculated (*dashed curve*) and observed (*solid curve*) temperature profiles for the present Venusian atmosphere (Kasting and Pollack 1983)

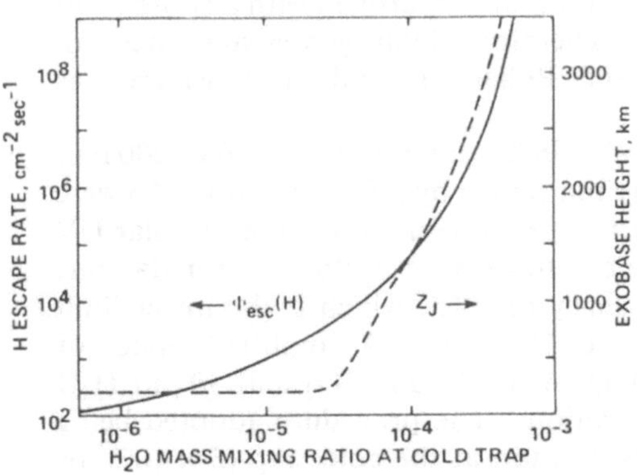

Fig. 199. Variation of the Jeans' escape rate (*solid curve*) and exobase height (*dashed curve*) as a function of the H_2O mass mixing ratio at the cold trap (Kasting and Pollack 1983)

mass mixing ratio at the lower boundary (the cold trap). In the present atmosphere $\varrho_0 \simeq 10^{-6}\,g\,cm^{-3}$ at 90 km because the hygroscopic trap by sulfuric acid near the cloud tops rather than condensation at 90 km determines $C_0(H_2O)$. However, this hygroscopic trap should be less effective with increasing water vapor amounts. The position of the cold trap was approximately evaluated for different $C_0(H_2O)$ to be in the range of 90 to 132 km for $C_0(H_2O)$ varying from 5×10^{-7} to 0.5.

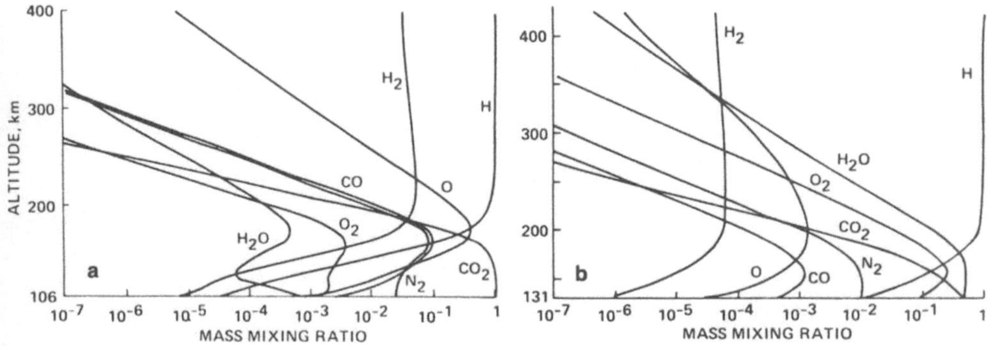

Fig. 200 a, b. Mass mixing ratio profiles for the H_2O mass mixing ratio at the cold trap equal to (a) 7×10^{-4} and (b) 0.46 (Kasting and Pollack 1983)

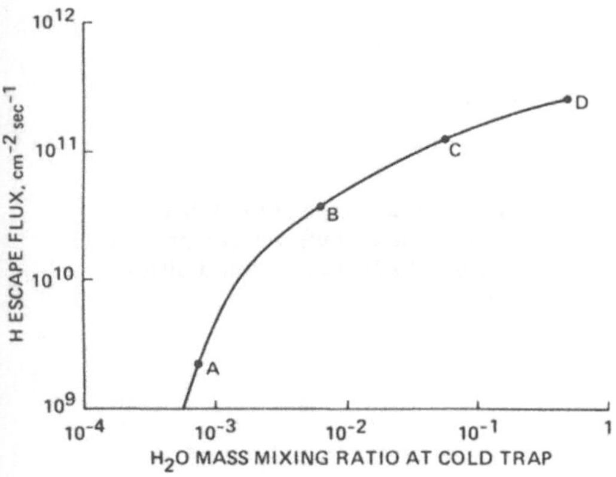

Fig. 201. Hydrodynamic escape flux of H normalized to the ground level as a function of the H_2O mass mixing ratio at the cold trap. The points *A*, *B*, *C*, and *D* correspond to the four cases described in the text (Kasting and Pollack 1983)

Variation of the Jeans' escape rate normalized to the planet radius and the altitude of the exobase, z_J, as a function of $C_0(H_2O)$ are shown in Fig. 199. z_J is determined by the atomic oxygen and remains constant up to $C_0(H_2O) = 3 \times 10^{-5}$. Then the atomic hydrogen determines the exobase height which results in a steep increase of both the escape rate and the exobase height. For $C_0(H_2O) \gtrsim 5 \times 10^{-4}$ the Jeans' escape formulation gives the exospheric temperatures of thousands of degrees and the exobase height moving out to infinity. Therefore, in this case the hydrodynamic approach should be used. Figure 200 demonstrates the calculated atmospheric composition for $C_0(H_2O) = 7 \times 10^{-4}$ and 0.46, and Fig. 201 shows the hydrodynamic escape flux versus $C_0(H_2O)$. Temperature, mass density, and bulk velocity profiles for the points *A, B, C, D* (see Fig. 201) are given in Fig. 202.

The atmospheric composition for case *A* significantly differs from the present atmosphere in that the atomic oxygen amount decreases with the increase of those of H and O_2. This is the result of the reactions

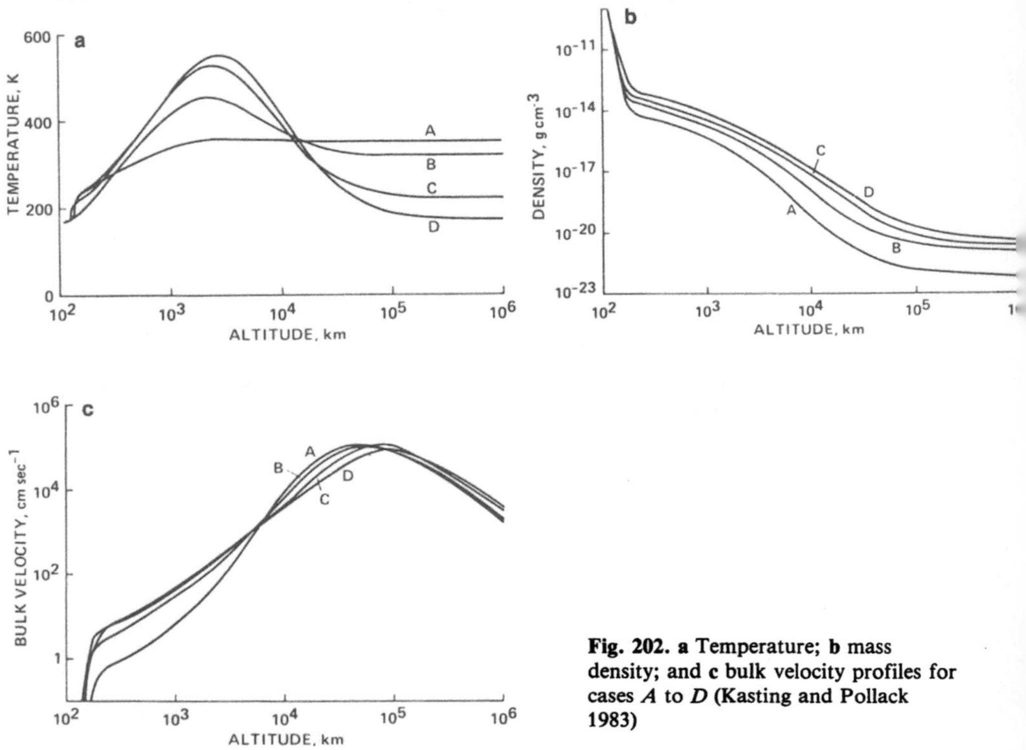

Fig. 202. **a** Temperature; **b** mass density; and **c** bulk velocity profiles for cases *A* to *D* (Kasting and Pollack 1983)

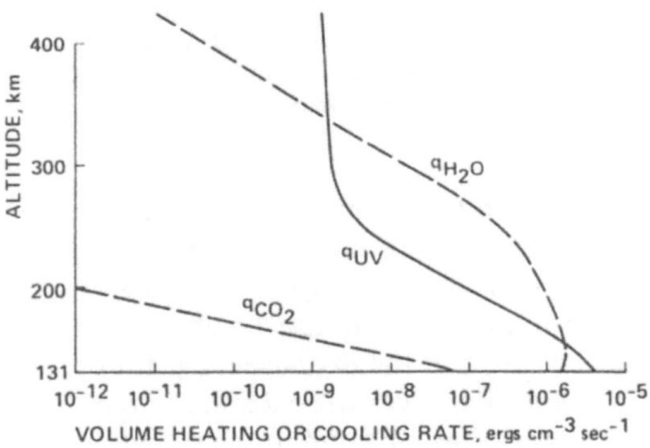

Fig. 203. Heating (*solid curve*) and cooling (*dashed curves*) rates for case *D* (Kasting and Pollack 1983)

$$H_2O + h\nu \rightarrow H + OH$$
$$OH + O \rightarrow O_2 + H\ .$$

For larger $C_0(H_2O)$ corresponding to case *D* the atomic oxygen amounts become very low and the enhanced OH formation results in a decrease of $[H_2]$ due to the reaction

$$H_2 + OH \rightarrow H_2O + H\ .$$

Table 41. Effect of enhanced EUV heating for case D (Kasting and Pollack 1983)

Relative EUV flux	$\phi_{esc}(H)$ ($cm^{-2}\,s^{-1}$)	F ($g\,ster^{-1}\,s^{-1}$)	f
1	2.7×10^{11}	1.6×10^5	4.9×10^{-7}
2	6.6×10^{11}	4.0×10^5	6.2×10^{-5}
4	1.3×10^{12}	8.0×10^5	3.5×10^{-2}
8	2.3×10^{12}	1.4×10^6	0.54
16	3.5×10^{12}	2.1×10^6	0.65

Atmospheric heating and cooling rates for case *D* are depicted in Fig. 203. The heating rate obtained should be considered as minimal due to low heating efficiency, 0.15, and solar minimum EUV fluxes. Table 41 demonstrates the effect of the enhanced heating rate.

The D/H fractionation factor is defined as the ratio of deuterium and hydrogen escape fluxes divided by the ratio of their total atmospheric amounts, $f = \phi_D/\phi_H \times (H/D)$. In the case of hydrodynamic flow the fractionation factor equals the ratio of *D/H* at infinity to that at the cold trap assumed to be 5×10^{-5}. For a low deuterium escape probability this factor is small. Calculations show that a considerable fractionation occurs with practically no deuterium escape ($f = 8 \times 10^{-9}$ and 5×10^{-7} for cases *A* and *D*, respectively). However, for a larger heating rate this fractionation becomes less pronounced (Table 41).

Dependence Between the Water Vapor Mixing Ratios at the Cold Trap and in the Lower Atmosphere. This problem is related to atmospheric dynamics and thermal regime. It was considered approximately by Kasting and Pollack (1983) and in more detail by Watson et al. (1983). Figure 204 reproduces this dependence for three values of the pressure, p_b, at the bottom of the moist convective layer. The curves do not differ significantly. Data of Fig. 204 result in a dependence of a hydrogen thermal escape rate on the H_2O mass mixing ratio in the lower atmosphere, $C_1(H_2O)$ (Fig. 205). Hence the thermal escape rate is large only for $C_1(H_2O) > 0.1$.

Nonthermal escape of hydrogen and deuterium was considered by Kumar et al. (1983).

First of all they calculated the escape rates in the present atmosphere produced by the charge exchange reactions

$$H^+ + H \rightarrow H^* + H^+$$
$$H^+ + O \rightarrow H^* + O^+ .$$

Their rate coefficients equal $2.2 \times 10^{-11}\, T_i^{1/2}\, cm^3\, s^{-1}$. Hot hydrogen atoms formed have energies equal to those of H^+ ions. Calculating the escape rates by these reactions McElroy et al. (1982a, see Chap. 4.4) assume $T_i \simeq 1000 - 1500$ K which results in an escape fraction of 3%. However, measurements show $T_i \simeq 2300$ K on the day side and ~5000 K near the mid-

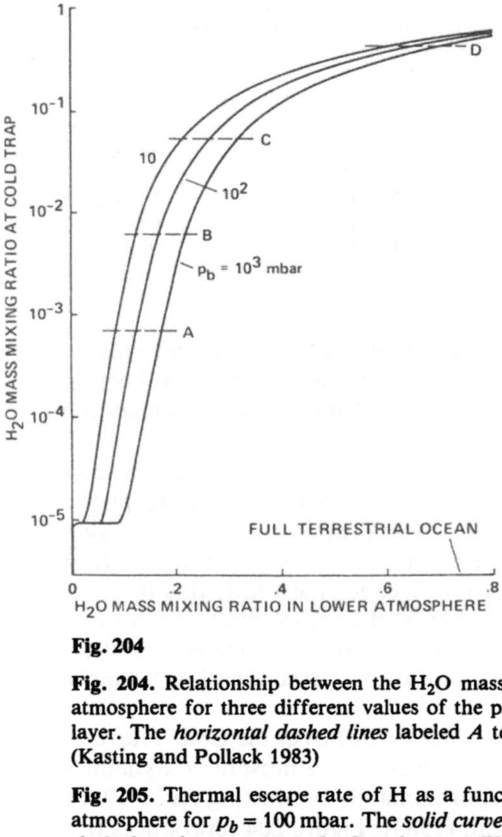

Fig. 204

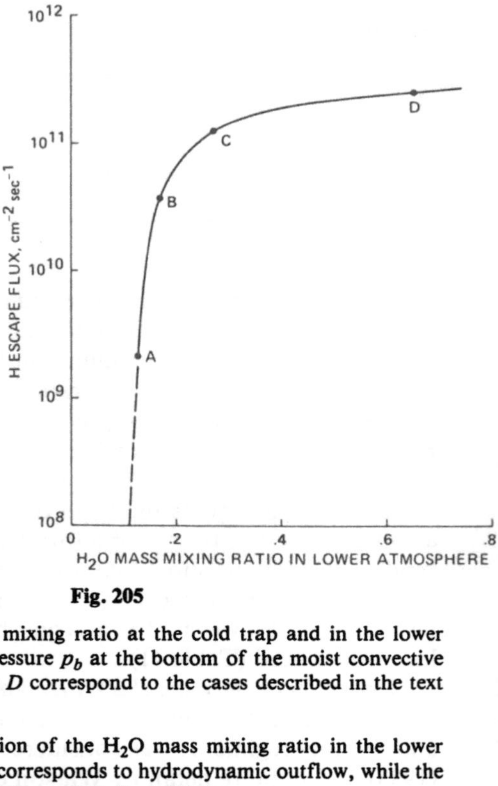

Fig. 205

Fig. 204. Relationship between the H_2O mass mixing ratio at the cold trap and in the lower atmosphere for three different values of the pressure p_b at the bottom of the moist convective layer. The *horizontal dashed lines* labeled *A* to *D* correspond to the cases described in the text (Kasting and Pollack 1983)

Fig. 205. Thermal escape rate of H as a function of the H_2O mass mixing ratio in the lower atmosphere for $p_b = 100$ mbar. The *solid curve* corresponds to hydrodynamic outflow, while the *dashed* portion represents the Jeans' escape (Kasting and Pollack 1983)

night region (Fig. 206). So the escape rate was recalculated by Kumar et al. based on the PV data on H, O, H^+ number densities and T_i and their diurnal variations. The escape probabilities of H and D atoms for the first reaction versus ion temperatures are depicted in Fig. 207. Figure 208 demonstrates diurnal variations of hot hydrogen production and escape rate. The double peak is due to some difference in local time between maximum of [H] and T_i. The global average escape rates are 1.2×10^7 and 4.2×10^3 cm^{-2} s^{-1} for *H* and *D*, respectively.

Contrary to [H] the atomic oxygen number densities reveal a strong night side minimum. Therefore, diurnal variations of $H^+ - O$ charge exchange rate are not so pronounced. Kumar et al. evaluate the escape probability as very low, ~0.3%, for this reaction, and its contribution, 2×10^4 cm^{-2} s^{-1}, can be neglected.

As a next step, calculations of the escape rates produced by the $H^+ - H$ reaction were performed for hydrogen-rich thermospheres which could be

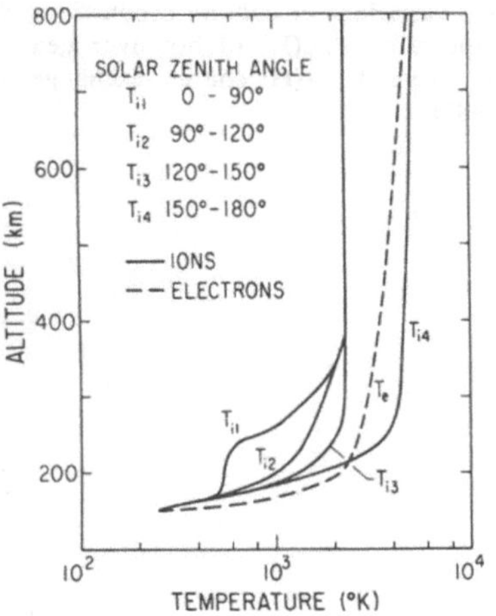

Fig. 206

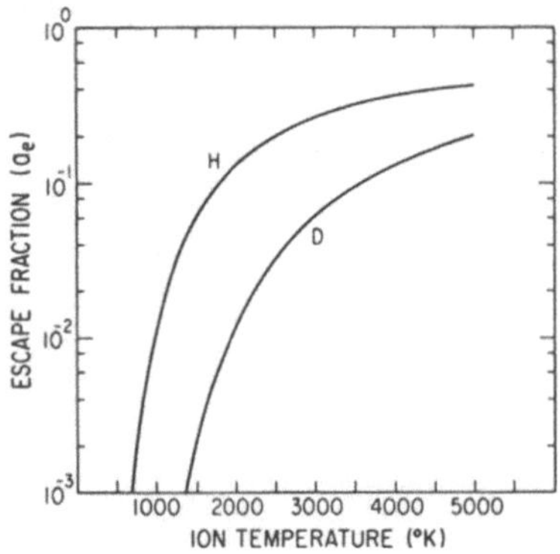

Fig. 207

Fig. 206. The ion and electron temperatures in the Venusian ionosphere measured by the PV orbiter (from Kumar et al. 1983). Electron temperature profile is uniform over the globe

Fig. 207. The escape fraction, a_e, for hydrogen and deuterium atoms formed by the reaction $H^+ + H$ with respect to the incident ion temperature (Kumar et al. 1983)

representative for Venus in the past. A method developed was checked by a comparison with the present atmosphere having the predawn bulge (Fig. 162). Figure 209 demonstrates the day side ionospheric composition calculated for the neutral atmosphere with the hydrogen mixing ratio at the homopause exceeding the present value by a factor of 1000. Here hydrogen atoms determine the exobase altitude where their column density, $N_c(\mathrm{H})$, equals $\sigma^{-1} = 4.5 \times 10^{14}\ \mathrm{cm}^{-2}$ with σ being the collisional cross-section. Then the maximal day side escape rate for $T_i = 2300$ K is

$$\phi_e^{\max}(\mathrm{H}) = kN_c(\mathrm{H})[\mathrm{H}^+]\, a_e(\mathrm{H}) = 1.5 \times 10^{10}\,\mathrm{cm}^{-2}\,\mathrm{s}^{-1}\,.$$

Here $a_e(\mathrm{H})$ is the escape fraction equal to 0.175 for $T_i = 2300$ K. At the night side, the maximal ion number densities at the exobase level are at least by an order of magnitude less than on the day side and with H^+ as a dominant ionospheric species here the maximal night side escape rate constitutes 1.4×10^9 $\mathrm{cm}^{-2}\,\mathrm{s}^{-1}$.

Calculations were made with the hydrogen mixing ratio at the homopause varying from 2.5 ppm to 2%. Therefore, the thermospheric thermal regime, dynamics, and circulation should be similar to those in the present thermosphere, e.g., the hydrogen night side bulge should form. Figure 210 shows the results obtained. The behavior of the $H^+ - H$ escape rate dependences on the day and night sides is explained by the saturation of $[H^+]$ at the exobase followed by that of [H].

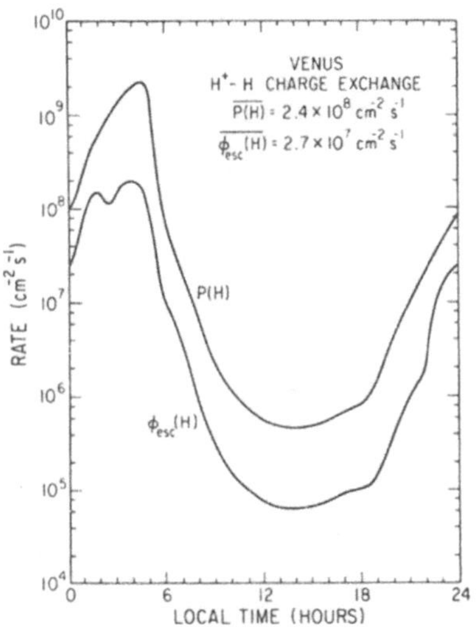

Fig. 208. Diurnal variations of column production, P(H), and escape rates, ϕ_{esc}(H) of hot hydrogen atoms resulting from $H^+ - H$ charge exchange (Kumar et al. 1983)

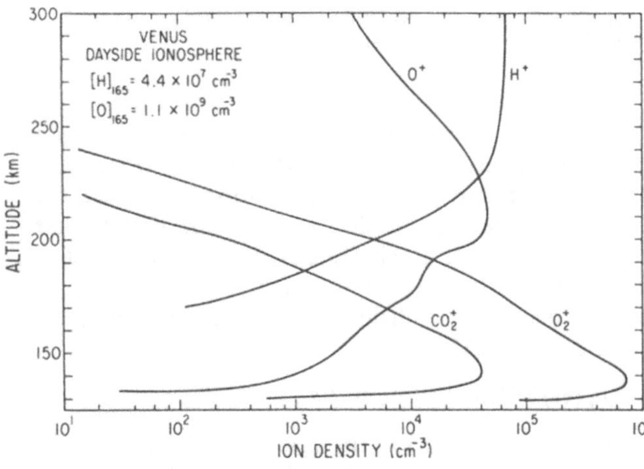

Fig. 209. A model of the Venusian day side ionosphere at noon in the past when the H mixing ratio at the homopause was 1000 times the present value (Kumar et al. 1983)

Then the hot oxygen impact mechanism

$$O_2^+ + e \rightarrow O^* + O^*$$
$$O^* + H \rightarrow O^* + H^*$$

was studied. According to McElroy et al. (1982a, see Chap. 4.4) it yields a global mean hydrogen escape rate of 8×10^6 cm^{-2}s^{-1} produced mainly on the predawn hydrogen bulge in the present atmosphere. However, with only

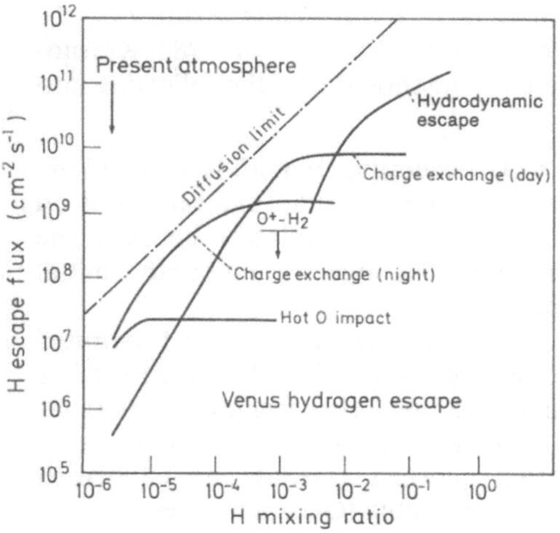

Fig. 210. Atomic hydrogen escape flux on Venus as a function of the elemental hydrogen mixing ratio at the homopause (Kumar et al. 1983, see also the text)

twice the current mixing ratio, $f_H = 5$ ppm, the exobase level would be determined by atomic hydrogen here and shifted from 152 to 160 km. One may expect the linear increase of the escape rate up to $f_H = 8$ ppm. With the further increase of f_H the exobase should be much higher than the O_2^+ layer and the hot H* atoms formed could not escape. Therefore, the escape region moves to the periphery of the predawn bulge where the exobase altitude is 160 – 170 km. The radius of this ring-shaped region would increase, while its area and the total escape flux would remain essentially constant and equal 2.5×10^7 $cm^{-2}s^{-1}$ up to $f_H \simeq 10^{-3}$ when the exobase level is high enough over the whole planet and this mechanism would cease.

The contribution of ion-molecular reactions

$$O^+ + H_2 \rightarrow OH^+ + H^*$$
$$OH^+ + e \rightarrow O + H^*$$

to the hydrogen escape rate was studied by Kumar et al. (1981, see Chap. 4.4), who identified the ion mass 2 as H_2^+. This identification results in $f_{H_2} \simeq 10$ ppm at the homopause and a hydrogen escape flux of $10^8\ cm^{-2}s^{-1}$ on the day side. $[H_2]$ is about 200 times larger and $[O^+]$ is ~20 times smaller in the predawn bulge than in the day side atmosphere, and the predawn bulge covers ~0.3 of the night side, therefore, the night side mean escape flux is equal to $2 \times 10^8\ cm^{-2}s^{-1}$ for $f_{H_2} = 10$ ppm. For larger f_{H_2} this mechanism is saturated due to O^+ not being the major ionospheric constituent at the exobase level. Though large amounts of H_2 in the lower atmosphere are confirmed by the V13, 14 gas chromatographic measurements (see Chap. 3.6) the identification of mass 2 as D^+ seems real (see Chap. 4.4) and a value of the hydrogen escape rate of $5 \times 10^8\ cm^{-2}s^{-1}$ which is due to the ion-molecular reaction should be considered as the upper limit (Fig. 210).

Initial Amount of Water Vapor and D/H Ratio. Time History of the Venusian Atmosphere (Kasting and Pollack 1983, Kumar et al. 1983, Krasnopolsky 1985b). Considering these problems one should take the following into account:

1. comparable amounts of N_2 and CO_2 generating materials received by Venus and the Earth and even larger amounts of primordial rare gases on Venus presume a substantial exchange of planetesimals between the Earth's and Venusian zones of accretion. This suggests comparable initial endowments of water on both planets;
2. the D/H ratio equals 1.5×10^{-4} in the terrestrial seawater and hydrated minerals in meteorites, 5×10^{-5} for outer planets, 1.8×10^{-5} in interstellar clouds. The equal ratios for seawater and meteorites suggests a relatively high and similar D/H ratio for bound water throughout the inner solar system due to isotopic fractionation between gaseous and condensed phases of the interstellar medium or solar nebula (Geiss and Reeves 1981). This implies that the initial D/H ratio on Venus was close to the present ratio for seawater and equals 1.5×10^{-4};
3. total time of water loss from Venus should not exceed the age of the solar system, 4.5×10^9 yr (4.5 b.y.).

   Kasting and Pollack (1983) and Kumar et al. (1983) try to reconcile these assertions with the measured $D/H = 1.6 \times 10^{-2}$ and the calculated dependence of escape rates on the water vapor amount on Venus. The simplest approach is to assume a time-independent fractionation factor f and all water vapor being in the atmosphere and not in subsurface reservoirs.

   Then

$$D/H = (D/H)_0 \left[\frac{P_0(H_2O)}{P(H_2O)} \right]^{1-f} .$$

McElroy et al. (1982a) considered collisions with hot oxygen atoms formed by O_2^+ recombination as a source of hydrogen escape which do not produce deuterium escape, i.e., $f = 0$. They assumed the escape rate to be proportional to the total amount of water vapor which results in $(D/H)_0 = 8 \times 10^{-5}$ and $P_0(H_2O) = 1.1$ bar (in Chap. 4.4 values for $D/H = 10^{-2}$ were given). Here the present H mixing ratio is adopted to be equivalent to ~ 100 ppm of H_2O with allowance for 20 – 200 ppm water vapor and possible contribution of H_2 and H_2S (Chap. 3.10).

Calculations of hydrogen and deuterium escape by a charge exchange process in the present atmosphere give

$$f = \frac{4.2 \times 10^3}{1.2 \times 10^7 \times 1.6 \times 10^{-2}} = 2.2\%$$

which may be assumed throughout all phases of the nonthermal escape. This factor includes a diffusive separation at the exobase level equal to $10^{-2}/$

$1.6 \times 10^{-2} \simeq 0.6$, a scale height ratio of 0.5, and a ratio of escape probabilities for a charge exchange process of 0.07 (Fig. 207). Unfortunately, Kumar et al. (1983) define f as $\frac{\phi_{esc}(D)\,P(H)}{\phi_{esc}(H)\,P(D)} = 8\%$ where $P(H)$, $P(D)$ are the mean rates of the charge exchange process with H and D, respectively.

However, this correction does not drastically change the situation, and starting from the present $\alpha = D/H = 1.6 \times 10^{-2}$ and $(D/H)_0 = 1.5 \times 10^{-4}$, one finds the initial H_2O amount to be 0.17% of a full terrestrial ocean (FTO). On the other hand, extrapolation of the present amounts of H and D to the beginning of the nonthermal escape phase when $C_1(H_2O) = 0.14$, i.e., $[H_2O]/[CO_2] = 0.4$ or 6% FTO, results in $\alpha = 5 \times 10^{-6}$ at that moment. Further extrapolation of FTO in the Venusian atmosphere gives $\alpha_0 = 7 \times 10^{-7}$ and 2×10^{-6} for $f = 0$ and 0.5, respectively, for the hydrodynamic flow phase. Besides, the time to remove a full terrestrial ocean appears to equal ~20 by.

Both possibilities, the comparatively low initial H_2O amount and the very small initial D/H ratio, seem to be unfavorable. To avoid them Kumar et al. (1983) propose a continuous long-term H_2O outgassing which can lower the present D/H ratio and for which the simple relation (*) appears invalid.

This idea was studied by Krasnopolsky (1985b). He argued that extrapolation of the measured noon hydrogen number density, $5 \times 10^4\,cm^{-3}$ (see Chap. 4.4) at 165 km, down to the homopause results in $f_H = 0.6$ ppm instead of 2.5 ppm adopted by Kumar et al. (1978, 1983). The value obtained agrees well with that from theory (Fig. 177). However, the dependence $\phi_H(f_H)$ derived by Kumar et al. (1983) (Fig. 210) can be valid if f_H is taken as the elemental hydrogen mixing ratio at the homopause and not the atomic hydrogen ratio. Then the sulfuric acid trap function, $f_H(f_{H_2O})$, was calculated. With the downward flux of sulfuric acid, $\phi_{H_2SO_4}$ independent of f_{H_2O} and equal to $-3 \times 10^{13}\,cm^{-2}\,s^{-1}$ (Chap. 3.10), the gradient of f_{H_2O} in the upper cloud layer can be calculated by the equation

$$(n+1)\,\phi_{H_2SO_4} + K[CO_2]\,\frac{df_{H_2O}}{dz} = 0$$

which for $K = 2 \times 10^4\,cm^2\,s^{-1}$ (Chap. 3.13) results in

$$f_{H_2O}(55\text{ km}) - f_{H_2O}(63\text{ km}) = 10^{-4}(n+1)\,.$$

Here n is the number of H_2O molecules bound with H_2SO_4, f_{H_2O} (63 km) corresponds to water vapor pressure determined by the concentration of sulfuric acid solution, i.e., n. Therefore this equation gives the unambiguous dependence of f_{H_2O}(63 km) on f_{H_2O}(55 km) the latter assumed equal to the lower atmospheric f_{H_2O}. Further decrease of f_{H_2O} above 63 km to 75 km was taken into account to yield $f_H(f_{H_2O})$ for the sulfuric acid trap (Fig. 211). The time of this trap operation appears much shorter than proposed by Kumar et al. (1983).

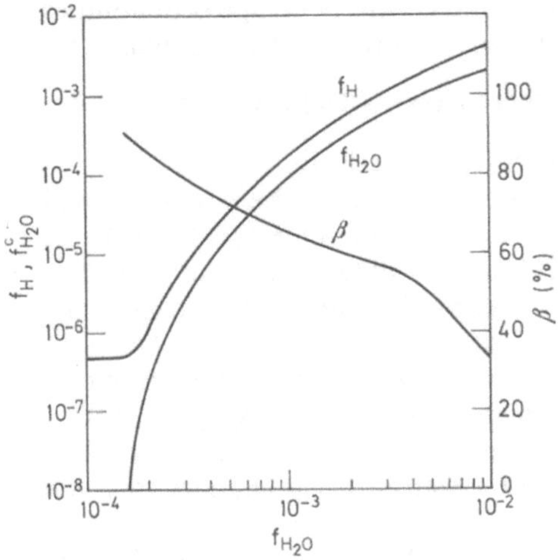

Fig. 211. Water vapor and elemental hydrogen mixing ratios, $f^c_{H_2O}$ and f_H, at 90 km and the concentration of sulfuric acid solution, β, in the upper cloud layer vs the lower atmospheric mixing ratio, $f_{H_2O} \cdot f_{HCl} = 5 \times 10^{-7}$ is taken into account (Krasnopolsky 1985b)

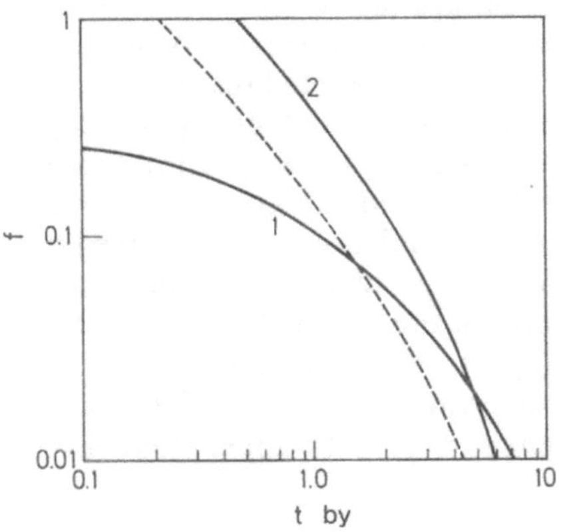

Fig. 212. Nonthermal escape fractionation factor vs time for $\phi'_H < \phi_{max}$ (*curve 1*) and $\phi'_H > \phi_{max}$ (*curve 2*) the latter for $T_c = 170$ K. *Dashed curve* shows ϕ'_H / ϕ_{HP} (Krasnopolsky 1985b)

Zahnle and Walker (1982) demonstrate that solar EUV radiation ($\lambda < 100$ nm) was more intense in the past by a factor of $\gamma = 4.5$ b.y./t, though the radiation along 200 nm was almost at the present level. This should increase the hydrogen escape rate up to the diffusion limit. The increase of ϕ_H and ϕ_D is also attributable to the increase in ionic temperature assumed to be $T_i = T_{i0}\,\gamma^{0.5}$. On reaching the diffusion limit for hydrogen ϕ_D continues to grow thus giving large fractionation factors, f (Fig. 212). Krasnopolsky (1985b) argues that water vapor mixing ratio at the cold trap should be increased by a factor of 7 due to supersaturation.

The corrections for supersaturation at the cold trap, for the sulfuric acid trap function, and the effects of the enhanced EUV radiation on the nonthermal escape rate are quite instrumental in solving the problem of time to remove a full terrestrial ocean (fto). Furthermore, it is not necessary to vary the cold trap temperature and the pressure at the bottom of the moist convective layer which were considered by Kasting and Pollack (1983) as poorly known values and appear now to be adequate.

Krasnopolsky (1985b) studied various arbitrary outgassing scenarios to yield the minimum deuterium enrichment factor starting with a full terrestrial ocean. The obtained minimum corresponds to the measured *H/D* ratio and can exist for the initial injection of 1 fto, the outgassing rate of the order of 10^9 cm^{-2} s^{-1} or less during the first 2 billion years, a subsequent steep increase of the outgassing rate up to 10^{10} cm^{-2} s^{-1}, and a gradual fall-off to the present value of the escape rate, 2×10^7 cm^{-2} s^{-1}. This outgassing history seems very artificial, and the total injection of water of the order of 1% fto is more probable.

The discussion above is based on the assumption that the total amount of planetary carbon dioxide was present in the atmosphere throughout the whole history of the planet.

An alternative approach was proposed by Kasting et al. (1984). They consider the primitive atmosphere of Venus to be similar to the present terrestrial atmosphere with a bulk of carbon dioxide in carbonaceous rocks and an ocean of liquid water. Kasting et al. predict a surface temperature of 76 °C and a partial pressure of water vapor of 0.4 bar for a solar flux smaller than the present one by a factor of 1.3. In this case, vapor mixing ratio at the cold trap is 0.07, the diffusion-limited escape flux is 5×10^{12} cm^{-2} s^{-1}, and the flux of the order of 10^{12} cm^{-2} s^{-1} is intense enough to lose a full terrestrial ocean in 0.5 b.y. No deuterium fractionation at that time and no subsequent deuterium escape were assumed. In this case the predicted deuterium enrichment is in accord with the measured one. Advantages and disadvantages of this approach have been discussed by Kasting et al. (1984).

We would like to emphasize two problems arising here. First of all the calculations were performed for the terrestrial atmosphere having ~1 bar of nitrogen and 0.3 mbar of carbon dioxide. Kasting et al. evaluate the amount of atmospheric carbon dioxide for Venus as 0.3 bar, which results in the increase of the Venusian atmospheric temperature by 35 °C. Furthermore they find that if *e*-type absorption of water vapor outside the 8 to 12 μm region is taken into account the surface temperature should be much higher. It is difficult to understand how the ocean of liquid water could form at such a high temperature.

Second, the assumption of no deuterium fractionation during the hydrodynamic flow phase is not supported by calculations. The nitrogen amount on Venus is larger by a factor of 3 than on Earth, therefore, for Venus, the water vapor mixing ratio at the cold trap and the escape flux should be lower than calculated for the Earth. This tends to increase deuterium fractionation also.

So the result obtained shows that the initial material of the Earth and Venus was not fully similar. This is consistent with the present amounts of nonradiogenic argon ^{36}Ar on the planets which is by two orders of magnitude more abundant on Venus than on the Earth. Therefore the inner part of the primitive solar nebula was not mixed well. In this case the hydrated silicates thermochemically stable at low temperatures could exist near the Earth's orbit and not the Venus' one (Prinn 1984b) that is why the initial endowment of water on Venus was only a percentage of that on the Earth. Prinn argues also that the present state of the oxidation of Venus is incompatible with the large amount of oxygen absorbed by rocks after escape of a full terrestrial ocean.

With the equal contribution of H_2O and HCl to the present f_H at the homopause the former will be negligible in 0.6 b.y. At that time $O^* - H$ process will dominate in the hydrogen escape rate equal to $3 \times 10^6\,cm^{-2}\,s^{-1}$. The deuterium escape will be negligible. Free chlorine formed by decomposition of HCl and the escape of hydrogen will reduce the water amount in the lower atmosphere by the thermochemical reaction

$$Cl_2 + H_2O + CO \rightarrow 2HCl + CO_2$$

while the HCl amount will remain the same. Further loss of hydrogen will slow down due to the decrease of the solar EUV radiation.

4.17 Conclusions

A large scope of experimental data on the structure and chemical composition of the Venusian atmosphere has been obtained in the last years. This promotes theoretical studies of processes determining the chemical composition and development of photochemical models of atmospheric composition. The models concerning general problems of atmospheric photochemistry on Venus, as well as analyses of some partial problems, show substantial progress made recently in the field. They consider initial data for noncontradictory models of atmospheric neutral composition and basic photochemical reactions which define number densities of CO_2, H_2O, HCl, SO_2 photolysis and further chemical transformation products as well as nitrogen compounds. Lists of components and reactions have been greatly enlarged which made it possible to consider chlorine and sulfur compounds simultaneously. A calculation technique for the photochemical formation of the cloud layer by an approximate description of radiative and particle transfer was developed, as were more universal methods of prescribing boundary conditions which are valid for both photochemical and diffusive equilibrium and intermediate cases. Numerical techniques used consider not only eddy and molecular diffusion, but particle sedimentation and some heterophase equilibriums as well.

Below are some results of the papers discussed.

1. The ionospheric models developed demonstrate the ionospheric composition measured to be consistent with the observed neutral composition of the

upper atmosphere. The role of metastable species as well as some minor components like H, H_2, N, NO, C has been studied.

2. A mean ionization rate for the night side ionosphere at high solar activity is equal to $\sim 2 \times 10^8 \, cm^{-2} s^{-1}$, which is maintained mostly by plasma flow from the day to the night side across the terminator with some contribution of electron fluxes. A variety of nighttime ionospheric phenomena (established, depleted, and disappearing ionospheres) can be attributed to variations in the magnetic field strength and direction, as well as in the height of the ionopause which in turn depends on the solar wind pressure.

3. The influence of ionospheric reactions on the neutral composition of the upper atmosphere, including formation of some minor components, has been investigated.

4. With mean temperatures of 280 and 120 K on the day and night sides, respectively, the daytime number densities of heavy thermospheric constituents (O, CO, N_2, CO_2) exceed the nighttime ones by two orders of magnitude at 180 km. The flow formed has a speed close to the speed of sound and results in formation of the night side bulges of light components (H, H_2, He) shifted to the predawn region.

5. The mean hydrogen escape rate equals $10^7 \, cm^{-2} s^{-1}$ with the $H^+ - H$ charge exchange process contributing 60% and the rest being due to the collisions of hydrogen atoms with hot oxygen atoms formed by dissociative recombination of O_2^+ ions. The mean escape rate of hot oxygen atoms equals $3 \times 10^6 \, cm^{-2} s^{-1}$.

6. The calculated neutral composition of the upper atmosphere agrees with measurements for the eddy mixing coefficient equal to $\sim 10^{13} M^{-1/2} \, cm^2 s^{-1}$.

7. Almost all atomic oxygen produced by CO_2 photolysis transforms into molecular oxygen by the three body association and the reactions of $O + ClO$, $O + OH$ and $NO_2 + O$. O_2 destruction occurs mainly via reactions of $ClCO_3 + Cl$ and $S + O_2$. For large H_2 abundance, $f_{H_2} \gtrsim 1$ ppm in the lower atmosphere which could have been valid in the past, $SO + HO_2$ and $NO + HO_2$ are also important as destruction processes. Molecular oxygen amounts above 60 km depend strongly on the efficiency of these reactions and atmospheric mixing. With a high efficiency an O_2 sink forms at 60 to 80 km, and almost all O_2 molecules are concentrated in the layer at 80 to 100 km. The O_2 amount is determined by atmospheric mixing and may agree with the spectroscopic limit. With a low efficiency the O_2 layer is expanded to 65 km, and the O_2 amount exceeds the spectroscopic limit.

8. The O_2 produced is consumed for CO_2 and H_2SO_4 formation and by a downward flux into the lower atmosphere. The net of sulfuric acid formation can be given as

$$CO_2 + SO_2 + 2H_2O + h\nu \rightarrow CO + H_2SO_4 \cdot H_2O$$

with the inverse reaction in the lower atmosphere. With a high efficiency of the COCl cycle carbon dioxide is rapidly reconstructed after its photolysis, and photochemical transformation of SO_2 to more oxidized and reduced forms occurs:

$$3SO_2 + 4H_2O + h\nu \rightarrow 2H_2SO_4 \cdot H_2O + S_a .$$

Some low oxidized acids like $H_2S_2O_2$ and $H_2S_2O_3$ and even H_2S become possible rather than elemental sulfur, S_a.

9. The atmospheric photochemistry is subject to amounts of H_2 in the lower atmosphere. For $f_{H_2} < 10^{-7}$ odd chlorine and its reactions are predominant, which obviously occurs in the present atmosphere. Then transformation of HCl to H_2O and Cl_2 and perhaps $COCl_2$ proceeds in the mesosphere:

$$2HCl + CO_2 + h\nu \rightarrow H_2O + CO + Cl_2$$

with a total rate of $3 \times 10^9\ cm^{-2}\ s^{-1}$. Following the transfer of products the inverse reaction should proceed in the lower atmosphere. For $f_{H_2} > 10^{-6}$, odd hydrogen prevails and a transformation of H_2 to H_2O occurs:

$$H_2 + CO_2 + h\nu \rightarrow H_2O + CO$$

with a rate of $\sim 10^{11}\ cm^{-2}\ s^{-1}$. Large H_2 abundance should be maintained by some compensation of this sink and of a possible thermochemical sink in the lower atmosphere near the surface, where H_2 should transform to H_2O via a reaction with CO_2. The necessary reactions are unknown, but they can be stimulated by visible radiation in the cloud layer and the lower atmosphere. In this case a clearer situation could have existed in the past when a large H_2 mixing ratio was formed by thermochemical equilibrium near the surface with large water vapor amounts.

10. The sulfur dioxide mixing ratio has a steep decrease near the cloud tops due to SO_2 photolysis and subsequent formation of sulfuric acid. Number densities and the scale height of SO_2 depend on the eddy mixing coefficient, K, and the quantum yield of H_2SO_4 formation by SO_2 photolysis which is determined by the ratio of rates of the reactions $R68$ $SO_2 + O + M$ to $R72$ $SO + O + M$ and $R111$ $SO + ClO$. The results of calculations adequately reproduce the measured SO_2 profile.

11. As a consequence of sulfuric acid photochemical formation the drying of the atmosphere occurs in the upper cloud layer. The calculated solution concentration equals 85% which agrees with measurements and corresponds to monohydrate $H_2SO_4 \times H_2O$. Therefore, H_2O and SO_2 should feature similar altitude profiles inside the cloud layer and the ratio of their number densities should be ~ 2, in accord with the optical measurements of H_2O by Venera probes. Even a very small variation of this ratio maintained at the lower cloud

boundary during the mixing time ($\sim$1 yr) entails substantial changes in atmospheric humidity and sulfuric acid solution concentration in the upper cloud layer.

12. The observed behavior of the water vapor mixing ratio ($\sim$200 ppm in the lower cloud layer and several ppm in the upper one) can be explained if a significant part or a total mode 3 of the cloud particles consists of sulfuric acid or some products of its further conversion. Apart from this, such behavior is in accordance with an eddy mixing coefficient of 2×10^4 cm^2 s^{-1} at the heights of the cloud layer.

13. Models of thermospheric and ionospheric composition have been developed for various water vapor mixing ratios in the lower atmosphere. These models were adequate in the past and dependences of thermal and non-thermal escape rates on the atmospheric water vapor amount could be obtained.

14. These data on the escape rates as well as the D/H ratio equal to 1.6×10^{-2} and the age of the planet are consistent with the total release of water into the atmosphere of the order of percent of a full terrestrial ocean starting with the initial isotopic ratio of seawater, 1.5×10^{-4}.

Much may be done in the study of the atmospheric chemical composition. First of all this means further progress in measuring the lower atmospheric composition and cloud layer particle composition. Main problems to be solved are possible presence of molecular oxygen, explanation of the water vapor altitude profile, molecular hydrogen amounts, search for sulfur compounds (COS, H_2S), etc. It is desirable to confirm by other techniques some results which have been obtained by gas chromatographs and seem to be inconsistent with the theory. These problems also necessitate laboratory measurements of rate coefficients of some reactions and new theoretical investigations. The COCl cycle may serve as an example. After the importance of this cycle was shown it was studied more carefully in the laboratory, and this study demonstrated that the formation of $ClCO_3$ is an intermediate step in the cycle. Rate coefficients of the main phases of this cycle have been measured. Then the data obtained have been used to refine the photochemical theory of the Venusian atmosphere which also utilizes reactions of $ClCO_3$ with O, H, Cl with unknown rate coefficients. Further progress may be due to investigations of reactions of $ClCO_3$ with NO, SO, and its photolytic and thermal collision destruction.

Data on the chemical composition of cloud particles are still very scarce. Though the presence of sulfuric acid and, perhaps, $FeCl_3$, seems reliable, new measurements are needed. Improved photochemical models should take into account additional sulfur compounds like S_n, S_2O, $(SO)_2$, H_xS_y, $H_xS_yO_z$ as well as compounds of sulfur with chlorine and other elements. The development in these directions is very promising since no photochemical models exist now for the atmosphere below 50 km and calculations of the thermochemical equilibrium are not satisfactory at present. Development of two- and three-dimensional models accounting for diurnal and latitudinal variations may also give important and interesting results.

References

Anderson DE (1974) Mariner 6, 7, and 9 ultraviolet spectrometer experiment: analysis of hydrogen Lyman alpha data. J Geophys Res 79:1513

Anderson DE (1976) The Mariner 5 ultraviolet photometer experiment: analysis of hydrogen Lyman alpha data. J Geophys Res 81:1213

Anderson DE, Hord CW (1971) Mariner 6 and 7 ultraviolet spectrometer experiment: analysis of hydrogen Lyman alpha data. J Geophys Res 76:6666

Anderson E, Leovy CB (1978) Mariner 9 television limb observations of dust and ice hazes on Mars. J Atm Sci 35:723

Anderson LG (1976) Atmospheric chemical kinetics data survey. Rev Geophys Space Phys 14:151

Andreichikov BM (1978) Water vapor distribution in the upper troposphere of Venus measured by Venera 4, 5, 6 space probes. Geokhimia 11 (in Russian)

Aumann HH, Martonchik JV, Orton GS (1982) Airborne spectroscopy and spacecraft radiometry of Venus in the far infrared. Icarus 49:227

Avduevsky VS, Akim EL, Aleshin VI et al. (1975) Martian atmosphere at the location of Mars 6 landing. Cosmic Res 13:4

Avduevsky VS, Borodin NF, Vasilyer UN et al. (1979) Analysis of parameters of the Venus atmosphere measured by Venera 11 and 12 descent probes. Cosmic Res 17:655

Barin I, Knacke O (1973) Thermodynamic properties of inorganic substances. Springer, Berlin Heidelberg New York Tokyo

Barker ES (1972) Detection of molecular oxygen in the Martian atmosphere. Nature (Lond) 238:447

Barker ES (1975) Observations of Venus water vapor over the disk of Venus: the 1972 – 1974 data using the H_2O lines at 8197 Å and 7176 Å. Icarus 25:268

Barker ES (1976) Martian atmospheric water vapor observations: 1972 – 1974 apparition. Icarus 28:247

Barker ES (1979) Detection of SO_2 in the spectrum of Venus. Geophys Res Lett 6:117

Barker ES, Woodman JH, Perry MA, Hapke B, Nelson R (1975) Relative spectrophotometry of Venus from 3067 to 5960 Å. J Atm Sci 32:1205

Bar-Nun A (1980) Production of nitrogen and carbon species by thunderstorms on Venus. Icarus 42:338

Barsukov VL, Khodakovsky IL, Volkov VP, Florensky KP (1980) The geochemical model of the troposphere and lithosphere of Venus based on new data. Space Res 10:197

Barth CA (1968) Interpretation of the Mariner 5 Lyman-measurements. J. Atm Sci 25:564

Barth CA, Dick ML (1974) Ozone and the polar hood of Mars. Icarus 22:205

Barth CA, Hord CW (1971) Mariner 6 and 7 ultraviolet spectrometer: topography and polar cap. Science (Wash DC) 173:197

Barth CA, Hord CW, Pearce JB, Kelly KK, Anderson GP, Stewart AI (1971) Mariner 6 and 7 ultraviolet spectrometer experiment: upper atmosphere data. J Geophys Res 76:2213

Barth, CA, Stewart AI, Hord CW, Lane AL (1972) Mariner 9 ultraviolet spectrometer experiment: Mars airglow spectroscopy and variations in Lyman alpha. Icarus 17:457

Barth CA, Hord CW, Stewart AI, Lane AL, Dick ML, Anderson GP (1973) Mariner 9 ultraviolet spectrometer experiment: seasonal variations of ozon on Mars. Science (Wash DC) 179:795

Bauer SH, Jeffers P, Lifshitz A., Yavada B (1971) Reaction between CO and SO_2 at elevated temperatures: a shock-tube investigation. Symp Int Conbust 13:417

Bauer SJ, Donahue TM, Hartle RE, Taylor HA (1979) Venus ionosphere: photochemical and thermal diffusion control of ion composition. Science (Wash DC) 205:109

Bauer SJ, Taylor HA (1981) Modulation of Venus ion densities associated with solar variations. Geophys Res Lett 8:840

Baulch DL, Drysdale DD, Duxbury J, Grant S (1976) Evaluated kinetic data for high temperature reactions. Butterworth, London

Baulch DL, Cox RA, Hampson RF, Kerr JA, Troe J, Watson RT (1980) Kinetic and photochemical data for atmospheric chemistry. J Phys Chem Ref Data 9:295

Belton MJS (1982) An interpretation of the near-ultraviolet absorption spectrum of SO_2: implications for Venus, Io, and laboratory measurements. Icarus 52:149

Bertaux JL, Blamont J, Babichenko SI et al. (1975) Measurements of intensity and spectral properties of Lyman-α in the upper atmosphere of Mars. Cosmic Res 13:42

Bertaux JL, Blamont J, Marcelin M, Kurt VG, Romanova NN, Smirnov AS (1978) Lyman alpha observations of Venera 9 and 10. I. The nonthermal hydrogen population in the exosphere of Venus. Planet Space Sci 26:817

Bertaux JL, Blamont JE, Lepine VM, Kurt VG, Romanova NN, Smirnov AS (1981) Venera 11 and Venera 12 observations of EUV emissions from the upper atmosphere of Venus. Planet Space Sci 29:149

Bertaux JL, Lepine VM, Kurt VG, Smirnov AS (1982) Altitude profile of H in the atmosphere of Venus from Lyman-α observations of Venera 11 and Venera 12 and origin of the hot exospheric component. Icarus 52:221

Borucki WJ (1982) Comparison of Venusian lightning observations. Icarus 52:354

Brace LM, Theis RF, Niemann HB, Mayz HG, Haegy WR, Nagy AF (1979) Empirical models of the electron temperature and density in the night side Venus ionosphere. Science 205:102

Brace LH, Taylor HA, Gombosi TI, Kliore AJ, Kundsen WC, Nagy AF (1983) The ionosphere of Venus: observations and their interpretations. In: Hunten DM, Colin L, Donahue TM, Moroz VI (eds) Venus. Univ Arizona Press, p 779

Brinton HC, Taylor HA, Niemann HB, Mayr HG, Nagy AF, Cravens TE, Strobel DF (1980) Venus night time hydrogen bulge. Geophys Res Lett 7:865

Broadfoot AL (1972) The solar spectrum 2100 – 3200 Å. Astrophys J 173:681

Broadfoot AL, Kumar S, Belton MJS, McElroy MB (1974) Ultraviolet observations of Venus from Mariner 10: preliminary results. Science (Wash DC) 183:1315

Burgin MS, Sholokhov VS (1981) Scattering of resonance lines of HeI 584 Å and OII 834 Å in the upper atmosphere of Venus. Cosmic Res 19:797

Carleton NP, Traub WA (1972) Detection of molecular oxygen on Mars. Science (Wash DC) 177:988

Chamberlain JW (1963) Planetary coronae and atmospheric evaporation. Planet Space Sci 11:901

Chameides WL, Walker JCG, Nagy AF (1979) Possible chemical impact of planetary lightning in the atmospheres of Venus and Mars. Nature (Lond) 280:820

Cieslik S, Nicolet M (1973) The aeronomic dissociation of nitric oxide. Planet Space Sci 21:925

Cimino J (1982) The composition and vertical structure of the lower cloud deck on Venus. Icarus 51:334

Clancy RT, Muhleman DO, Jakosky BM (1983) Variability of carbon monoxide in the Mars atmosphere. Icarus 55:282

Coffeen DL, Gehrels T (1969) Wavelength dependence of polarization. XV. Observations of Venus. Astron J 74:433

Connes J, Connes P, Maillard JP (1969) Atlas des spectres infrarouges de Venus, Mars, Jupiter et Saturne. CNRS, Paris

Connes P, Connes J, Benedict WS, Kaplan LD (1967) Traces of HCl and HF in the atmosphere of Venus. Astrophys J 147:1230

Connes P, Connes J, Benedict WS, Kaplan LD (1968) Carbon monoxide in the Venus atmosphere. Astrophys J 152:731

Connes P, Noxon JF, Traub WA, Carleton NP (1979) $O_2(^1\Delta g)$ emission in the day and night airglow of Venus. Astrophys J 233:L29

Conrath B, Curran R, Hanel R et al. (1973) Atmospheric and surface properties of Mars obtained by infrared spectroscopy on Mariner 9. J Geophys Res 78:4267
Conway RR, McCoy RP, Barth CA, Lane AL (1979) IUE detection of sulfur dioxide in the atmosphere of Venus. Geophys Res Lett 6:629
Counselman CC III, Gourevitch SA, King RW, Loriot GB, Ginsberg ES (1980) Zonal and meridional circulation of the lower atmosphere of Venus determined by radio interferometry. J Geophys Res 85:8026
Craig RA, Reynolds RT, Ragent B, Carle GC, Woeller F, Pollack JB (1983) Sulfur trioxide in the lower atmosphere of Venus? Icarus 53:1
Cravens TE, Kliore AJ, Kozyra JU, Nagy AF (1981) The ionospheric peak on the Venus dayside. J Geophys Res 86:11323
Cravens TE, Crawford SL, Nagy AF, Gombosi TI (1983) A two-dimensional model of the ionosphere of Venus. J Geophys Res 88:559
Dalgarno A, McElroy MB, Stewart AI (1969) Electron impact excitation of dayglow. J Atm Sci 26:753
Dalgarno A, McElroy MB (1970) Mars: is nitrogen present? Science (Wash DC) 170:167
Danielson RE, Moore DR, van de Hulst HC (1969) The transfer of visible radiation through clouds. J Atm Sci 26:1078
Davies DW (1979) The vertical distribution of Mars water vapor. J Geophys Res 84:2875
Davies DW (1981) The Mars water cycle. Icarus 45:398
Dementyeva NN, Kurt VG, Smirnov AS, Titarchuk LG, Chuvahin SD (1972) Preliminary results of measurements of UV emissions scattered in the Martian upper atmosphere. Icarus 17:475
Deming D, Espenak F, Jennings D, Kostiuk T, Mumma M (1982) Evidence for high-altitude haze thickening on the dark side of Venus from 10-micron heterodyne spectroscopy of CO_2. Icarus 49:35
Dickinson RE, Ridley EC (1977) Venus mesosphere and thermosphere temperature structure. II. Day-night variations. Icarus 30:162
Dollfus A, Coffeen DL (1970) Polarization of Venus I. Disk observations. Astron Astrophys 8:251
Dollfus A, Auriere M, Santer R (1979) Wavelength dependence of polarization. XXXVII. Regional observations of Venus. Astron J 84:1419
Donahue TM, Hoffman JH, Hodges RR (1981) Krypton and xenon in the atmosphere of Venus. Geophys Res Lett 8:513
Donahue TM, Hoffman JH, Hodges RR, Watson AJ (1982) Venus was wet: a measurement of the ratio of *D* to *H*. Science (Wash DC) 216:630
Dostovalov SB, Chuvahin SD (1973) Distribution of neutral hydrogen in the Martian upper atmosphere. Cosmic Res 11:767
Durrance ST (1981) The carbon monoxide fourth positive bands in the Venus dayglow. I. Synthetic spectra. J Geophys Res 86:9115
Durrance ST, Barth CA, Stewart AIF (1980) Pioneer Venus observations of the Venus dayglow spectrum 1250 – 1430 Å. Geophys Res Lett 7:222
Ekonomov AP, Moshkin BE, Golovin YM, Parfentyev NA, Sanko NF (1979) Spectrophotometric experiment at the Venera 11 and 12 lander probes. I. Method, results, and preliminary analysis of measurements. Cosmic Res 17:714
Ekonomov AP, Moshkin BE, Moroz VI, Golovin YM, Gnedykh VI, Grigoryev AV (1983) UV photometry at the Venera 13 and 14 lander probes. Cosmic Res 21:254
Ertel D, Moroz VI, Linkin VM et al. (1984) Venera 15, 16: first results of the infrared spectrometer experiment. Pisma Astron Zh 10:243
Esposito LW (1984) Sulfur dioxide shows evidence for Venus vulcanism. Science (Wash DC) 223:1072
Esposito LW, Gates LJ (1981) Horizontal and vertical distribution of sulfur dioxide on Venus. EOS 62:321
Esposito LW, Travis LD (1982) Polarization studies of the Venus UV contrasts: cloud height and haze variability. Icarus 51:374
Esposito LW, Winick JR, Stewart AI (1979) Sulfur dioxide in the Venus atmosphere: distribution and implications. Geophys Res Lett 6:601

Evans WF, Hunten DM, Llewellin EJ, Vallance Jones A (1968) Altitude profile of the infrared atmospheric system of oxygen in the dayglow. J Geophys Res 73:2885
Farmer CB (1976) Liquid water on Mars. Icarus 28:279
Farmer CB, Davies DW, Holland AL, La Porte DD, Docus PE (1977) Mars: water vapor observations from the Viking orbiters. J Geophys Res 82:4225
Feldman PD, Moos HW, Clark JT, Lane AL (1979) Identification of the UV nightglow from Venus. Nature (Lond) 279:221
Fink U, Larson HP, Kuiper GP, Poppen RF (1972) Water vapor in the atmosphere of Venus. Icarus 17:617
Fjeldbo G, Eshleman VR (1969) Atmosphere of Venus as studied with Mariner 5 dual radio-frequency occultation experiment. Radio Sci 4:879
Fjeldbo G, Kliore A, Seidel B (1970) The Mariner 1969 occultation measurements of the upper atmosphere of Mars. Radio Sci 5:381
Fjeldbo G, Kliore A, Eshleman VR (1971) The neutral atmosphere of Venus as studied with the Mariner 5 radio occultation experiment. Astron J 76:123
Fjeldbo G, Seidel B, Sweetman D, Howard T (1975) The Mariner 10 radio occultation measurements of the ionosphere of Venus. J Atm Sci 32:1232
Flasar FM, Goody RM (1976) Diurnar behavior of water on Mars. Planet Space Sci 24:269
Florensky CP, Volkov VP, Nikolaeva OV (1978) A geochemical model of the Venus troposphere. Icarus 33:537
Florensky KP, Nikolaeva OV, Volkov VP, Kudrjashova AF, Pronin AA, Goktin YM, Chikina EA, Bachkirova AS (1983) On the oxidation conditions on the surface of Venus based on the data of the geochemical indicator "Contrast" on board Venera 13 and 14. Cosmic Res 21:351
Fox JL (1982a) The chemistry of metastable species in the Venussian ionosphere. Icarus 51:248
Fox JL (1982b) Atomic carbon in the atmosphere of Venus. J Geophys Res 87:9211
Fox JL, Dalgarno A (1979a) Electron energy deposition in carbon dioxide. Planet Space Sci 27:491
Fox JL, Dalgarno A (1979b) Ionization, luminosity, and heating of the upper atmosphere of Mars. J Geophys Res 84:7315
Fox JL, Dalgarno A (1981) Ionization, luminosity, and heating of the upper atmosphere of Venus. J Geophys Res 86:629
Fox JL, Dalgarno A (1983) Nitrogen escape from Mars. J Geophys Res 88:9027
Fox JL, Victor GA (1981) O^{++} in the Venussian ionosphere. J Geophys Res 86:2438
Geiss J, Reeves H (1981) Deuterium in the solar system. Astron Astrophys 93:189
Gelman BG, Zolotukhin VG, Lamonov NI et al. (1979) An analysis of the chemical composition of the atmosphere of Venus by Venera 12 gas chromatograph. Cosmic Res 17:708
Gerard JC, Stewart AIF, Bougher SW (1981) The altitude distribution of the Venus ultraviolet nightglow and implications on vertical transport. Geophys Res Lett 8:633
Germogenova TA (1962) Solution of radiative transfer equation for a plane parallel layer. J Comp Math Phys 1:1168
Gierasch PJ, Goody PM (1968) A study of the thermal and dynamical structure of the Martian lower atmosphere. Planet Space Sci 16:615
Glushko VP, Gurvich LV, Bergman GA (eds) (1978) Thermodynamical properties of individual substances. Nauka, Moscow
Gmelin Handbuch der Anorganischen Chemie. Chlor. Springer, Heidelberg, 1969
Golovin YM, Moshkin BE, Ekonomov AP (1981) Aerosol component properties as measured by the Venera 11, 12 spectrophotometer. Cosmic Res 19:421
Golovin YM, Ustinov EA (1982) Subcloud aerosol in the atmosphere of Venus. Cosmic Res 20:104
Golovin YM, Moroz VI, Moshkin BE, Sanko NF, Ekonomov AP (1982) Atmosphere of Venus as measured by the Venera 11, 12 spectrophotometers. Cosmic Res 20:572
Good JC, Schloerb FP (1981) Martian CO abundance from the $J = 1 \rightarrow 0$ rotational transition: evidence for temporal variation. Icarus 47:166
Good JC, Schloerb FP (1983) Limits on Venus' SO_2 abundance profile from interferometric observations at 3.4 mm wavelength. Icarus 53:538

Griggs N (1968) Absorption coefficients of ozone in the ultraviolet and visible regions. J Chem Phys 49:857
Gringauz KI, Verigin MI, Breus TK, Gombosi T (1977) Electron fluxes measured by Venera 9, 10 in the optical umbra of Venus – the main source of ionization in the nightside ionosphere of Venus. Dokl Akad Nauk SSSR 232:1039
Gringauz KI, Verigin MI, Breus TK, Gombosi T (1979) The interaction of the solar wind electrons in the optical umbra of Venus with the planetary atmosphere – the origin of the nighttime ionosphere. J Geophys Res 84:2123
Gringauz KI, Verigin MI, Breus TK, Ivanova SV (1981) Low energy electrons in the optical umbra of Venus measured by Venera 9, 10 orbiters – the origin of the nighttime ionosphere. Comparison with the Pioneer Venus orbiter results. Cosmic Res 19:430
Gringauz KI, Verigin MI, Breus TK, Schwachunova LA (1983) On the main source of ionization at the main peak at the nighttime ionosphere of Venus. Cosmic Res 21:746
Hanel P, Conrath B, Hovis W, Kunde V, Lowman P, Maguire W, Straat P, Burke T (1972) Investigation of the Martian environment by infrared spectroscopy on Mariner 9. Icarus 17:423
Hansen JE, Hovenier JW (1974) Interpretation of the polarization of Venus. J Atm Sci 31:1137
Hansen JE, Travis LD (1974) Light scattering in the planetary atmospheres. Space Sci Rev 16:527
Hanson WB, Sanatani S, Zuccaro DR (1977) The Martian ionosphere as observed by Viking retarding potential analyzers. J Geophys Res 82:4351
Hapke B, Nelson R (1975) Evidence for an elemental sulfur component of the clouds from Venus spectroscopy. J Atm Sci 32:1212
Hartle RE, Taylor HA (1983) Identification of deuterium ions in the ionosphere of Venus. Geophys Res Lett 10:965
Hartle RE, Mayr HG, Bauer SJ (1978) Global circulation and distribution of hydrogen in the thermosphere of Venus. Geophys Res Lett 5:719
Hess SL (1976) The vertical distribution of water vapor in the atmosphere of Mars. Icarus 28:269
Higgins JE (1976) The solar EUV flux between 230 and 1220 Å on November 9. J Geophys Res 81:1301
Hill RD (1979) A survey of lightning energy estimates. Rev Geophys Space Phys 17:155
Hochanadel CJ, Gromley JA, Ogren PJ (1972) Absorption spectrum and reaction kinetics of the HO_2 radical in the gas phase. J Chem Phys 56:4426
Hoffman JH, Hodges RR Jr, McElroy MB, Donahue TM (1979) Composition and structure of the Venus atmosphere: Results from Pioneer Venus. Science 205:49
Hoffman JH, Hodges RR Jr, Donahue TM, McElroy MB (1980) Composition of the Venus lower atmosphere from the Pioneer Venus mass spectrometer. J Geophys Res 85:7882
Hudson RD (1971) Critical review of ultraviolet photoabsorption cross sections for molecules of astrophysical and aeronomic interest. Rev Geophys Space Phys 2:305
Hulst van de HC (1957) Light scattering by small particles. Wiley, New York
Hunten DM (1974) Aeronomy of Mars lower atmosphere. Rev Geophys Space Phys 12:529
Hunten DM (1982) Thermal and nonthermal escape mechanisms for terrestrial bodies. Planet Space Ser 30/773:1
Hunten DM, McElroy MB (1970) Production and escape of hydrogen on Mars. J Geophys Res 75:5989
Intrilligator DS, Collard HR, Mihalov JD, Whitten RC, Wolfe JH (1979) Electron observations and ion flows from the Pioneer Venus orbiter plasma analyzer experiment. Science (Wash DC) 205:116
Irvine WM (1968) Monochromatic phase curves and albedos for Venus. J Atm Sci 25:610
Istomin VG, Grechnev KV, Kochnev VA, Ozerov HN (1979) Venus lower atmospheric composition as measured by mass spectrometers. Cosmic Res 17:703
Istomin VG, Grechnev KV, Kochnev VA (1980) Mass spectrometry of the Venus lower atmosphere: krypton isotopes and other recent results of processing the data of measurements on board the Venera 11, 12 spacecraft. Preprint D-298, Space Res Inst, Moscow
Istomin VG, Grechnev KV, Kochnev VA (1983) Venera 13 and Venera 14: mass spectrometry of the atmosphere. Cosmic Res 21:410

Ivanov-Kholodny GS, Kolosov MA, Savich NA, Alexanrov YM, Vasiljev MB, Vyshlov AS, Dubrovin VM, Zaitsev AL, Samovol VA, Samoznaev LN, Sidorenko AI, Hasjanov AF, Stern DY (1977) Ionosphere of Venus as measured by the Veneras 9, 10 orbiters and some features of its formation. Usp Phys Nauk 123:694

Ivanov-Kholodny GS, Kolosov MA, Savich NA et al. (1979) Daytime ionosphere of Venus as studied with Venera 9 and 10 dual frequency radio occultation experiments. Icarus 39:209

Izakov MN (1978) The Martian upper atmosphere structure from the Viking spacecraft experiments. Icarus 36:189

Izakov MN, Krasitzki OP (1977) Mars model atmospheric composition. Cosmic Res 15:455

Izakov MN, Morozov SK (1975) Venus thermospheric structure and dynamics. Cosmic Res 13:404

Izakov MN, Morozov SK (1976) Mars equatorial thermospheric structure and dynamics. Cosmic Res 14:476

Izakov MN, Krasitzki OP, Pavlov AV (1981) Venus model day side ionospheric composition. Cosmic Res 19:739

Jackson GE, Calvert JG (1971) The triplet sulfur dioxide-carbon monoxide reaction excited within the $SO_2(^1A_1) - SO_2(^3B_1)$ forbidden band. J Am Chem Soc 93:2593

Janssen MA, Klein MJ (1981) Constraints on the composition of the Venus atmosphere from microwave measurements near 1.35 cm wavelengths. Icarus 46:58

Kaplan LD, Munch G, Spinrad H (1964) An analysis of the spectrum of Mars. Astrophys J 139:237

Kaplan LD, Connes J, Connes P (1969) Carbon monoxide in the Martian atmosphere. Astrophys J 157:L187

Kasprzak WT, Hedin AE, Niemann HB, Spencer NW (1980) Atomic nitrogen in the upper atmosphere of Venus. Geophys Res Lett 7:106

Kasting JF, Pollack JB (1983) Loss of water from Venus. I. Hydrodynamic escape of hydrogen. Icarus 53:479

Kasting JF, Pollack JB, Ackerman TP (1984) Response of the Earth's atmosphere to increase in solar flux and implications for loss of water from Venus. Icarus 57:335

Kaufman F (1975) Hydrogen chemistry: perspective on experiment and theory. In: Atmospheres of the Earth and the planets. Astrophysics and Space Science Library, vol 51, p 219

Kawabata K et al. (to be published 1984)

Kawabata K, Coffeen DL, Hansen JE, Lane WA, Sato M, Travis LD (1980) Cloud and haze properties from Pioneer Venus polarimetry. J Geophys Res 85:8129

Keating GM, Nicholson JM, Lake LR (1980) Venus upper atmosphere structure. J Geophys Res 85:7941

Kenner RD, Ogryzlo EA, Turley S (1979) On the excitation of the night airglow on Earth, Venus, and Mars. J Photochem 10:199

Kerzhanovich VV (1977) Mars 6: improved analysis of the descent module measurements. Icarus 30:1

Kerzhanovich VV, Makarov YF, Marov MY et al. (1979) Estimates of wind speed and turbulence in the Venus atmosphere based on mutual Doppler measurements from Veneras 11 and 12. Cosmic Res 17:690

Khodakovsky IL (1982) Atmosphere-surface interactions on Venus and implications for atmospheric evolution. Planet Space Sci 30:803

Khodakovsky IL, Volkov VP, Sidorov YI, Dorofeeva VA, Borisov MV, Barsukov VL (1979) Geochemical model of the Venus troposphere and crust according to new data. Geokhimia 12:1747

Kieffer HH, Chase SC, Miner E, Munch G, Neugebauer G (1973) Preliminary report of infrared radiometric measurements from the Mariner 9 spacecraft. J Geophys Res 78:4291

Kieffer HH, Martin TZ, Peterfreund AR, Jakosky BM, Miner ED, Palluconi FD (1977) Thermal and albedo mapping of Mars during of Viking primary mission. J Geophys Res 82:4229

Kliore AJ, Cain DL, Levy GS (1967) Radio occultation measurements of the Martian atmosphere over two regions with Mariner 4 space probe. Space Research VII, Moon and Planets

Kliore AJ, Cain DL, Fjeldbo G, Seidel BL, Sykes MJ, Rasool SI (1972) The atmosphere of Mars from Mariner 9 occultations. Icarus 17:484

Kliore AJ, Patel IR, Nagy AF, Cravens TE, Gombosi TI (1979) Initial observations of the night side ionosphere of Venus from Pioneer Venus orbiter radio occultations. Science (Wash DC) 205:99

Knollenberg RG (1984) A reexamination of the evidence for large, solid particles in the clouds of Venus. Icarus 57:161

Knollenberg RG, Hunten DM (1980) The microphysics of the clouds of Venus: results of the Pioneer Venus particle size spectrometer experiment. J Geophys Res 85:8039

Knudsen WC, Spenner K, Whitten RC, Spreiter JR, Miller KL, Novak V (1979) Thermal structure and energy influx to the day and night side Venus ionosphere. Science (Wash DC) 205:105

Knudsen WC, Spenner K, Miller KL, Novak V (1980) Transport of ionospheric O^+ ions across the Venus terminator and implications. J Geophys Res 85:7803

Kolosov MA, Yakovlev OI, Matyugov SS, Timofeev TS, Chub EV, Yakovleva GD (1978) Day side atmosphere of Venus as measured by the Veneras 9 and 10 orbiter radio occultations. Cosmic Res 16:278

Kondratjev VN (1971) Rate coefficients of gas phase reactions. Nauka, Moscow

Kong TY, McElroy MB (1977a) The global distribution of O_3 on Mars. Planet Space Sci 25:839

Kong TY, McElroy MB (1977b) Photochemistry of the Martian atmosphere. Icarus 32:168

Krasitzky OP (1978) Model of diurnal variations of the Martian atmospheric composition. Cosmic Res 16:434

Krasnopolsky VA (1973) Atmospheric ionization rates by solar EUV radiation. Geomagn i aeronomia 13:882

Krasnopolsky VA (1975) On the structure of Mars' atmosphere at 120 – 220 km. Icarus 24:28

Krasnopolsky VA (1978) Upper limits of abundances of some components of Mars' and Venus' atmospheres based on the twilight airglow spectroscopy on board the Mars 5 and Veneras 9 and 10 orbiters. Cosmic Res 16:895

Krasnopolsky VA (1979a) Vertical distribution of water vapor and Mars model lower and middle atmosphere. Icarus 37:182

Krasnopolsky VA (1979b) On the dependence of the thermospheric nitric oxide and atomic nitrogen number densities on solar activity. Cosmic Res 17:413

Krasnopolsky VA (1979c) Veneras 9 and 10: spectroscopy of scattered radiation in overcloud atmosphere. Preprint No 523, Space Res Inst, Moscow; 1980. Cosmic Res 18:899

Krasnopolsky VA (1979d) Night side ionosphere of Venus. Planet Space Sci 27:1402

Krasnopolsky VA (1980) On lightnings in the Venus' atmosphere according to the Venera 9 and 10 data. Cosmic Res 18:429

Krasnopolsky VA (1981a) Spectroscopic estimate of the carbon monoxide amount in the upper atmosphere of Mars. Cosmic Res 19:902

Krasnopolsky VA (1981b) Excitation of oxygen emissions in the night airglow of the terrestrial planets. Planet Space Sci 29:925

Krasnopolsky VA (1982a) Atomic carbon in atmospheres of Mars and Venus. Cosmic Res 20:595

Krasnopolsky VA (1982b) Composition of the Venus day side atmosphere at 110 – 200 km. Cosmic Res 20:647

Krasnopolsky VA (1982c) 5577 Å night airglow and electron fluxes in the nighttime atmosphere of Venus. Cosmic Res 20:742

Krasnopolsky VA (1983a) Venus spectroscopy in the 3000 – 8000 Å region by Veneras 9 and 10. In: Hunten DM, Colin L, Donahue TM, Moroz VI (eds) Venus. Univ Arizona Press, p 459

Krasnopolsky VA (1983b) Lightnings and nitric oxide on Venus. Planet Space Sci 31:1363

Krasnopolsky VA (1985a) Chemical composition of Venus clouds. Plant Space Sci 33:109

Krasnopolsky VA (1985b) Total injection of water vapor into the Venus' atmosphere. Icarus 62:221

Krasnopolsky VA, Krysko AA (1976) On the night airglow of the Martian atmosphere. Space Res 16:1005

Krasnopolsky VA, Parshev VA (1977) Altitude profile of water vapor on Mars. Cosmic Res 15:776

Krasnopolsky VA, Parshev VA (1979a) Ozone and photochemistry of the Martian lower atmosphere. Planet Space Sci 27:113

Krasnopolsky VA, Parshev VA (1979b) On the chemical composition of the troposphere and cloud layer of Venus based on Veneras 11 and 12 and Pioneer Venus measurements. Cosmic Res 17:763

Krasnopolsky VA, Parshev VA (1980a) Initial data of calculations of chemical composition of the Venus atmosphere down to 50 km. Preprint No 590, Space Res Inst, Moscow; 1981. Cosmic Res 19:87

Krasnopolsky VA, Parshev VA (1980b) Photochemistry of the Venus atmosphere down to 50 km: results of calculations. Preprint No 591, Space Res Inst, Moscow; 1981. Cosmic Res 19:261

Krasnopolsky VA, Parshev VA (1981) Chemical composition of the atmosphere of Venus. Nature (Lond) 282:610

Krasnopolsky VA, Parshev VA (1983) Photochemistry of the Venus atmosphere. In: Hunten DH, Colin L, Donahue TM, Moroz VI (eds) Venus. Univ Arizona Press, p 431

Krasnopolsky VA, Tomashova GV (1980) Venus night airglow variations. Cosmic Res 18:766

Krasnopolsky VA, Krysko AA, Rogachev VN, Parshev VA (1976) Venus night airglow spectroscopy by Veneras 9 and 10. Cosmic Res 14:1789

Krasnopolsky VA, Parshev VA, Krysko AA, Rogachev VN (1979) Structure of the lower and middle atmosphere of Mars based on the data of ultraviolet photometry on board the Mars 5 orbiter. Preprint, Space Res Inst, Moscow. Cosmic Res 18:120

Ksanfomaliti LV (1979) Lightnings in the Venus cloud layer. Cosmic Res 17:747

Ksanfomaliti LV (1983) The electrical activity of the Venus' atmosphere. I. Descent probe measurements. Cosmic Res 21:279

Ksanfomaliti LV, Scarf FL, Taylor WWL (1983) The electrical activity of the atmosphere of Venus. In: Hunten DM, Colin L, Donahue TM, Moroz VI (eds) Venus. Univ Arizona Press, p 565

Kuiper GP (1968) Identification of the Venus cloud layers. Comm Lunar Planet Lab 6/100:229

Kumar S, Broadfoot AL (1975) He 584 Å airglow emission from Venus: Mariner 10 observations. Geophys Res Lett 2:357

Kumar S, Hunten DM, Broadfoot AL (1978) Non-thermal hydrogen in the Venus exosphere: the ionospheric source and the hydrogen budget. Planet Space Sci 26:1063

Kumar S, Hunten DM, Taylor HA (1981) H_2 abundance in the atmosphere of Venus. Geophys Res Lett 8:237

Kumar S, Hunten DM, Pollack JB (1983) Nonthermal escape of hydrogen and deuterium from Venus and implications for loss of water. Icarus 55:1369

Kurt VG, Dostovalov SB, Sheffer EK (1968) The Venus far UV observations with Venera 4. J Atm Sci 25:668

Kurt VG, Smirnov AS, Titarchuk LG, Chuvahin SD (1974) Observation of OI 1300 Å radiation in the Martian atmosphere. Icarus 21:35

Kurt VG, Smirnov AS, Titarchuk LG, Bertaux JL, Lepine VM (1983) Light scattering in resonant lines in the upper atmosphere of Venus according to the Veneras 11 and 12 measurements. Cosmic Res 21:545

Kuzmin AD (1983) Radio astronomical studies of Venus. In: Hunten DM, Colin K, Donahue TM, Moroz VI (eds) Venus. Univ Arizona Press, p 36

Kuzmin AD, Marov MY (1974) Physics of the Planet Venus. Nauka, Moscow

Lane AL, Barth CA, Hord CW, Stewart AI (1973) Mariner 9 ultraviolet spectrometer experiment: observations of ozone on Mars. Icarus 18:102

Lane WA, Opstbaum R (1983) High altitude Venus haze from Pioneer Venus limb scans. Icarus 54:48

Lawrence GM, Barth CA, Argabright V (1977) Excitation of the Venus night airglow. Science (Wash DC) 195:573

Lee LC, Carlson RW, Judge DL, Ogawa M (1973) The absorption cross sections of N_2, O_2, CO, NO, CO_2, N_2O, CH_4, C_2H_6 and C_4H_{10} from 180 to 700 Å. J Quant Spectrosc Radiat Transfer 13:1023

Levine JS, Gregory GL, Harvey GA, Howell WE, Boruck WJ, Orville RE (1982) Production of nitric oxide by lightning on Venus. Geophys Res Lett 9:893

Lewis JS (1970) Venus: atmospheric and lithospheric composition. Earth Planet Sci Lett 10:73

Lewis JF, Kreimendal (1980) Oxidation state of the atmosphere and crust of Venus from Pioneer Venus observations. Icarus 42:330

Lindal GF, Hotz HB, Sweetnam DN, Shippony Z, Brenkle JP, Hartsell GV, Spear RT (1979) Viking radio occultation measurements of the atmosphere and topography of Mars: Data acquired during 1 Martian year of tracking. J Geophys Res 84:8443

Liu SC, Donahue TM (1976) The regulation of hydrogen and oxygen escape from Mars. Icarus 28:231

Lyot B (1929) Recherches sur la polarisation de la lumière des planets et de quelques substances terrestres. Ann Observ Paris (Meudon) 8:161

Makarova EA, Kharitonov AV (1972) Energy distribution in solar spectrum and solar constant. Nauka, Moscow

Mantas GP, Hanson WP (1979) Photoelectron fluxes in the Martian atmosphere. J Geophys Res 84:369

Marov MY, Ryabov OL (1972) Venus model atmosphere. Preprint No 39, Inst Appl Mathematics, Moscow

Marov MY, Byvshev BV, Baranov BP, Lebedev VN, Lukashevich NL, Maximov AV, Manuilov K, Frolov AM, Shari VP (1983) Study of Venus cloud structure by the Venera 13 and 14 nephelometers. Cosmic Res 21:269

Marov MY, Lystsev VE, Lebedev VN (1978) The structure and microphysical properties of the Venus clouds. Preprint No 144, Inst Appl Mathematics, Moscow

Marov MY, Lystsev VE, Lebedev VN, Lukashevich NL, Shari VP (1980) The structure and microphysical properties of the Venus clouds: Venera 9, 10 and 11 data. Icarus 44:608

Marov MY, Galtsev AP, Shari VP (1984) H_2O profile in the lower atmosphere of Venus deduced from the net flux measurements. Cosmic Res 22:267

Martonchik JV, Beer R (1975) Analysis of spectrophotometric observations of Venus in the 3 – 4 micron region. J Atm Sci 32:1151

Massie ST, Hunten DM, Sowell DR (1983) Day and night models of the Venus thermosphere. J Geophys Res 88:3955

Masurzky H, Eliason E, Ford PG, McGill GE, Pettengill GH, Schaber GG, Schubert G (1980) Pioneer Venus radar results: geology from images and altimetry. J Geophys Res 85:8233

Mayr HG, Harris I, Hartle RE, Hoegy WR (1978) Diffusion model for the upper atmosphere of Venus. J Geophys Res 83:4411

Mayr HG, Harris I, Niemann HB, Brinton HC, Spencer NW, Taylor HA, Hartle RE, Hoegy WR, Hunten DM (1980) Dynamic properties of the thermosphere inferred from Pioneer Venus mass spectrometer measurements. J Geophys Res 85:7841

McElroy MB (1969) Structure of the Venus and Mars atmospheres. J Geophys Res 74:29

McElroy MB (1972) Mars: an evolving atmosphere. Science (Wash DC) 175:443

McElroy MB, Donahue TM (1972) Stability of the Martian atmosphere. Science (Wash DC) 177:986

McElroy MB, McConnell JC (1971) Atomic carbon in the atmospheres of Mars and Venus. J Geophys Res 76:6674

McElroy MB, Kong TY, Lung YL, Nier AO (1976a) Composition and structure of the Martian upper atmosphere: analysis of results from Viking. Science (Wash DC) 144:1295

McElroy MB, Yung YL, Nier AO (1976b) Isotopic composition of nitrogen: implications for the past history of Mars' atmosphere. Science (Wash DC) 194:70

McElroy MB, Kong TY, Yung YL (1977) Photochemistry and evolution of Mars' atmosphere: a Viking pespective. J Geophys Res 82:4379

McElroy MB, Prather MJ, Rodriguez JM (1982a) Escape of hydrogen from Venus. Science (Wash DC) 215:1614

McElroy MB, Prather MJ, Rodriquez JM (1982b) Loss of oxygen from Venus. Geophys Res Lett 9:649

Meier RR, Anderson DE, Stewart AIF (1983) Atomic oxygen emissions observed from Pioneer Venus. Geophys Res Lett 10:214

Mikhailov AV, Ivanov-Kholodny GS (1975) Model of Mars' upper atmosphere based on ionospheric data. Geomagn i aeronomia 15:29

Mills KC (1974) Thermodynamic data for inorganic sulfides, selenides, and tellurides. Butterworths, London

Moroz VI (1964) New observations of Venus infrared spectrum (1.2 – 3.8 μm). Astron Zh 41:711

Moroz VI (1967) Physics of planets. Nauka, Moscow

Moroz VI (1971) Atmosphere of Venus. Usp Phys Nauk 104:255

Moroz VI (1978) Physics of Mars. Nauka, Moscow

Moroz VI (1981a) The atmosphere of Venus. Space Sci Rev 29:3

Moroz VI (1981b) Albedo of Venus in the range from 0.2 to 4.0 μm. Cosmic Res 19:591

Moroz VI, Nagip AE (1975a) Mars 3: water vapor in the atmosphere of the planet. Cosmic Res 13:738

Moroz VI, Nagip AE (1975b) Preliminary results of water vapor measurements by Mars 5. Cosmic Res 13:38

Moroz VI, Parfentiev NA, Sanko NF, Jegulev VS, Zasova LV, Ustinov EA (1976) Preliminary results of narrow band photometric sounding of the Venus cloud layer at 0.8 – 0.87 μm by the Veneras 9 and 10 descent probes. Cosmic Res 14:743

Moroz VI, Parfentiev NA, Sanko NF (1979) Veneras 11 and 12 spectrophotometric experiments. II. Analysis of the Venera 11 spectral data by the layer addition technique. Cosmic Res 17:727

Moroz VI, Golovin YM, Moshkin BE, Ekonomov AP (1981) Veneras 11 and 12 spectrophotometric experiment. III. Results of the photometric measurements. Cosmic Res 19:599

Moroz VI, Moshkin BE, Ekonomov AP, Grigoriev AV, Gnedykh VI, Golovin YM (1983) Veneras 13 and 14 spectrophotometric experiment. II. Preliminary analysis of H_2O absorption bands in spectra. Cosmic Res 21:246

Moshkin BE, Ekonomov AP, Moroz VI, Golovin YM, Gnedykh VI, Grigoriev AV, Djachkov AV (1983) Veneras 13 and 14 spectrophotometric experiment. I. Methodics, results, and preliminary analysis of measurements. Cosmic Res 21:236

Mount GH, Rottman GJ, Timothy JG (1980) The solar spectral irradiance 1200 – 2550 Å at solar maximum. J Geophys Res 85:427

Mukai S, Mukai T (1981) Infrared properties of haze particles of Venus. Icarus 48:482

Mukhin LM, Gelman BG, Lamonov NI, Melnikov VV, Nenarokov DF, Okhotnikov BP, Rotin VA, Khohlov VN (1983) Gas chromatographic analysis of the Venus atmospheric chemical composition by the Veneras 13 and 14 descent probes. Cosmic Res 21:225

Mumma MJ, Morgan HD, Mentall JE (1975) Reduced absorption of the nonthermal $CO(A^1\pi - X^1\Sigma^+)$ fourth positive group by thermal CO and implications for the Mars upper atmosphere. J Geophys Res 80:168

Nagy AF, Cravens TE, Smith SG, Taylor HA, Brinton HC (1980) Model calculations of the day side ionosphere of Venus: ionic composition. J Geophys Res 85:7795

Nagy AF, Cravens TE, Yee JH, Stewart AIF (1981) Hot oxygen atoms in the upper atmosphere of Venus. Geophys Res Lett 8:629

Nagy AF, Cravens TE, Gombosi TI (1983a) Basic theory and model calculation of the Venus ionosphere. In: Hunten DM, Colin L, Donahue TM, Moroz VI (eds) Venus. Univ Arizona Press, p 841

Nagy AF, Cravens TE, Paxton LJ, Stewart AI (1983b) Nonthermal oxygen, carbon and nitrogen in the Venus exosphere. Paper presented at XVIII Assembly of IUGG, Hamburg, FRG

Nakata RS, Watanabe K, Matsunaga FM (1965) Absorption and photoionization coefficients of CO_2 in the region 580 – 1970 Å. Sic Light 14:54

Nicolet M, Cieslik S (1980) The photodissociation of nitric oxide in the atmosphere and stratosphere. Planet Space Sci 26:105

Niemann HB, Kasprzak WT, Hedin AE, Hunten DM, Spencer NW (1980) Mass spectrometric measurements of the neutral gas composition of the thermosphere and exosphere of Venus. J Geophys Res 85:7817

Nier AO, McElroy MB (1977) Composition and structure of Mars' upper atmosphere: results from the neutral mass spectrometers on Viking 1 and 2. J Geophys Res 82:4341

Nier AO, McElroy MB, Yung YL (1976) Isotopic composition of the Martian atmosphere. Science 194:68

Noxon JF, Traub WA, Carleton NP, Connes P (1976) Detection of O_2 dayglow emission from Mars and the Martian ozone abundance. Astrophys J 207:1025

Ogawa T, Kondo Y (1977) Diurnal variability of thermospheric N and NO. Planet Space Sci 25:735

Okabe H (1971) Fluorescence and predissociation of sulfur dioxide. J Am Chem Soc 93:7095

O'Leary B (1975) Venus: vertical structure of stratospheric hazes from Mariner 10 pictures. J Atm Sci 32:1091

Oran ES, Julienne PS, Strobel DF (1975) Aeronomy of odd nitrogen in the thermosphere. J Geophys Res 80:3068

Owen TS (1964) A determination of the Martian CO_2 abundance. Comm Lunar Planet Lab 2:133

Owen TS, Kuiper G (1964) A determination of the composition and surface pressure of the Martian atmosphere. Comm Lunar Planet Lab 2:113

Owen T, Sagan C (1972) Minor constituents in planetary atmospheres: ultraviolet spectroscopy from the Orbiting Astronomical Observatory. Icarus 16:557

Owen TS, Biemann K, Rushneck DR, Biller JE, Homarth DW, Lafleur AL (1977) The composition of the atmosphere at the surface of Mars. J Geophys Res 82:4635

Oyama VI, Carle GC, Woeller F, Pollack JB (1979) Venus lower atmosphere composition: analysis by gas chromatography. Science (Wash DC) 203:802

Oyama VI, Carle GC, Woeller F, Pollack JB, Reynolds RT, Craig RA (1980) Pioneer Venus gas chromatography of the lower atmosphere of Venus. J Geophys Res 85:7891

Parisot JP, Moreels G (1984) Photochemistry of Fluorine in the atmosphere of Venus. Icarus 59:69

Parkinson TD, Hunten DM (1972) Spectroscopy and aeronomy of O_2 on Mars. J Atm Sci 29:1380

Petryanov IV, Andreichikov BM, Korchuganov BM et al. (1981a) Application of the FP filter to study aerosol of the Venus clouds. Dokl Akad Nauk SSSR 258:57

Petryanov IV, Andreichikov BM, Korchuganov BM et al. (1981b) Iron in the Venus clouds. Dokl Akad Nauk SSSR 260:834

Phillips LF, Schiff HI (1962) Mass spectrometric studies of atomic reactions. III. Reactions of hydrogen atoms with nitrogen dioxide and with ozone. J Chem Phys 37:1233

Pollack JB, Yung YL (1980) Origin and evolution of planetary atmospheres. Annu Rev Earth Planet Sci 8:425

Pollack JB, Stecker DW, Witteborn FC, Erickson EF, Baldwin BJ (1978) Properties of the clouds of Venus as inferred from airborne observations of its near infrared reflectivity spectrum. Icarus 34:28

Pollack JB, Ragent B, Boese R (1979) Nature of the ultraviolet absorber in the Venus clouds: interferences based on Pioneer Venus data. Science (Wash DC) 205:76

Pollack JB, Toon OB, Whitten RC et al. (1980) Distribution and source of the UV absorption in Venus atmosphere. J Geophys Res 85:8141

Prather MJ, McElroy MB (1983) Helium on Venus: Implications for uranium and thorium. Science 220:410

Prinn RG (1975) Venus: chemical and dynamical processes in the stratosphere and mesosphere. J Atm Sci 32:1237

Prinn RG (1978) Venus: chemistry of the lower atmosphere prior to the Pioneer Venus mission. Geophys Res Lett 5:973

Prinn RG (1979) On the possible roles of gaseous sulfur and sulfanes in the atmosphere of Venus. Geophys Res Lett 6:807

Prinn RG (1984a) The sulfur cycle and clouds of Venus. In: Hunt GE (ed) Recent advances in planetary meteorology. Cambridge Univ Press, England

Prinn RG (1984b) The photochemistry of the atmosphere of Venus. In: Levine JS (ed) The photochemistry of atmospheres. Academic Press, New York

Ragent B, Blamont J (1980) The structure of the clouds of Venus: results of the Pioneer Venus nephelometer experiment. J Geophys Res 85:8089

Rao TN, Collier SS, Calvert JG (1969a) Primary photophysical processes in the photochemistry of sulfur dioxide. J Am Chem Soc 91:1609

Rao TN, Collier SS, Calvert JG (1969b) The quenching reactions of the first excited singlet and triplet states of sulfur dioxide with oxygen and carbon dioxide. J Am Chem Soc 91:1616

Reed RA, Forrest WJ, Houck JR, Pollack JB (1978) Venus: the 17 to 38 micron spectrum. Icarus 33:554

Rohatgi NK, DeMore WB (1978) Ultraviolet absorption cross sections of hydrogen peroxide. Geophys Res Lett 5:113

Romand J, Vodar B (1948) Spectra d'absorption de l'acide chlorydrique gaze aux dans la région le Schumann. Compt Rend Acad Sci Paris 226:238

Rottman SJ, Moos HW (1973) The ultraviolet (1200 – 1900 Å) spectrum of Venus. J Geophys Res 78:8033

Rusch DW, Cravens TE (1979) A model of the neutral and ion nitrogen chemistry in the daytime thermosphere of Venus. Geophys Res Lett 6:791

Sagan C, Hanst PL, Young AT (1965) Nitrogen oxides on Mars. Planet Space Sci 13:73

Samuelson RE, Hanel RA, Herath LW, Kunde VG, Maguire WC (1975) Venus cloud properties: infrared opacity and mass mixing ratio. Icarus 25:49

Sanko NF (1980) Gaseous sulfur in the Venus atmosphere. Cosmic Res 18:600

Savich NA, Samovol VA (1976) The nighttime ionosphere of Mars from Mars 4 and Mars 5 dual-frequency radio occultation measurements. Space Res 17:1009

Sawada T, Strickland DJ, Green AES (1972) Electron energy deposition in CO_2. J Geophys Res 77:4812

Scarf FL, Russell CT (1983) Measurements of Venus lightnings from the Pioneer orbiter. Paper presented at IUGG Assembly, Hamburg

Scarf FL, Taylor WWL, Russell CT, Brace LH (1980) Lightning in Venus: orbiter detection of whistler signals. J Geophys Res 85:8158

Schofield JT, Taylor FW (1982) Net global thermal emission from the Venusian atmosphere. Icarus 52:245

Schofield JT, Taylor FW, McCleese DJ (1982) The global distribution of water vapor in the middle atmosphere of Venus. Icarus 52:263

Schubert G, Covey C, Del Genio A, Elson LS, Keating G, Seiff A, Young RE, Apt J, Connselman III CC, Kliore AJ, Limaye SS, Revercomb HE, Smorovsky LA, Suomi VE, Taylor F, Woo R, von Zahn U (1980) Structure and circulation of the Venus atmosphere. J Geophys Res 85:8007

Schunk RW, Nagy AF (1980) Ionospheres of the terrestrial planets. Rev Geophys Space Phys 18:813

Seery DI, Britton D (1964) The continuous absorption spectra of chlorine, bromine, bromine chloride, iodine chloride, and iodine bromide. J Phys Chem 68:2263

Seiff A, Kirk DB (1977) Structure of the atmosphere of Mars in summer at mid-latitudes. J Geophys Res 84:4364

Seiff A, Kirk DB, Young RE, Blanchard RC, Findlay JT, Kelly GM, Sommer SC (1980) Measurements of thermal structure and thermal contrasts in the atmosphere of Venus and related dynamical observations: results from the four Pioneer Venus probes. J Geophys Res 85:7903

Shemansky DE (1972) CO_2 extinction coefficient 1700 – 3000 Å. J Chem Phys 56:1582

Shimazaki T (1981) A model of temporal variations in ozone density in the Martian atmosphere. Planet Space Sci 29:21

Shimazaki T, Shimizu M (1979) The seasonal variation of ozone density of the Martian atmosphere. J Geophys Res 84:1269

Shimizu M, Ashihara O (1972) Electron temperature in the Martian ionosphere. Publ Astron Soc Japan 24:201

Sholokhov VS, Burgin MS (1982) Atomic oxygen in the upper atmosphere of Venus. Cosmic Res 20:312

Sill GT (1972) Sulfuric acid in the Venus clouds. Comm Lunar Planet Lab 171:191

Sill GT (1983) The clouds of Venus: sulfuric acid by the lead chamber process. Icarus 53:10

Sinton WM (1962) Trans IAU IXB:246. Academic Press, New York

Slanger TG (1978) Generation of $O_2(c^1\Sigma_u^-, C^3\Delta_u, A^3\Sigma_u^+)$ from oxygen atom recombination. J Chem Phys 69:4779

Slanger TG (1979) The $O_2(C^3\Delta_u - a^1\Delta_g)$ band system. Chem. Phys Lett 66:344
Slanger TG, Black G (1978) The $O_2(C^3\Delta_u - a^1\Delta_g)$ bands in the nightglow spectrum of Venus. Geophys Res Lett 5:947
Slanger TG, Huestis DL (1981) $O_2(c^1\Sigma_u^- - X^3\Sigma_g^-)$ emission in the terrestrial nightglow. J Geophys Res 86:3551
Sobolev VV (1972) Light scattering in planetary atmospheres. Nauka, Moscow
Spenner K, Kundsen WC, Whitten RC, Michalson PF, Miller KL, Novak V (1981) On the maintenance of the Venus nighttime ionosphere: electron precipitation and plasma transport. J Geophys Res 86:9170
Steffes PG, Eshleman VR (1982) Sulfuric acid vapor and other cloud related gases in the Venus atmosphere: abundances inferred from observed radio opacity. Icarus 51:322
Stewart AI (1972) Mariner 6 and 7 ultraviolet spectrometer experiment: implications of the CO_2^+, CO and O airglow. J Geophys Res 77:54
Stewart AI, Barth CA (1979) Ultraviolet night airglow of Venus. Science (Wash DC) 204:59
Stewart AI, Barth CA, Hord CW, Lane AL (1972) Mariner 9 ultraviolet spectrometer experiment: structure of Mars' upper atmosphere. Icarus 17:469
Stewart AI, Anderson DE, Esposito LW, Barth CA (1979) Ultraviolet spectroscopy of Venus: initial results from the Pioneer Venus orbiter. Science (Wash DC) 203:777
Stewart AI, Gerard J, Rush DW, Bougher SW (1980) Morphology of the Venus ultraviolet night airglow. J Geophys Res 85:7861
Strickland DJ, Thomas GE, Sparks PR (1972) Mariner 6 and 7 ultraviolet spectrometer experiment: analysis of the OI 1304 and 1356 Å emissions. J Geophys Res 77:4052
Strickland DJ, Stewart AI, Barth CA, Hord CW, Lane AL (1973) Mariner 9 ultraviolet spectrometer experiment: Mars atomic oxygen 1304 Å emission. J Geophys Res 78:4547
Strickler SJ, Vikesland JP, Bier HD (1974) $^3B_1 - ^1A_1$ transition of SO_2 gas. II. Radiative lifetime and radiationless processes. J Chem Phys 60:664
Strobel DF, Oran ES, Feldman PD (1976) The aeronomy of odd nitrogen in the thermosphere. II. Twilight emissions. J Geophys Res 81:3745
Suomi VE, Smorovsky LA, Revercomb LA (1980) Net radiation in the atmosphere of Venus: measurements and interpretation. J Geophys Res 85:8200
Surkov YA (1983) Studies of Venus rocks by Veneras 8, 9 and 10. In: Hunten DM, Colin L, Donahue TM, Moroz VI (eds) Venus. Univ Arizona Press, USA, p 154
Surkov YA, Kirnozov FF, Gurianov VI et al. (1981) A study of the Venus cloud aerosol by Veneral 12 (preliminary data), 1982. Geochimia 1:3; Cosmic Res 20:435
Surkov YA, Kirnozov FF, Glazov VN, Dunchenko AG, Atrasrashkevich (1982) New data on the Venus cloud aerosol (preliminary results of measurements by Venera 14). Pisma v Astron Zh 8:700
Surkov YA, Ivanova VF, Pudov AN, Pavlenko VA, Davydov NA, Sheinin DM (1983a) Water vapor mixing ratio in the Venus atmosphere as measured by Veneras 13 and 14. Cosmic Res 21:231
Surkov YA, Moskaleva LP, Sheglov OP, Kharjukova VP, Manveljan OS (1983b) Elemental composition of Venus' rocks. Cosmic Res 21:308
Sze ND, McElroy MB (1975) Some problems in Venus aeronomy. Planet Space Sci 23:763
Takacz PZ, Broadfoot AL, Smith GR, Kumar S (1980) Mariner 10 observation of hydrogen Lyman alpha emission from the Venus exosphere: evidence of the complex structure. Planet Space Sci 28:687
Taylor FW, Beer R, Chanine MT, Diner DJ, Elson LS, Haskins RD, McCleese DJ, Martonchik JV, Reichley PE, Bradley SP, Delderfield J, Schofield JT, Farmer CB, Froidevaux L, Leung J, Coffey MT, Gille JC (1980) Structure and meteorology of the middle atmosphere of Venus: infrared remote sensing from the Pioneer orbiter. J Geophys Res 85:7963
Taylor HA, Brinton HC, Bauer SJ, Hartle RE (1980) Global observations of the composition and dynamics of the ionosphere of Venus: implications for the solar wind interaction. J Geophys Res 85:7765
Thomas GE (1971) Neutral composition of the upper atmosphere of Mars as determined from the Mariner 6 and 7 UV spectrometer experiment. J Atm Sci 28:859

Thompson BA, Harteck P, Reeves RR (1963) Ultraviolet absorption coefficients of CO_2, CO, O_2, H_2O, NH_3, NO, SO_2 and CH_4 between 1850 and 4000 Å. J Geophys Res 68:6431
Timofeeva TS, Yakovlev OI, Efimov AI (1978) Radiowave fluctuations and turbulence of the Venus nighttime atmosphere from the Venera 9 orbiter radio occultations. Cosmic Res 16:285
Titov DV (1983) On the possibility of aerosol formation by the reaction between SO_2 and NH_3 in the Venus atmosphere. Cosmic Res 21:401
Titov DV (1984) Microstructure of the Venus cloud layer based on results of the Venera 11 spectrophotometric experiments (in Russian). Cosmic Res 2:120
Tomasko MG, Doose LR, Smith PH, Odell AP (1980) Measurements of the flux of sunlight in the atmosphere of Venus. J Geophys Res 85:8167
Toon OB, Turco RP, Pollack JB (1982) The ultraviolet absorber on Venus: amorphous sulfur. Icarus 51:358
Toon OB, Ragent B, Colburn D, Blamont J, Cot C (1984) Large, solid particles in the clouds of Venus: Do they exist? Icarus 57:143
Torr MR, Torr DG, Oug RA, Hinteregger HE (1979) Ionization frequencies for major thermospheric constituents as a function of solar cycle 21. Geophys Res Lett 6:771
Traub WA, Carleton NP (1974) A search for H_2O and O_2 on Venus. In: Woszcyk A, Iwaniszewska C (eds) Exploration of planetary atmospheres. Reidel, Boston, p 223
Traub WA, Carleton NP, Connes P, Noxon JF (1979) The latitude variation of O_2 dayglow and O_3 abundance on Mars. Astrophys J 229:846
Trauger JT, Lunine JI (1983) Spectroscopy of molecular oxygen in the atmospheres of Venus and Mars. Icarus 55:272
Tull RG (1970) High-dispersion spectroscopic observations of Mars: IV. The latitude distribution of atmospheric water vapor. Icarus 13:43
Turco RP, Toon OB, Whitten RC, Keesee RG (1983) Venus: mesospheric hazes of ice, dust, and acid aerosols. Icarus 33:18
Ustinov EA (1977) Inverse problem of multiple scattering theory and interpretation of measurements of radiation scattered in the Venus cloud layer. Cosmic Res 15:768
Ustinov EA, Moroz VI (1978) An attempt to specify the H_2O mixing ratio in the Venus atmosphere from the narrow band photometry by Venera 9 and 10. Cosmic Res 16:127
Veverka J (1971) A polarimetric search for a Venus halo during the 1959 inferior conjunction. Icarus 14:282
Vikesland JP, Strickler SJ (1974) $^3B_1 - {}^1A_1$ transition of SO_2 gas. I. Franck-Condon treatment and transition moments. J Chem Phys 60:660
Vinogradov AP, Surkov YA, Andreichikov BM, Kalinkina OM, Glechisheva IM (1970) Chemical composition of the Venus atmosphere. Cosmic Res 8:578
Vinogradov AP, Surkov YA, Kirnozov FF (1973) The content of uranium, thorium, and potassium in the rocks of Venus as measured by Venera 8. Icarus 20:253
Volkov VP, Sidorov YI, Khodakovsky IL, Barsukov VL (1982) On the possible condensates in the Venus main cloud layer. Geokhimia 3
Walker JCG (1982) The earliest atmosphere of the Earth. Precambrian Res 17:147
Wallis MK (1978) Exospheric density and escape fluxes of atomic isotopes on Venus and Mars. Planet Space Sci 26:949
Warneck P, Marmo FF, Sullivan JO (1964) Ultraviolet absorption of SO_2: dissociation energies of SO_2 and SO. J Chem Phys 40:1132
Watson AJ, Donahue TM, Stedman DH, Knollenberg RG, Ragent B, Blamont J (1979) Oxides of nitrogen and the clouds of Venus. Geophys Res Lett 6:743
Watson AJ, Donahue TM, Walker JCG (1981) The dynamics of a rapidly escaping atmosphere: applications to the evolution of Earth and Venus. Icarus 48:150
Watson AJ, Donahue TM, Kuhn WR, Walker JCG (1983) Temperatures in a runaway greenhouse on the prehistoric Venus: implications for water loss. Science (Wash DC)
Wehrbein WH, Hord CW, Barth CA (1979) Mariner 9 ultraviolet spectrometer experiment: vertical distribution of ozone on Mars. Icarus 38:288
Widing KG, Purcell JD, Sandlin GD (1970) The UV continuum 1450 – 2100 Å and the problem of solar temperature minimum. Sol Phys 12:52

Williams MA, Krider EP, Hunten DM (1983) Planetary lightning: Earth, Jupiter, and Venus. Rev Geophys Space Phys 21:892
Wilson WJ, Klein MJ (1981) Venus: observed variations in the carbon monoxide distribution. Paper presented at Int Conf Venus Environment, Pal Alto
Wilson WJ, Klein MJ, Kakar RK, Gulkis S, Olsen ET, Ho PTP (1981) Venus. I. Carbon monoxide distribution and molecular-line searches. Icarus 45:624
Winick JR, Stewart AI (1980) Photochemistry of SO_2 in Venus' upper cloud layers. J Geophys Res 85:7849
Wofsy SC, Sze ND (1975) Venus cloud models. In: McCormac BM (ed) Atmospheres of Earth and Planets. Reidel, Dordrecht, p 369
Woichshyn PM (1974) Global seasonal atmospheric fluctuations on Mars. Icarus 22:325
Woo R, Ishimaru A (1981) Eddy diffusion coefficient for the atmosphere of Venus from radio scintillation measurements. Nature (Lond) 289:383
Woo R, Armstrong JW, Kliore AJ (1982) Small-scale turbulence in the atmosphere of Venus. Icarus 52:335
Wraight PC (1982) Association of atomic oxygen and airglow excitation mechanism. Planet Space Sci 30:251
Yakovlev OI, Efimov AI, Matjugov SS, Timofeeva TS, Chub EV, Yakovleva GD (1978) Night-time atmosphere of Venus as measured by radio occultations of the Venera 9 and 10 orbiters. Cosmic Res 16:113
Young AT (1973) Are the clouds of Venus sulfuric acid? Icarus 18:564
Young AT (1977) An improved Venus cloud model. Icarus 32:1
Young AT (1983) Venus cloud microphysics. Icarus 56:568
Young LDG (1971) Interpretation of high resolution spectra of Mars. J Quant Spectrosc Radiat Transfer 11:385
Yung YL, Strobel DF, Kong TY, McElroy MB (1977) Photochemistry of nitrogen in the Martian atmosphere. Icarus 30:26
Yung YL, DeMore WB (1982) Photochemistry of the stratosphere of Venus: implications for atmospheric evolution. Icarus 51:199
von Zahn U, Fricke KH, Hoffman HJ, Pelka K (1979) Venus: eddy coefficient in the thermosphere and the inferred helium content of the lower atmosphere. Geophys Res Lett 6:337
von Zahn U, Fricke KH, Hunten DM, Krankowsky D, Mauersberger K, Nier AO (1980) The upper atmosphere of Venus during morning conditions. J Geophys Res 85:7829
von Zahn U, Kumar S, Niemann H, Prinn R (1983) Composition of the Venus atmosphere. In: Hunten DM, Colin L, Donahue TM, Moroz VI (eds) Venus. Univ Arizona Press, p 299
Zahnle KJ, Walker JCG (1982) The evolution of solar ultraviolet luminosity. Rev Geophys Space Phys 20:280
Zasova LV, Krasnopolsky VA, Moroz VI (1981) Vertical distribution of SO_2 in upper cloud layer of Venus and origin of UV absorption. Adv Space Res 1:13
Zeldovich YB, Raizer YB (1966) Physics of shockwaves and high temperature hydrodynamic phenomena. Academic, New York
Zipf EC, McLaughlin RW (1978) On the dissociation of nitrogen by electron impact and by EUV photoabsorption. Planet Space Sci 26:449
Zurek RW (1976) Diurnal tide in the Martian atmosphere. J Atm Sci 33:321

Subject Index[1]

Pages less than 99 refer to Mars, more than 98 refer to Venus

1 See also Table of Contents